theory and design

of digital machines

LINCOLN LABORATORY PUBLICATIONS

theory and design
of digital machines

Thomas C. Bartee

Lincoln Laboratory
Massachusetts Institute of Technology

Irwin L. Lebow

Lincoln Laboratory
Massachusetts Institute of Technology

Irving S. Reed

RAND Corporation
(formerly with Lincoln Laboratory
Massachusetts Institute of Technology)

McGraw-Hill Book Company

New York   San Francisco   Toronto   London

Preface

This book has been written to serve as a text for a course in digital-machine design fundamentals at the senior or first-year-graduate level. It is therefore suitable for mature students and practicing engineers. While the treatment of the material begins with fundamental notions and builds systematically from these fundamentals, it is felt that some knowledge of or "feeling" for digital computers is a desirable prerequisite, although prior experience in actual machine design is not required. Problems are included at the end of each chapter. These problems, of varying difficulty and complexity, are meant to illustrate and amplify the material presented in the text.

The book should also prove useful to such people as mathematicians, physicists, and programmers, that is, to people who are interested in or acquainted with the use of computers and who desire more familiarity with the internal structure of computing machines.

The primary purpose of this book is the presentation of techniques for digital-machine design. It is based, to a great extent, on the experience of the authors in designing special-purpose and general-purpose computers at the Massachusetts Institute of Technology Lincoln Laboratory.

The material covered falls in the general category of what has sometimes been called *logical design*, a term which has come to mean different things to different people. In this book the process called logical design has been subdivided, and the different aspects of the design process have been defined somewhat more precisely. The term logical design is used here to attempt to distinguish in the broadest of terms between what has and has not been included in this book. The book does not describe electronic-circuit design, and specific electronic components are mentioned only as examples; nor is there any but the barest reference to computer programming. It is both the theoretical and practical aspects of much of the material lying between circuit design and programming that form the subject matter for this book.

The treatment of the material is as general as seemed reasonable. From a design viewpoint, there is little or no essential difference between general-purpose and special-purpose machines, and the two classes are treated together until specific illustrative examples are discussed. In the same spirit, there is no attempt at completeness in describing particular characteristics of machines, such as arithmetic algorithms and numerical representations. We have selected only that material which seemed most useful, omitting many specialized areas. Extensive lists of references to supplementary material are provided at the end of most of the chapters.

The subject matter of the book may be divided into three parts.

Chapters 1 through 3 and the Appendix present the necessary foundations for machine design. Definitions and descriptions of elementary computer components are included together with the mathematics which best describes their operational use.

Chapters 6 to 10 cover the various aspects of machine design. Chapters 6 to 8 are quite general and introduce the concepts and notation used in Chaps. 9 and 10, which describe the design of a general-purpose and several special-purpose computers, respectively.

The remaining three chapters, 4, 5, and 11, discuss material of a more theoretical nature, which is somewhat peripheral to the main course of the book. All the necessary mathematical tools for the machine-design section (Chaps. 6 to 10) are presented in Chaps. 3 and 7. Chapter 4 presents an extension of the algebraic techniques for the design of logic circuits introduced in Chap. 3, while Chap. 5 essentially repeats much of the material in Chap. 3 but with considerable mathematical rigor. It is because of their close relationship in content to Chap. 3 that Chaps. 4 and 5 were placed after Chap. 3 rather than farther back in the book.

A complete treatment of the material in this book would take Chaps. 3, 4, 5, and 6 in order, while an abbreviated study would either omit Chaps. 4 and 5 or include only selected material from these chapters.

Chapter 11 contains some advanced material in the theory of sequential machines, finite automata, etc. In fact, Chaps. 3, 4, 5, 7, and 11 taken together comprise much of the material usually classified under the heading of switching circuits and can be used for a short course in this subject.

The following members of Lincoln Laboratory read sections of the book and made helpful suggestions during its preparation: M. I. Schneider, D. I. Schneider, E. Weiss, P. E. Wood, and N. Zierler. We would also especially like to thank W. W. Peterson of the University of Florida, E. J. McCluskey of Princeton University, and D. L. Epley of Stanford University for their assistance.

Finally, we should like to mention the following for their careful work in the preparation of the final manuscript: Mrs. Irene Alderson, Mrs. Jean Coombs, Mrs. Helen Hennessey, Mrs. Joyce Kresa, and Mrs. Sheila Toomey

Thomas C. Bartee
Irwin L. Lebow
Irving S. Reed

Contents

Chapter 9   General-purpose Computers   202

Chapter 10   Special-purpose Computers   236

Chapter 11   Sequential Machines   270

Appendix   Number Representation Systems   *305*

Index   *319*

1

Digital Machines and Systems

1-1. Definitions. It is difficult to describe precisely the subject matter of any particular area of science or technology. We generally discover the problems, results, and techniques in a given area gradually, and the extent of a given discipline is fully realized only after considerable specific knowledge has been accrued. In attempting to introduce such terms as *digital machine, digital calculator,* and *digital computer,* we run a further risk: the material being discussed is in a state of rapid expansion, and any but the broadest of definitions is sure to be subject to change.

Nevertheless, first definitions of some of the subjects discussed in this book must be attempted, although the definitions and descriptions in this chapter will be subject to later refinement. These earlier definitions must be general of necessity, and generality is often a synonym for vagueness.

Let us define a digital machine as a machine which represents information in a discrete manner. A digital machine is therefore contrasted to an *analog machine,* which represents information in a continuous manner.

Insight into the fundamental difference between analog and digital machines will be gained if we consider the problem of physically representing the values of a continuous independent variable x in first an analog and then a digital machine. If the variable x is constrained to range through a closed interval $[A,B]$, an analog machine will represent the value of the variable as a physical position, a voltage, a current, etc., which is continuously variable through some interval, say $[V_-,V_+]$, where V_- and V_+ are two voltage values with $V_- < V_+$. We can place the least value in this interval, V_-, in correspondence with the least possible value for x, which is A, and the upper value in the interval, V_+, in correspondence with the maximum value which can be taken by the variable x, which is B. The variable can then be assumed to be capable of taking on any value in the interval $[A,B]$, and there will be a corresponding value of the voltage in the analog system.

The devices used to represent information in a digital machine are

1

operated in a discrete manner.   That is, a given device is either constrained in or naturally disposed to be in only one of some finite number
of *states* at a given time t.   Suppose we are given a device with n states,
which we designate $q_1, q_2, \ldots, q_n$.   In order to represent the values of
our continuous variable x in the interval $[A,B]$, we partition the interval
$[A,B]$ into n subintervals $[X_1,X_2), [X_2,X_3), \ldots, [X_n,X_{n+1}]$, where
$A = X_1 < X_2 < \cdots < X_{n+1} = B$, and place subinterval $[X_i,X_{i+1})$
in correspondence with state q_i of our physical device.   Notice that if
the variable x was not continuous, but was used to represent an integer
in the range $[A,B]$, where $A \leq B$ and $B - A \leq n$, we could assign the
integral values in this interval to the states of our physical device in a
perfectly natural and obvious way.

We defined a digital machine as a machine which represents information in a discrete manner.   This makes an ordinary two-position switch a
digital machine, for it represents the information ON or OFF by means of
an open or closed connection.   It also makes a set of digital machines
which are interconnected a digital machine.   Notice that if m storage
devices are used to represent information, and if each storage device can
be in one of n states at a given time t, then the total number of different
internal states the machine can have is n^m.   Similarly, for s machines
$M_1, M_2, \ldots, M_s$ which are interconnected, the total state of the
resulting machine at time t must be one of $S_1 S_2 \cdots S_s$ states, where S_i
is the number of states which can be taken by machine M_i.   In our
usage, the term digital machine is synonymous with the term *finite-state
automaton*, which also refers to a machine (automaton) with a finite
number of internal states.

Given enough relays and switches and a few push buttons and levers,
we can construct an electromechanical calculator, which is a digital
machine, but which requires another level of refinement.   A digital
calculator is defined as a digital machine which can perform arithmetic
calculations on data stored in (represented by) the machine.   A digital
calculator is therefore a digital machine, but with the further properties
that the calculator must be able to store or represent information and
must also be able to perform arithmetic calculations on the information
stored.

A common form of digital calculator is the adding machine, which
stores numerical data introduced via the keys of the calculator and
performs the arithmetic operation of addition.   A desk calculator is a
more complicated device which performs the arithmetic operations of
addition, subtraction, multiplication, and sometimes division.   Many of
the early electromechanical business-data-processing machines were
referred to as *automatic calculators;* their operation generally consisted of
reading data from a set of perforated cardboard cards and performing a

sequence of operations on this data, the sequence of operations being determined by a wired patch board electrically connected to the internal circuitry of the calculator.

This type of machine is a step advanced from the manually operated desk calculator, for the operations performed are sequenced automatically. Once the cards are introduced, the machine determines its own sequence of operations according to the pattern wired into the patch board, and the intermediate results are also stored in the machine during computation.

Two important points are involved here. The sequence of operations wired into the patch board is generally referred to as the *program* for the machine. A program is defined as the sequence of instructions to the machine which will cause it to perform a desired set of calculations. Now a calculator of this type is capable of storing the program. Suppose that a digital machine stores both its program and the data to be used and also executes the program automatically; a machine with these capabilities is generally referred to as an *automatic digital* computer.

The set of physical devices which is used to retain the information being processed is referred to as the *memory* of the machine. Most machines use several different types of devices in their memories, and the different devices are often classified according to their speed of operation or the particular use for which they are intended.

Many of the memory devices used in a computer are such that information can be read from or stored in them at the operating speeds of the machine. That is, the machine can both read information from and store information in these devices during operation and can "write over" information previously stored. The set of devices with these characteristics is called the *internal memory* and is thereby distinguished from input storage media such as perforated paper cards and tape and output storage media such as printed rolls of paper. A machine which can store both its program and the data which is used in its internal memory is called a *stored-program* computer.

A *digital system* is a data- or information-processing system which performs a major portion of its processing operations using a digital machine or a set of interrelated digital machines. When we speak of the system we include not only the machine but also the operators and all the peripheral equipment associated with the system.

1-2. General-purpose and Special-purpose Computers. Automatic digital computers are quite often separated into two general classes: general-purpose computers and special-purpose computers. A general-purpose computer is a computer designed to solve a wide variety of problems, and a special-purpose computer is one which is designed to perform a specific function. The typical general-purpose computer is also a stored-program computer in that it stores the sequences of instructions comprising a given program in its internal memory.   This makes it possible to read a particular program and the associated data into the computer's memory, have the computer perform the calculations specified by the program, and then to read in and operate another, and perhaps quite different, program.   General-purpose computers which are not stored-program computers often use patch boards or plug boards into which a given program is wired (often these patch boards are changeable so that a computer of this type can perform a number of different programs, each with its own wired patch board).

A stored-program computer which stores both the program and the data on which the program operates in the same memory will have the ability to modify its program during operation, thus making the computer extremely flexible.   Also, this makes use of the computer memory in an efficient fashion, for long programs with little data or short programs with considerable data can be accommodated by a memory which need only be able to contain the sum of the program and data.   A general-purpose stored-program computer may be used to solve a problem in lens design, may then process a payroll, and may later solve a problem in astronomy.

The special-purpose computer is designed to solve a specific problem or a restricted class of problems.   It is usually designed to perform its special task efficiently and rapidly.   Accordingly, it is relatively inflexible and unsuitable for tasks other than the one for which it was constructed.   Thus, for example, a special-purpose missile-guidance computer is generally unsuitable for payroll processing or matrix inversion.

In theory, a given general-purpose computer can be programmed to perform any data-processing job that a special-purpose computer can perform.   In this sense the general-purpose computer possesses a form of universality in that it is capable of simulating any other computer.   The special-purpose computer is useful, therefore, when efficiency is more important than flexibility.   As a general rule the more special-purpose the machine, the more efficient, but, of course, the less versatile.   At present we would not use a typical general-purpose computer in a missile-guidance system, for instance, for the weight, power consumption, etc., would be prohibitive, even though the general-purpose computer would perform all the calculations necessary to guidance.†   Often it would

† Some of the computers used in guidance systems are very similar to small general-purpose computers. Computers constructed along the lines of general-purpose computers—stored-program, automatic digital computers with a reasonable set of instructions—but which are designed for specific functions, are often capable of performing any data-processing task, except for speed and memory limitations. These computers perform general tasks very inefficiently, however, but perform the specific task for which they were designed very efficiently. Such computers are perhaps equally worthy of the titles general-purpose and special-purpose.

simply be bad economics to use a general-purpose computer in a specific function; for instance it would be foolish to tie up a large general-purpose computer in order to encode messages for transmission over a communications channel, and here a small special-purpose computer would come into play.

The special-purpose computer is especially useful in *real-time control systems*, systems which process data and perform control functions during actual operation of a system. The digital computer used in this type of control system must process the data rapidly in order not to fall behind in assessing the present situation, for the computer will be called on to make effective decisions in real time. Special-purpose computers are especially applicable to these systems, because the data-processing calculations to be performed are often quite specific and a given machine can be designed for a specific task. Quite often a real-time control system will contain several special-purpose machines, each performing a particular task. In this way, complex data-processing calculations can be performed in parallel, thus keeping the operating speed of specific machines within reasonable bounds. In several large real-time systems, for instance in the Air Force SAGE system† and in other similar defense systems, a number of special-purpose computers are used to perform many of the routine, highly repetitious operations which are necessary; these smaller machines transmit the results of their calculations to a central general-purpose machine which then integrates, correlates, and calculates, using the data from the special-purpose machines. In this way the total data from several machines are organized and used to control the air situation. Other real-time control systems of a similar nature are being used in the process-control area, which deals with the control of manufacturing processes such as oil refining and the dynamic operation of chemical reactors. The real-time control systems engaged in the controlling of manufacturing processes are forerunners of the automatic factory, in which the operation will be completely controlled by digital machines.

Workers in the real-time control-system area tend to classify systems as *fully automatic* or *semiautomatic*, the latter term designating systems in which certain of the decisions are made by human operators, using data supplied by the computer(s) involved.

1-3. Classification of Subject Matter and a Brief Outline. By custom, the subject matter of digital machines has been divided into several general study areas. As might be expected, these areas overlap considerably, and in actual practice the subject matter is irrevocably tied together. Nevertheless, divisions are made, and specialists in the various areas are given specific titles such as *programmer*, *system designer*, and *circuit designer*. In general, then, the literature and most organizations

† SAGE is a contraction of "Semiautomatic Ground Environment."

divide the work areas concerned with digital machines into:

Circuits and components

Programming and programming techniques, including automatic programming systems and numerical analysis

Logic design, switching circuits, and number systems

System design and structural design of machines

Theoretical studies, automata (finite-state machines), Turing machines, and compatability

The first subjects listed above, circuits and components, involve the actual hardware used from the detailed viewpoint of, for instance, magnetic-core operating characteristics, transistor switching-circuit design, and cryotron development. The research here is concerned with developing new devices and designing circuitry, tape transports, etc., to utilize most efficiently the basic devices. The great amount of research in these areas has led to faster, more reliable, smaller, lighter, less power-consuming, more economical devices and circuitry, and improvement will doubtless continue. This book does not deal with this subject in any detail. The existence of good transistor circuits, memory devices, magnetic tape transports, etc., is assumed, and only the characteristics of these devices which affect the system design will be treated. Similarly, the second set of topics listed above will not be treated in detail. Several books have been published describing programming, along with the associated disciplines of numerical analysis, machine languages, and automatic programming, and we have included references to several books on these subjects at the end of this chapter.

The third set of topics listed above—logic design, switching circuits, and number systems—is treated in some detail in Chaps. 2 to 5 and the Appendix. The subject matter of these chapters is closely bound to such mathematical subjects as Boolean algebra, symbolic logic, algebraic logic, number representation systems, and set theory, and we have included some introductory material in these areas. Chapter 2 introduces several physical storage devices, the concept of storage cell and register, and some material on representation of data in registers. Chapters 3 and 4 introduce the basic concepts of Boolean algebra and switching-circuit theory, presenting the mathematical material in parallel with the description of digital-machine gating elements. Chapter 5, Mathematical Foundations, describes the mathematical material in a more rigorous way, introducing Boolean algebra via rings and fields, and introducing the concept of Boolean functions of time. Chapters 4 and 5 are to be considered optional, although knowledge of this material is felt to be of importance to anyone interested in digital systems, and the concepts in these chapters are certainly of considerable value.   The chapters which follow do not assume a knowledge of the material in Chaps. 4 and 5.

In Chaps. 6 to 10, a systematic procedure is developed for designing digital machines.   The design process is divided into three categories: system design, structural design, and logic design.   By system design we mean the planning and specification of the gross characteristics of the machine.† The structural design then consists of describing the machine as a set of interacting registers directed by a control unit.   Logic design consists of using the discipline of switching theory to derive the details of the machine.   Chapter 6 introduces some of the basic concepts and notation for the structural design, defining the transfer operation as the fundamental interaction between registers.

Chapter 7 deals with sequential networks, which are sometimes called sequential circuits or finite-state automata, from the viewpoint of the logical configuration.   The subject matter of Chap. 7 is generally included under the title "switching circuits," which indicates some of the overlap inherent in any division of material (Chap. 7 deals with *sequential networks*, Chaps. 3 and 4 with *combinational networks*).

Chapters 8, 9, and 10 develop the systematic design procedures; Chap. 8 includes the elements of structural design and introduces the control unit.   Chapter 9 describes the design of a small general-purpose computer from the system design through the structural design and including the logic design.

Chapter 10 treats special-purpose computers.   This chapter may be divided into two sections.   The first section describes the design of a special-purpose computer which processes radar returns in a real-time control system and then transmits target coordinates over a communications channel.   The particular machine described here is a simplified version of the AN/FST-2 which was used in the SAGE Air Defense System; this particular machine was designed at Lincoln Laboratory.   The second section of Chap. 10 describes the digital differential analyzer.   An introduction to the digital differential analyzer is followed by some design details, after which programming, or the setting up of problems for this type of digital machine, is described.

Chapter 11 presents some theoretical material and very briefly treats

† The term "system design" is used in several ways, and common usage is somewhat ambiguous.   Originally, system design referred to the planning and design of a machine from a broad viewpoint, but including the specification of the actual machine construction.   At the time, system design referred to about everything save the design of specific circuitry, gears, switches, etc.   For digital systems this included what is generally called logical design.   There is a tendency in very large systems to use the term system design to refer to the specification and planning of a system from a very broad viewpoint, quite often not including the logical design phase.

such subjects as minimal-state machines, Turing machines, probabilistic machines, finite-state automata, and information processing from a more theoretical, viewpoint.   Again there is considerable overlap with the subject matter in previous chapters, and as in previous chapters the list of references can be used to supplement the material presented.

REFERENCES

1. Alt, F. L.: "Electronic Digital Computers," Academic Press, Inc., New York, 1958.
2. Bartee, T. C.: "Digital Computer Fundamentals," McGraw-Hill Book Company, Inc., New York, 1960.
3. Berkeley, E. C., and L. Wainwright: "Computers, Their Operation and Applications," Reinhold Publishing Corporation, New York, 1956.
4. Caldwell, S. H.: "Switching Circuits and Logical Design," John Wiley & Sons, Inc., New York, 1958.
5. Engineering Research Associates, Inc.: "High-speed Computing Devices," McGraw-Hill Book Company, Inc., New York, 1950.
6. Flores, I.: "Computer Logic," Prentice-Hall, Inc., Englewood Cliffs, N.J., 1960.
7. Hollingdale, S. H.: "High Speed Computing," The Macmillan Company, New York, 1959.
8. Humphrey, W. S., Jr.: "Switching Circuits," McGraw-Hill Book Company, Inc., New York, 1958.
9. Ivall, T. E.: "Electronic Computers," Philosophical Library, Inc., New York, 1956.
10. Jeenel, J.: "Programming for Digital Computers," McGraw-Hill Book Company, Inc., New York, 1959.
11. McCracken, D. D.: "Digital Computer Programming," John Wiley & Sons, Inc., New York, 1959.
12. Montgomerie, G. A.: "Digital Calculating Machines," D. Van Nostrand Company, Inc., Princeton, N.J., 1956.
13. Phister, M.: "Logical Design of Digital Computers," John Wiley & Sons, Inc., New York, 1958.
14. Pressman, A. I.: "Design of Transistorized Circuits for Digital Computers," John Francis Rider, Publisher, Inc., New York, 1959.
15. Richards, R. K.: "Arithmetic Operations in Digital Computers," D. Van Nostrand Company, Inc., Princeton, N.J., 1955.
16. Richards, R. K.: "Digital Computer Components and Circuits," D. Van Nostrand Company, Inc., Princeton, N.J., 1957.
17. Scott, N. R.: "Analog and Digital Computer Technology," McGraw-Hill Book Company, Inc., New York, 1960.
18. Staff of the Harvard Computation Laboratory: "Synthesis of Electronic Computing and Control Circuits," Harvard University Press, Cambridge, Mass., 1951.
19. Wilkes, M. V.: "Automatic Digital Computers," John Wiley & Sons, Inc., New York, 1956.

2

Binary Cells and Registers

2-1. Introduction. A digital machine is one in which the data are represented discretely. This is accomplished by the use of physical devices exhibiting a finite number of states which are placed in correspondence with the values of the variables to be represented. The basic physical devices used for this purpose are almost exclusively two-state or binary devices which we call *binary cells*. The types of binary cells in use differ widely in their behavior. Even when the physical phenomena associated with a given device are basically continuous, a binary mode of operation is usually employed for reasons of simplicity and reliability.

Since a two-state device can represent only a single binary digit, it is apparent that to represent the values of continuous variables with any amount of precision or to represent integers of any magnitude, the bistable devices must be handled in sets and a given set of binary values used to represent the value of a particular variable. In actual practice the data stored in digital machines are divided into sets of ordered characters called *words;* in a machine a word is therefore an ordered set of characters which has meaning. Now the term "word" refers to the information stored and not to the physical devices in which the information is stored. We will refer to an ordered set of physical devices used to store a specific piece of data as a *register*.

This chapter is divided into three basic parts. First, some of the characteristics of the devices used to store information in digital machines are examined from a systems-design viewpoint. Second, the concept of a register as an ordered set of cells is introduced and some of the systems whereby information is represented in registers are examined. Finally, the idea of a *function of a register* is introduced as the means by which functions of variables stored in registers are computed.

The material on representation systems for digital machines in this chapter assumes a knowledge of the binary number system, other number representation systems such as octal and ternary, and such related topics as positional notation and binary-coded decimals. The Appendix of this book contains an introduction to number representation systems which furnishes background material in this area.

2-2. The Binary Storage Cell.   In order to make the notions introduced in the previous section more precise, we shall consider the properties of storage devices, first from an idealized picture and subsequently from the point of view of realistic storage devices.

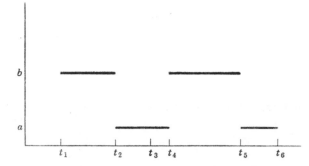

Fɪɢ. 2-1. State value versus time for an idealized binary storage cell.

Let us first describe the characteristics of an idealized primitive storage element which we call a binary storage cell or simply a binary cell.   This is defined to be a physical device with the following properties:

1. It possesses two possible stable configurations or states which we label state A and state B.

2. Any changes of state (from A to B or from B to A) occur instantaneously, i.e., at the discrete times $t_1, t_2, \ldots, t_j, \ldots$.

3. A measurement of some physical property of the cell such as residual magnetism, voltage, or current at some point always results in a number a if the cell is in state A, or a number b if the cell is in state B.   This definition of measurement implies that the measuring device does not disturb the cell, or equivalently, extracts no energy from the cell.   Since by property 2 the cell is assumed to change state instantaneously, a measurement is meaningless at the switching times $t_1, t_2, \ldots, t_j, \ldots$.

If we perform an experiment in which we measure the physical state of a binary cell at all times $t \neq t_j$ and plot the results of these measurements as a function of time, we obtain a two-valued function as shown in Fig. 2-1.   For the purpose of drawing the curve we have assumed with no loss of generality that $a < b$.

A binary cell is, then, an idealized physical device which can store a two-valued function of time.   The determination of the state of the cell at some meaningful time t is equivalent to the evaluation of the function at time t and yields one *bit* of information in the sense of Shannon.   The binary cell is the fundamental storage element of the digital data processor.

Correspondingly, the two-valued function is the fundamental mathematical function for the representation of physical variables in the machine. That the fundamental storage element is defined to be binary, and correspondingly that the fundamental functions are two-valued, is due to the fact that real physical storage elements are most easily constructed as two-state devices. (In the next section several types of actual binary cells will be described.)

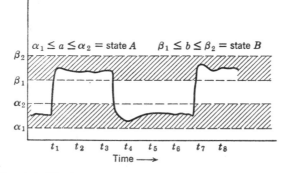

Fig. 2-2. Physical measurements of a binary storage cell.

There is a natural one-to-one correspondence between a cell and the function which it stores. We shall often have occasion to use this correspondence and consider the cell itself as a two-valued function of time. It follows, then, that if at some time t the cell X stores a function with value a, the cell can be treated as a function X with value $X(t) = a$.

2-3. Examples of Binary Storage Cells. An idealized binary cell was described in the preceding section. Indeed, no real device can possess properties 2 and 3, i.e., switch instantaneously and yield a precise value from an electrical measurement. Nevertheless, these conditions can be relaxed so that imperfect physical devices can store two-valued functions.

A physical two-state device will not yield precise values as the result of measurements of its physical state, but rather it will yield two narrow ranges of values, as shown in Fig. 2-2. The spread in value is due to departures of both the cell and the measuring device from the idealized property 3. Any real measuring device absorbs a small amount of energy from the cell; this in turn affects the value of the measurement. If the spread in values is small enough so that the ranges of the values for each state do not overlap, then property 3 is effectively met. We call a measurement which does not affect the state of a cell *nondestructive*.

Similarly, any real device requires a finite time to change states. If the interval of time during which a measurement is made does not overlap this switching interval, then property 2 is effectively met. Indeed we can even go one step further and restrict the measurement time to an interval of any desired duration so long as it does not overlap a switching interval.

Perhaps the most common example of a binary storage cell is the *static flip-flop* of Fig. 2-3.   This is a circuit generally consisting of two active

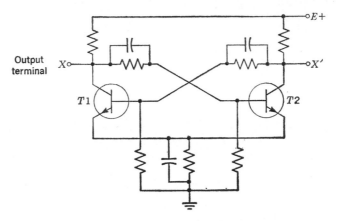

FIG. 2-3. Transistor static flip-flop.

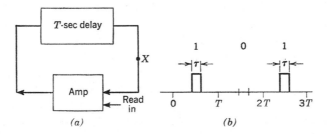

FIG. 2-4. Dynamic flip-flop.

circuit elements such as transistors or vacuum tubes, together with several passive elements, and having the steady-state property that only one transistor or vacuum tube can conduct current at any time.   In Fig. 2-3, state A occurs when transistor T_1 is conducting current, state B when T_2 is conducting current.   A measurement of the output voltage at the output terminal X will give a value close to some value a determined by the circuit parameters in state A and close to some value b in state B.   This measurement will be meaningful provided that it is taken when the flip-flop is not switching from one state to the other.

Figure 2-4a illustrates another example of a binary cell, called a *dynamic flip-flop*.   This cell consists of a delay element and an amplifier connected in a loop. The two states are represented by the presence or absence of a pulse in the loop. Suppose that the delay element produces a delay of T seconds and a pulse, if present, is of length τ seconds. Then at point X one can observe a τ-second pulse every T seconds. The state measurement consists in observing at point X the presence or absence of the pulse during the given τ-second interval (Fig. 2-4b).

Magnetic materials form the basis for another class of binary cells. In Fig. 2-5 is a section of magnetic tape in which the two states are represented by differing directions of the magnetic lines of force (flux).

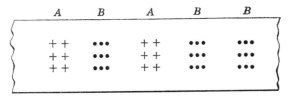

FIG. 2-5. Magnetic-tape storage cells.

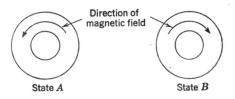

FIG. 2-6. Magnetic-core storage cell.

The cell is in state A when the flux lines point into the plane of the page and in state B when the lines point out of the page. The states are measured by observing the polarity of the voltage induced in the detecting element (reading head) when the tape is moved past it. A different type of magnetic binary cell is the magnetic core shown in Fig. 2-6. The core is in state A when the magnetic flux lines point in the counterclockwise direction, and in state B when the lines point in the clockwise direction.

In this example of a cell, nondestructive measurement techniques are not always satisfactory. That is to say, it is not always possible to obtain a measurement of the state of the core by extracting an amount of energy sufficiently small to prevent disturbing the core. More often a large magnetic force is applied to the core, of a polarity which will switch the core from state B to state A, and the amount of flux change is measured when the magnetic force is applied. Such a measurement technique is called *destructive*. A destructive measurement corresponds to the hypothetical nondestructive measurement implied in the definition of the cell followed by an input to the cell causing it to assume state A. The equivalent of a nondestructive measurement of the core can be obtained by following a destructive measurement by an input causing the core to assume the state which existed before the measurement.

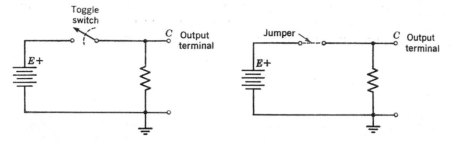

FIG. 2-7. Toggle-switch storage cell.　　　FIG. 2-8. Constant storage cell.

Another example of a cell is shown in Fig. 2-7. In this simple circuit the two states are provided by a toggle switch. The cell is in state A when the switch is closed and in state B when the switch is open. A measurement of the voltage at the output terminal C yields the battery voltage $E +$ in state A and 0 volts in state B, provided that this measurement is not made when the switch is being thrown.

Closely related to the cell just described is the *constant* cell shown in Fig. 2-8. The toggle switch is replaced by either a wire or no wire. In the former case the cell is in state A $(V_C = E +)$ for all time, and in the latter case the cell is in state B for all time. Such a cell stores a function of time which has the constant value a or b.

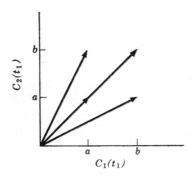

FIG. 2-9. Range of values of a two-cell register at a fixed time t_1.

All of the foregoing examples point to the facts that binary cells can have many diverse physical properties and that the representation of the states can be based upon many different physical phenomena. In their basic use, however, as elements for the storage of two-valued functions, they are equivalent.

2-4. The Register. As stated above, the binary cell is the basic storage element of the digital data-processing machine. A simple extension of the notion of the binary cell leads to the concept of the register.

We define a register to be an ordered set of binary cells. The properties of the register therefore follow immediately from the properties of the cell. Consider an n-cell register C, composed of the cells C_1, C_2, . . . , C_n. Each cell C_i ($i = 1, 2, . . . , n$) stores a two-valued function of time $C_i(t)$. Hence at any fixed time t_1, each $C_i(t_1)$ has the value a or b. The register C correspondingly stores the n two-valued functions of time $C_1(t)$, $C_2(t)$, . . . , $C_n(t)$, or in other words the *vector function* $C(t)$ with *components* $C_i(t)$, $i = 1, 2, . . . , n$. As in the case of the cell, we often use the correspondence between a register and the function it stores by saying that the register *is* a vector-valued function of time. Thus at some fixed time we say that the value of the register C is the value of the function it stores, that is, the n-tuple $[C_1(t), C_2(t), . . . , C_n(t)]$, $C_i(t) = a$ or b.

As an example consider the two-cell register $C = (C_1, C_2)$, storing the two-component vector function $C(t) = [C_1(t), C_2(t)]$. At some time t_1, C has the value $C(t_1) = [C_1(t_1), C_2(t_1)]$ where $C_i(t) = a$ or b. Figure 2-9 shows the four possible values of $C(t_1)$, the coordinate pairs (a,a), (a,b), (b,a), and (b,b).

As implied earlier in this chapter, the register is the physical device which stores representations of numerical values in the digital machine. We have just demonstrated that a register stores an n-component vector function, with each component a two-valued function of time. To complete the picture, we must define a means by which the values of a register are associated with a set of numerical values. The simplest and perhaps most natural association is between the 2^n different values of an n-cell register and the integers from 0 to $2^n - 1$, inclusive. If we associate the integers 0 and 1 with the numbers a and b, respectively, and imagine a binary point to be located to the right of the register, then the values of the register are n-tuples of 0's and 1's which correspond to the first 2^n nonnegative integers in the binary number system. We call this the *integer representation* of the register. Thus, for example, if the four-cell register $C = (C_1, C_2, C_3, C_4)$ has the value $bbab$, in the integer representation it is said to store the binary integer 1101.

The decimal equivalent $[\delta_I(C)]$ of an integer in the binary number system stored in register C is obtained by forming the polynomial

$$\delta_I(C) = \sum_{i=1}^{n} C_i 2^{n-i} \qquad C_i = 0, 1 \qquad i = 1, . . . , n$$

where the values of 2^{n-i} and the sum $\delta_I(C)$ are expressed as decimal integers. Thus in the above example the decimal equivalent of 1101 is 13. The octal number system is also often used. If the n cells of register C are grouped into the k three-cell subregisters

$$R^1 = \begin{cases} 0 & 0 & C_1 & n = 3k - 2 \\ 0 & C_1 & C_2 & n = 3k - 1 \\ C_1 & C_2 & C_3 & n = 3k \end{cases}$$

$$\cdots \cdots \cdots \cdots \cdots \cdots$$

$$R^{k-1} = C_{n-5}C_{n-4}C_{n-3}$$
$$R^k = C_{n-2}C_{n-1}C_n$$

then the octal equivalent of the integer representation of C is a k-tuple

$$\delta_I(R^1), \; \delta_I(R^2), \; \ldots, \; \delta_I(R^k)$$

Thus the octal equivalent of 1101 is 15.

Table 2-1 lists the values of a four-cell register together with its integer representation in the binary, octal, and decimal systems.

Table 2-1. Integer Representation of a Register in the Binary, Octal, and Decimal Number Systems

Register value	Binary	Octal	Decimal
aaaa	0000	0	0
aaab	0001	1	1
aaba	0010	2	2
aabb	0011	3	3
abaa	0100	4	4
abab	0101	5	5
abba	0110	6	6
abbb	0111	7	7
baaa	1000	10	8
baab	1001	11	9
baba	1010	12	10
babb	1011	13	11
bbaa	1100	14	12
bbab	1101	15	13
bbba	1110	16	14
bbbb	1111	17	15

From now on the values of a cell, a and b, will always be considered to represent the integers 0 and 1. Furthermore, when no other representation is stated explicitly, a register will be assumed to have the integer representation. In line with this we shall consider the integer representation (in any number system) of a register to be synonymous with the value of the register. Thus the four-cell register C with value *bbba* will be said to have the *binary value* 1110 or the *octal value* 16 or the *decimal value* 14.

Another representation of a register, closely related to the integer representation, is obtained by assuming the binary point to be located at the left end of the register. Thus a four-cell register with binary value 1110 has the *fractional representation* .1110. The decimal equivalent of the fractional representation of a register is given by the polynomial

$$\delta_F(C) = \sum_{i=1}^{n} C_i 2^{-i} = 2^{-n}\delta_I(C)$$

Similarly, the octal equivalent of this number is obtained by grouping the cells into three-cell subregisters beginning now at the left end of the register. Thus the binary fraction .1110 is equivalent to the decimal fraction $\frac{7}{8} = .875$ and the octal fraction .70.

Both the integer and fractional representations are so-called *binary* representations since the 2^n values of the n-cell register are made to correspond to 2^n numbers. Somewhat different are the so-called *binary-coded decimal* representations in which groups of cells are associated with decimal digits. For example, suppose that register C contains $n = 4k$ cells where k is an integer. Divide C into the k subregisters

$$R^1 = C_1 C_2 C_3 C_4$$
$$\cdots \cdots \cdots \cdots$$
$$R^k = C_{n-3} C_{n-2} C_{n-1} C_n$$

We restrict each subregister to take on the values $0, 1, \ldots, 9$ inclusive and assign a unique subregister value to each decimal digit. Then the binary-coded decimal representation of register C is the k-tuple

$$\delta_I(R^1),\ \delta_I(R^2),\ \ldots,\ \delta_I(R^k)$$

with decimal value

$$\delta_I(R^1)10^{k-1} + \delta_I(R^2)10^{k-2} + \cdots + \delta_I(R^k)10^{k-k}$$

In this representation the 10^k decimal integers are associated with 10^k of the 2^{4k} possible values of the register.

Clearly many other numerical representations, binary and nonbinary, are possible. The binary (or decimal) point, for example, may be located anywhere with respect to the cells of the register. Representations exist for both positive and negative numbers. Some of these are described further in the Appendix.

All the foregoing examples are of numerical representations of registers. Nonnumerical representations are also possible. Let C be a five-cell register with $2^5 = 32$ possible values. The 26 letters of the Roman alphabet may be represented by C by assigning to each letter one of the 32 possible values of the register, as in Table 2-2. In general, any list of k

things may be represented by a register with $\log_2 k$ cells or more by an association of each item with a unique value of the register.

Table 2-2

Value of register	Alphabet representation
00000	A
00001	B
00010	C
00011	D
00100	E
. . . .	. . .
11000	Y
11001	Z

2-5. Generalization of the Register. We have thus arrived at a definition of the register as the basic device in the digital computer for the storage of representations of values of variables and have shown how both numerical and nonnumerical data may be stored in a register by establishing correspondences between the numbers or data to be stored and the various values of the register. This definition, moreover, is basically independent of the physical implementation of the register.

A simple extension of the notion of register was alluded to in the previous section. Any part of a register (i.e., a subregister) is itself a register. In considering the octal and decimal integer representations, we assumed the register to be a set of subregisters in which each subregister had an octal or decimal value. In any n-cell register, each of the n cells is a one-cell register; there are $\binom{n}{2}^{\dagger} = n(n-1)/2$ different two-cell registers which preserve the original ordering, $\binom{n}{3}$ different three-cell registers, etc.

All told there are $\sum_{k=1}^{n} \binom{n}{k} = 2^n - 1$ different subregisters; we have included the entire n-cell register as the last term of the summation. All these subregisters are not in general meaningful for every register. It makes sense to consider a subset of cells of a register as a separate subregister only if at some time the subset is treated independently of the rest of the register. As a simple example suppose that a ten-cell register C

$\dagger$ The symbol $\binom{n}{k}$ is called a binomial coefficient and is defined by

$$\binom{n}{k} = \frac{n!}{k!(n-k)!}$$

It represents the number of distinct ways that k things may be selected from a total population of n things.

sometimes serves to store binary integers and sometimes stores two letters of the alphabet. The only meaningful subregisters in this case are the two five-cell "halves" of C and, of course, the entire register C itself.

The subregister is just a special case of the general class of functions of registers which we shall now discuss. To review briefly, a register is, by definition, an idealized physical device which stores a vector-valued function of time, each component of which can take on one of two values. The observation was also made that because of the one-to-one correspondence between the register and the function it stores, the register itself may be considered to be the function. We therefore arrive at a mathematical definition of an n-cell register as a function whose domain is the time axis and whose range is the n-dimensional vector space V_n, each component of which can take on one of two values. On the basis of this mathematical definition of the register, we can define a *function* or *transformation* of a register as follows: A function or transformation $D = f(C)$ of an n-cell register C is an m-component function, where each component D_i $(i = 1, 2, \ldots, m)$ of D is a two-valued function of, in general, the components C_j $(j = 1, 2, \ldots, n)$ of the register C. A function of a register is therefore an m-dimensional vector function with some of the mathematical properties of a register. A function of a register does not possess the physical properties of the register, for it does not exist as a set of storage cells, but the value of the function at time t is dependent upon the value of a physical register at time t. A register is therefore analogous to an independent variable and a function of a register to a dependent variable. Henceforth, when we desire to emphasize the difference between registers and their functions we shall call the former *independent registers* and the latter *dependent registers.*† We shall continue to use the unqualified word register to describe either a register or a function of a register where by context no ambiguity exists or where the difference between independent and dependent registers is irrelevant.

It is clear now that an m-cell subregister of an n-cell register is just a special case of a dependent register. Here the mapping from V_n into V_m is simply the projection of vectors in V_n onto the appropriate m-dimensional subspace V_m. In this special case, the dependent register does exist as a set of physical cells and may also be considered as an independent register. A less trivial example is that of the dependent register

$$D = (D_1, D_2, \ldots, D_n)$$

---

† A function of a register is physically realized by a set of *gates*. A network consisting only of gates and containing no memory cells is called a *combinational network* (Chap. 3 will describe this in detail). A network which contains both memory cells and gates is called a sequential network; these will be introduced in Chap. 7. Both combinational and sequential networks are often referred to as logical networks.

which is the function of the independent register

$$C = (C_1, C_2, \ldots, C_n)$$

where each component D_i is determined by

$$D_i = \begin{cases} 0 & \text{when } C_i = 1 \\ 1 & \text{when } C_i = 0 \end{cases}$$

As we shall see later, the dependent register D is called the *complement* of C and is a rather commonly used function. Another common dependent register is the n-cell register D defined by

$$D_i = C_{i+1} \qquad i = 1, 2, \ldots, n - 1$$
$$D_n = 0$$

This function is evidently equivalent to shifting each cell of C one place to the left and introducing the constant 0 into cell C_n.

In an analogous fashion we can introduce functions of two registers. Let C and D be independent registers with components given by

$$C = (C_1, C_2, \ldots, C_n)$$
$$D = (D_1, D_2, \ldots, D_r)$$

A function of C and D is an m-cell register $F(C,D)$ in which each component F_i ($i = 1, 2, \ldots, m$) is a two-valued function of all the C_j and D_k. The extension to functions of any finite number of registers follows in a similar manner.

A digital machine is essentially a collection of independent and dependent registers. In different machines the number of independent registers and the number and nature of the dependent registers can vary very widely according to the use of the machine. Regardless of this fact, any machine can be described by considering its registers, dependent and independent, and their interactions; this is the point of view that we shall adopt in the following chapters. But it is first necessary to treat functions of registers in much more detail. This we do in the following three chapters with a discussion of the algebra of two-valued functions.

PROBLEMS

2-1. A static flip-flop A is switched every τ seconds.
(*a*) Sketch the output voltage when the value of the cell alternates between 0 and 1.
(*b*) Sketch the output voltage representing the complement of A.
(*c*) A three-cell register $A_1 A_2 A_3$ assumes the integer values 0, 1, $\ldots$, 7. If the cells are static flip-flops switched every τ seconds, sketch the output voltage waveforms of A_1, A_2, and A_3.
2-2. Define a binary-coded ternary representation for an n-cell register.
2-3. An n-cell register has a binary-coded base r representation. Show that the

number of bits of information stored in the register is $n(\log_2 r)/p$, where p is the smallest integer greater than or equal to $\log_2 r$.

2-4. A computer has a total of ten instructions or operations which it can perform. Define a binary representation for the ten instructions using a minimum number of cells.

2-5. Suppose the computer operations in Prob. 2-4 are labeled $O_1, O_2, \ldots, O_{10}$. Define a binary representation for the operations such that the values for successive operations differ in one and only one digit. What is the minimum number of cells required for this so-called *Gray* code?

2-6. Define a binary representation for the computer operations of Prob. 2-4 such that each binary value contains two and only two 1's. What is the minimum number of cells required?

2-7. Let A designate a five-cell register. Let $W(A)$ be a dependent register storing the number of 1's in the binary value of A. What is the minimum number of cells in $W(A)$? Indicate a method of generating $W(A)$.

2-8. Let $P(A)$ be a dependent register designating whether $W(A)$ in Prob. 2-7 is odd or even. What is the minimum dimension of $P(A)$? Indicate how $P(A)$ might be generated.

2-9. Let A and B be five-cell registers with integer representations. Let $S(A,B)$ designate a dependent register, the integer representation of which is the sum of the integer representations of A and B. What is the minimum dimension of $S(A,B)$? Sketch how $S(A,B)$ might be generated (see Appendix).

3

Boolean Algebra and Logic Design

3-1. Introduction. The algebraic techniques presently used in the design of logical networks for digital machines were originally developed by nineteenth-century mathematicians to describe certain problems in set theory and in propositional logic. George Boole,[1] in 1854, published a work titled, "An Investigation of the Laws of Thought," in which he developed a symbolic notation which could be used in certain problems of formal logic and a calculus for handling the symbolic expressions. Subsequent mathematicians such as Schroder, Peano, Whitehead,[2] Russell,[3] and Huntington,[4] extended this work, developing an area of mathematics known as symbolic logic, which includes the algebra of Boole as an integral part. In 1938, Claude Shannon[5] described the use of the algebra proposed by Boole in designing switching circuits, and since that time Boolean algebra has been a major tool in the design of switching circuits for digital machines.

Boolean-algebra expressions may be used to represent the logical structure of a digital machine. A digital machine, from a circuit viewpoint, consists of a number of diodes, transistors, magnetic cores, etc., which are used to form digital-computer circuits. In designing a machine, it is expedient to consider these circuits according to the *logical* functions they perform rather than their electrical behavior. From this point of view, the basic machine will consist principally of two types of elements: memory or storage elements (i.e., cells and registers), and elements, called *gates*, which physically realize functions of these memory elements (i.e., elements for generating dependent registers).

With Boolean algebra it is possible to describe a set of digital computing circuits by means of a set of algebraic expressions. These mathematical expressions represent the physical circuits in an idealized form, and the behavior of the expressions is mirrored by the operation of the circuits. Further, with algebraic techniques it is not only possible to derive the expressions that describe a set of logical circuits which will physically realize the functions to be performed by a given section of the machine,

22

but it is also possible to simplify the expressions derived, thereby reducing the number of electrical components used and optimizing the design.

The algebra which will be described can be developed as an abstract mathematical entity and will be treated in this manner in Chap. 5.   This chapter will first introduce the basic concepts of the algebra; a set of basic postulates for the algebra will then be presented, followed by a list of useful theorems.   Electronic circuitry which physically realizes the relationships defined will be described in parallel with the development of the algebra, as will block-diagram representations of the logical circuitry.

3-2. Binary Boolean Algebra.   Chapter 2 presented the concepts of the binary storage cell and of n-cell registers constructed of such cells. Section 2-5 also defined functions or transformations of n-cell registers. Let us examine the possible functions of a register containing a single cell x,† letting $y = f(x)$ be a two-valued function of x.   There are four possible two-valued functions of x; Table 3-1 presents, in tabular form, the four functions.   We will designate the values taken by a binary cell with the symbols 0 and 1, as in Chap. 2.

Table 3-1. Functions of a Single Variable

Value of register x	Values for dependent register $y = f(x)$			
	$y = 1$	$y = 0$	$y = x$	$y = x'$
0	1	0	0	1
1	1	0	1	0

The first two of these functions, $f(x) = 1$ and $f(x) = 0$, are trivial; the third function, $f(x) = x$, is the identity function.   The fourth function is defined as the *complement* of x and is written $f(x) = x'$; the prime indicates that the variable x is in complement form.‡   If flip-flops are used as the binary storage cells x_i of a register x, both x_i and x_i' are usually available as outputs from each of the storage cells (refer to Fig. 2-5).   If other storage devices, such as the magnetic cores described in Chap. 2 (refer to Fig. 2-6), are used, it is necessary to form complements by using additional circuit elements.

† In Chaps. 3 to 5 lowercase letters will be used as independent variables in the Boolean expressions, these variables generally representing the state or *output value* of a memory cell.

‡ Sometimes a bar above a variable is used to indicate that a variable is a complemented form.   In this notation $x\bar{y}z$ corresponds to $xy'z$ in the notation of this book. Also, some authors refer to x' as *not x* or as the *negation* of x corresponding to our complement of x.

Figure 3-1 illustrates a *transistor inverter circuit* which inverts or complements a d-c-level signal. In many machines binary values are represented by d-c electrical signals which are, ideally, at one of two given logic levels at a given time. For example, the static flip-flop in Fig. 2-3 has d-c output levels. If the logic levels used are d-c signals, where 0 volts represents a 1 and -3 volts a 0 as in Fig. 3-1, an input to an inverter at

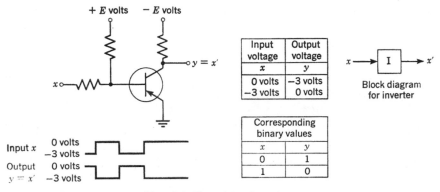

FIG. 3-1. Transistor inverter.

the 0-volt level (which represents a 1) will result in an output voltage of -3 volts (representing a 0), and a -3-volt level at the input will result in a 0-volt level at the output, so we may say that the circuit *complements* the input value. If we represent the input signal by the binary variable x, then the expression for the output function $y = f(x)$ may be written as $y = x'$. It is generally convenient to use block diagrams to represent the circuits used in digital machines instead of reproducing the actual circuit diagram. The block-diagram symbol for the inverter is shown in Fig. 3-1; it consists of a hollow block with an I inside.

Table 3-2. Three Binary Operations

x	y	$x + y$	$x \oplus y$	xy
0	0	0	0	0
0	1	1	1	0
1	0	1	1	0
1	1	1	0	1

Table 3-2 lists the values for three functions of a register containing two binary storage cells, designated x and y. The two storage cells of the register can take any one of four different combinations of values; for each combination the corresponding value for each function of the register is listed.   For instance, if $x = 0$ and $y = 1$, then the function $f(x,y) = xy$ has the value 0 and the function $g(x,y) = x + y$ has the value 1.   The values of the function $f(x,y) = x + y$ listed in the table define a binary operation called *logical addition;*† $x + y$ takes the value 1 if *either* x or y has the value 1.   In the next column the values of the function

$$g(x,y) = x \oplus y$$

define a binary operation called *sum modulo 2 addition;* $x \oplus y$ takes the value 1 only when x or y, but not both, takes the value 1 and is therefore often called the *exclusive-or operation.*

The values of the third function $h(x,y) = x \cdot y$ (where $x \cdot y$ is generally written xy) define a binary operation called *logical multiplication; xy* takes the value 1 only when both x *and* y are equal to 1.

3-3.   AND Gates and OR Gates.   There are a number of types of electronic circuits which physically realize the three binary operations between variables illustrated in Table 3-2.   The most used at this time, and probably the circuits which mirror most straightforward the logical addition and logical multiplication operations of the algebra, are constructed of diodes and resistors.   Figure 3-2 shows a two-input AND gate which performs the logical multiplication operation.   The inputs to the circuit consist of signals from two storage cells designated x and y. Several sets of input waveforms are shown, as are the corresponding output waveforms.   Notice that the circuit will perform the logical multiplication function on d-c-level inputs (in which case a $+5$-volt level represents a 1, a -5-volt level a 0), 5-volt positive pulses (here pulses represent 1's, the absence of pulses 0's), and combined pulse and d-c-level signals (in which case the d-c level must be at the $+5$-volt level, representing a 1, in order for the pulse to pass through the AND gate).   A table of d-c input voltages and corresponding output voltages is also shown, along with a table of the 0's and 1's represented by these signals.   (The values for the voltages given assume ideal diodes with infinite back resistance and zero forward resistance.)   Figure 3-2 also shows the block-diagram symbol for the AND gate, consisting of a hollow block with a dot inside.

The number of inputs to the AND-gate circuit may be increased, although because of the nonzero forward and noninfinite back resistance of actual diodes, there is generally some practical limit to the number of inputs to the same gate.   For each additional input another diode connected to the input is required, so that four diodes are required to construct a four-input AND gate.   If the inputs are designated a, b, c, and d,

---

† Throughout this book the symbol $+$ will be used to designate logical addition. The $+$ symbol will be used for "ordinary" addition of numbers stored in registers. Often the meaning of the symbol will be evident from the context, but when ambiguous interpretation is possible, written notice will be given of the meaning.

and the output z, the input-to-output relation for a four-input AND-gate circuit may be written as $z = abcd$, for the signal at the output of the gate represents a 1 only if all the inputs a, b, c, and d carry signals representing 1's simultaneously.

Many other types of circuitry may also be used in circuits for AND gates, including all-transistor AND gates, magnetic-core AND gates, and combined

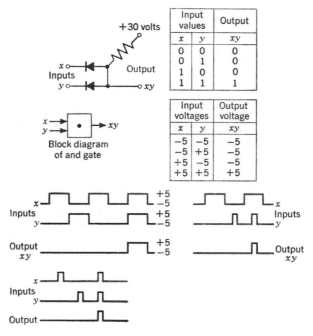

FIG. 3-2. Diode AND gate.

transistor-diode AND gates. From the point of view of logic design, the important point lies in the use of Boolean algebra to represent the AND gate symbolically.

Figure 3-3 shows a diode OR gate which physically realizes the logical addition operation. The two inputs are again designated as x and y and the output as z. If either of the input signals represents a 1, the output signal will represent a 1. A table of input- and output-signal potentials and a table of the corresponding 1's and 0's are also shown. If the circuit is extended to three inputs, another diode connected in the same manner will be required, and a five-input circuit will require five diodes. If the five inputs to an OR gate are designated by the letters a through e and the output as z, then the input-to-output relation may be written

$$a + b + c + d + e = z$$

The block-diagram symbol for the OR gate is also shown in Fig. 3-3; it consists of a hollow block with a plus symbol inside.

It is possible to connect the AND gates and OR gates together and to represent the outputs by means of logical expressions. In general, the circuits formed in this manner are referred to as *combinational logical networks* (a combinational network might consist of only a single AND gate or

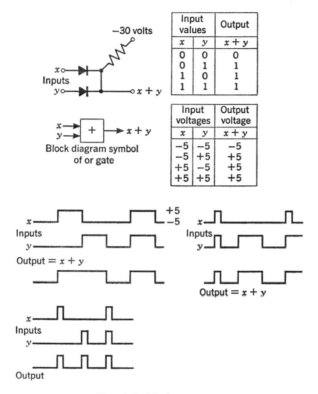

Input values		Output
x	y	$x + y$
0	0	0
0	1	1
1	0	1
1	1	1

Input voltages		Output voltage
x	y	$x + y$
−5	−5	−5
−5	+5	+5
+5	−5	+5
+5	+5	+5

FIG. 3-3. Diode OR gate.

OR gate). As before, combinational networks are defined as networks containing only gates; they contain no memory cells. Figure 3-4 shows the outputs from two AND gates connected as inputs to an OR gate. This particular configuration is referred to as an AND-to-OR network. There are six inputs, represented by variables a, b, c, d, e, and f, and a single output z; the input-output relation may be written $z = abc + def$. The expression $abc + def$ will be designated a *sum-of-products* expression. An examination of the circuit will indicate that the output signal will represent a 1 only when a and b and c, or d and e and f, represent 1's; at all other times the output will represent a 0.

This particular configuration is referred to as a two-level logical circuit because the longest path through which any signal must pass is through two diodes (logical elements) in the circuit illustrated, or through two

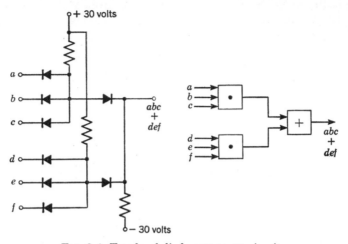

FIG. 3-4. Two-level diode AND-to-OR circuit.

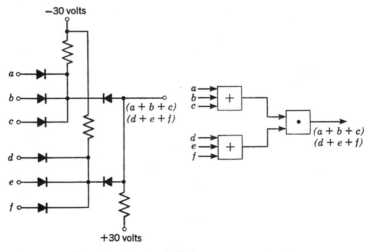

FIG. 3-5. Two-level diode OR-to-AND circuit.

blocks in the block diagram of the circuit.   There is a corresponding circuit consisting of two OR gates, the outputs of which are connected to an AND gate.   This configuration is referred to as an OR-to-AND circuit and is shown in Fig. 3-5.   This is also a two-level circuit; the expression relating the inputs to the output is shown in Fig. 3-5.   An expression such as $(a + b + c)(d + e + f)$ will be referred to as a *product-of-sums* expression.

The third operation described in Sec. 3-2, modulo 2 addition, can be physically realized by sets of AND gates and OR gates if flip-flops with both complemented and uncomplemented outputs are used as the binary storage cells, or by sets of AND gates, OR gates, and inverters if binary storage cells not having complemented outputs are used. Figure 3-6

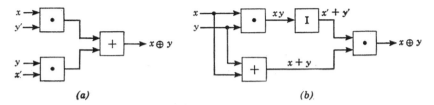

(a)   (b)

FIG. 3-6. Exclusive-OR circuits. (a) For storage cells with complemented outputs; (b) for storage cells with uncomplemented outputs.

shows the block diagrams for two configurations which realize this binary operation between variables. In each case the output signal will represent the value 1 only when either of the inputs, but not both, has signals representing the value 1. Often, the exclusive-or operation is defined by the relation $x \oplus y = xy' + x'y$.

3-4. Basic Rules for the Algebra. The list of relations below defines a binary Boolean algebra in terms of *complementation, logical addition,* and *logical multiplication.* The algebra may be derived using other operations (these will be introduced later in the chapter), but the use of these particular postulates leads more directly to the synthesis of actual digital-machine circuitry.

If $x \neq 1$, then $x = 0$. If $x \neq 0$, then $x = 1$.

$$0' = 1$$
$$1' = 0$$
$$0 + 0 = 0$$
$$0 + 1 = 1$$
$$1 + 0 = 1$$
$$1 + 1 = 1$$
$$0 \cdot 0 = 0$$
$$1 \cdot 0 = 0$$
$$0 \cdot 1 = 0$$
$$1 \cdot 1 = 1$$

There are two steps in the synthesis of a logical network which yields a given two-valued function of a register: first a Boolean-algebra expression is derived which describes the function, and second, calculations are performed with this Boolean-algebra expression in order to optimize in some way the circuit which physically realizes the expression. In most cases the original expression which is derived will contain redundancy, and by manipulating the expression it may be shortened so that when the circuitry which physically realizes the Boolean expression is constructed, a saving in circuit components will be realized. The following is a list of relations which are especially useful in calculating with Boolean expressions.

1. *Special properties of* 0 *and* 1 (the zero and unit rules)

$$0 + x = x \qquad 0 \cdot x = 0$$
$$1 + x = 1 \qquad 1 \cdot x = x$$

2. *The idempotence laws*

$$x + x = x \qquad x \cdot x = x$$

Boolean algebra is idempotent under both multiplication and addition; multiplying a quantity by itself or adding a quantity to itself leaves the quantity unchanged.

3. *Complementarity*

$$x + x' = 1$$
$$x \cdot x' = 0$$

Sometimes these relations are used to define 1 and 0. For instance, if we consider x to represent a class of objects and x' to represent all objects which are not members of the class x, the sum of the x and x' will form a *universal class* (1) and the product $x \cdot x'$ will define an *empty class* (0).

4. *Involution*

$$(x')' = x$$

A value twice complemented is left unchanged. Since x can take only the value 0 or 1, consider $0' = 1$ and $1' = 0$; therefore $(0')' = 0$. Also $1' = 0$ and $0' = 1$; therefore $(1')' = 1$.

5. *Commutativity*

$$x + y = y + x$$
$$xy = yx$$

The order of multiplication does not affect the product; nor does the order of addition affect the sum.

6. *Associativity*

$$x(yz) = (xy)z$$
$$x + (y + z) = (x + y) + z$$

This property was used where the notation for AND gates and OR gates was extended so that a four-input AND gate, for instance, could be represented by a product term containing no brackets. The meaning of the product term $abcd$ or the sum term $a + b + c + d$ is unambiguous.

7. *Distributive laws*

$$x(y + z) = xy + xz$$
$$x + yz = (x + y)(x + z)$$

The second of these may be clarified by noticing that $x \cdot x = x$ and then rewriting the theorem $(x \cdot x) + (y \cdot z) = (x + y)(x + z)$. Notice that a Boolean algebra is distributive for both the multiplication and addition operations.

8. *Laws of absorption*

$$x + xy = x$$
$$x(x + y) = x$$
$$x + x'y = x + y$$

These are very useful relations when expressions are to be simplified. The relation $x + x'y = x + y$ will be used to illustrate the *proof by perfect induction*, which will be explained following the listing of the basic relations. The first two of the absorption laws may be derived from each other, using the distributive laws: $x + xy = (x + x)(x + y) = x(x + y)$.

9. *De Morgan's theorems*

$$(x + y)' = x'y'$$
$$(xy)' = x' + y'$$

These theorems can be used to form the complement of a given expression. For instance, the complement of an expression consisting of a number of variables summed together can be formed by complementing each uncomplemented variable, uncomplementing each complemented variable, and replacing the plus signs with dots:

$$(x + y' + z')' = x' \cdot y \cdot z$$

If an expression consists of a number of product terms logically added together, each product term must be complemented, and the plus signs replaced with dots:

$$(x'y + xy'z' + yz)' = (x'y)' \cdot (xy'z')' \cdot (yz)'$$
$$= (x + y')(x' + y + z)(y' + z')$$

The theorem for converting expressions consisting of several variables added together may likewise be extended to cover sets of product terms added together:

$$(x + y' + z)' = x'yz'$$
$$[z(x + y' + z')(x' + y + z')]' = z' + (x + y' + z')' + (x' + y + z')'$$
$$= z' + x'yz + xy'z$$

De Morgan's theorem points out a basic duality which underlies all

Boolean algebra.   Each theorem has a *dual*, the basic form of which may be found by complementing each side of the equality expressed by the theorem.   As examples, the dual of $x + x = x$ is $x \cdot x = x$, the dual of $x + y = y + x$ is $xy = yx$, and the dual of $x + xy = x$ is

$$x(x + y) = x$$

Notice that a binary Boolean algebra can be derived from complementation and either logical multiplication or logical addition.   Logical addition can be expressed in terms of complementation and multiplication, $x + y = (x'y')'$; and logical multiplication can be formed by complementation and addition, $xy = (x' + y')'$.

The relations listed on the preceding pages may be proved either by deduction on the postulates of the algebra (Chap. 5 will follow this procedure) or by means of the proof by *perfect induction*.   Since the variables used in the algebra can take only one of two possible values at a given time, all possible combinations of values for the variables may be enumerated, and a given expression evaluated for each combination. Consider the rule $a(b + c) = ab + ac$, which indicates that the algebra is distributive over multiplication.   There are 2^n different ways of assigning binary values to n variables.   Since there are three variables a, b, and c in the rule, the rule may be checked by substituting each of the eight possible combinations of values for the variables into the equation and evaluating each side of the equation for each set of values.   If the resulting values correspond, the rule is proved.   A tabular representation of the values for the variables along with the resulting values for the expression being evaluated is called a *table of combinations*.   The following is a table of combinations for the expression $a(b + c) = ab + ac$.

Table of Combinations for $a(b + c) = ab + ac$

Values for variables			Left side of equation		Right side of equation		
a	b	c	$b + c$	$a(b + c)$	ab	ac	$ab + ac$
0	0	0	0	0	0	0	0
0	0	1	1	0	0	0	0
0	1	0	1	0	0	0	0
0	1	1	1	0	0	0	0
1	0	0	0	0	0	0	0
1	0	1	1	1	0	1	1
1	1	0	1	1	1	0	1
1	1	1	1	1	1	1	1

Since the values of both sides of the relation $a(b + c) = ab + ac$ agree for every possible combination of values which can be taken by the variables, the relation holds.

In order further to demonstrate use of the proof by perfect induction, and at the same time to illustrate the construction of other tables of combinations, proofs for the rules $a + a'b = a + b$ and $(a + b)' = a'b'$ are constructed below.

Table of Combinations for $a + a'b = a + b$

Values for variables		Left side of equation			Right side of equation
a	b	a'	$a'b$	$a + a'b$	$a + b$
0	0	1	0	0	0
0	1	1	1	1	1
1	0	0	0	1	1
1	1	0	0	1	1

Table of Combinations for $(a + b)' = a'b'$

Values for variables		Left side of equation		Right side of equation		
a	b	$a + b$	$(a + b)'$	a'	b'	$a'b'$
0	0	0	1	1	1	1
0	1	1	0	1	0	0
1	0	1	0	0	1	0
1	1	1	0	0	0	0

3-5. Algebraic Minimization of Boolean Expressions. When a function of a register is expressed using Boolean algebra, it is possible to perform calculations and derive other equivalent expressions. In general, if the number of occurrences of variables or terms in a given expression describing a function is reduced by these calculations, a corresponding reduction in the number of switching elements in the logical network required to realize the function will be made. The process of shortening an expression describing a specific function is called *minimizing* or *simplifying* the expression. The minimization problem is complicated by the different forms in which a given expression may be written, by the costs of the different types of elements required to construct networks, and by the electrical characteristics of the elements used in the networks, all of which must be taken into consideration.

In order to illustrate the minimization process we will initially limit the networks to diode AND gates and OR gates such as were shown in Figs. 3-2 and 3-3, and we will use flip-flops for the binary storage cells.   Consider the expression $z = abc + abc' + ab'c$.   Figure 3-7a shows a block-diagram drawing of the network for the expression in this form.   The expression $abc + abc' + ab'c$ is in sum-of-products form, and there are nine occurrences of variables and three product terms in the expression (twelve diodes would be required to construct the network).

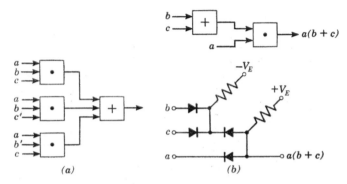

FIG. 3-7. Two logical networks for same function.   (a) Block diagram for $abc + abc' + ab'c$; (b) networks for $a(b + c)$.

According to the distributive law, the expression may be factored, giving $abc + abc' + ab'c = a(bc + bc' + b'c)$.   The product terms in parenthesis may be further factored to give $a[b(c + c') + b'c]$, and since $(c + c') = 1$, the expression may be shortened to $a(b + b'c)$, and finally $(b + b'c)$ may be shortened to $(b + c)$, using the absorption rule.   A final minimized expression is therefore $a(b + c)$, which is equivalent to the original expression $abc + abc' + ab'c$ but requires only four diodes if constructed as in Fig. 3-7b.

To demonstrate further the algebraic minimization of Boolean expressions, let us consider the expression $abcd + abcd' + a'bcd' + a'bc'd' + a'b'cd' + a'b'c'd'$.   This expression contains considerable redundancy, and would require 30 diodes to construct in its present form.   The steps which may be taken to minimize the expression are listed below:

$$abcd + abcd' + a'bcd' + a'bc'd' + a'b'cd' + a'b'c'd'$$
$$= abc(d + d') + a'bd'(c + c') + a'b'd'(c + c')$$
$$= abc + a'bd' + a'b'd'$$
$$= abc + a'd'(b + b')$$
$$= abc + a'd'$$

The expression $abc + a'd'$ contains five occurrences of variables and two product terms (seven diodes are required in the corresponding diode network).

The calculations performed above are straightforward, but as problems become larger, arriving at a minimal expression becomes increasingly difficult. As a result, it is desirable to use a well-defined procedure in minimizing larger expressions, if possible; descriptions of several systematic techniques will be presented in Chap. 4.

3-6. Canonical Expansions and Multilevel Networks. Before defining the term "canonical expansion," let us briefly review our terminology. Two types of expressions have been used in the preceding examples: sum-of-products expressions and product-of-sums expressions; these expressions are realized by two-level AND-to-OR and OR-to-AND networks, respectively. A product term consists of one or more variables multiplied together in a common term, and a sum term of one or more variables added together in a common term. A sum-of-products expression is defined as an expression consisting of two or more product terms logically added together, for instance, $xy + xz$ or $a + bc$. A product-of-sums expression consists of two or more sum terms multiplied together, for instance, $(a + b + c)(a + b' + c')$ or $(x + y')(x' + y)$.

A sum-of-products canonical expansion in n different variables is an expression consisting of a set of product terms logically added together in which each product term of the expression contains each of the n variables in the expression, with each variable either complemented or uncomplemented, and with no variable repeated in a given term, and no product term repeated. For instance, the sum-of-products expression in three variables $xyz' + x'y'z + xy'z$ and the expression in two variables $ab + ab'$ are both in canonical expansion form, but $x + yz$ is not. Some authors refer to canonical expansion expressions as *developed* or *expanded normal expressions*. The dual type of expression is defined as the product-of-sums canonical expansion, and is a set of sum terms logically multiplied together, where each sum term contains all the n variables in the expression, and where no term is repeated. For instance, $(a + b' + c)(a + b' + c')$ and $(x + y)(x + y')(x' + y')$ are product-of-sums canonical expansions.

Any expression can be converted into canonical form. For instance, $x + yz$ can be expanded, using the theorem $xy + xy' = x$, as follows:

$$x + yz = (xy + xy') + (xyz + x'yz)$$
$$= (xyz + xyz' + xy'z + xy'z') + xyz + x'yz$$
$$= xyz + xyz' + xy'z + xy'z' + x'yz$$

It is possible to represent any Boolean function in n variables as a sum-of-products canonical expansion expression with $\leq 2^n$ product terms and as a product-of-sums canonical expansion with $\leq 2^n$ sum terms. A systematic technique for deriving both types of canonical expressions, given a list of values for the function, will be described in Sec. 3-10.

It is obvious that other types of expressions can be used in the synthesis of logical networks. For instance, the expression $xw + xyz + xy'z'$ is a minimal expression in sum-of-products form but can be factored to yield $x(w + yz + y'z')$. The expression $xw + xyz + xy'z'$ can be constructed as a two-level AND-to-OR network and will require eleven logical elements

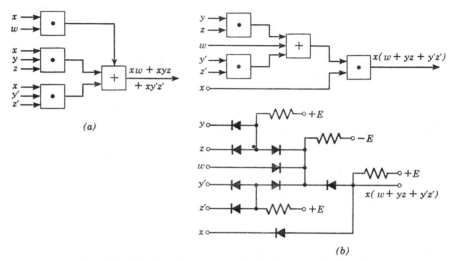

FIG. 3-8. (a) Two-level network; (b) three-level network.

(diodes, for instance) for construction while the expression $x(w + yz + y'z')$ can be constructed as a *three-level* AND-OR-AND network which will require only nine elements (diodes). Block diagrams for both circuits are shown in Fig. 3-8, as is a schematic of the three-level circuit. Although the three-level network requires fewer logical elements (in this case diodes), other considerations may lead to the use of the two-level network. For instance, from a circuit viewpoint the three-level network will require, in general, more input power than the two-level network, if resistances of comparable value are used. Further, the three-level network will be slower in responding to changes at the inputs than the two-level network, because each level through which the signal must pass will add an additional delay to the response time for the network. The choice of different networks, and in turn the Boolean expressions corresponding to these networks, is therefore complicated by physical considerations as well as economy in the number of elements used.

3-7. Functions of Two Variables. There are, in all, 16 functions of 2 variables. Several of these have been described and calculations performed using them.   Table 3-3 lists all 16 functions, in table-of-combinations form, designating each function from f_0 through f_{15}.   Functions f_0 and f_{15} are the 0 and 1 functions, respectively.   Function f_8 has been defined as logical multiplication, f_{14} as logical addition, f_3 as the complement of x, and f_5 as the complement of y.   The following is a list of all

Table 3-3. Functions of Two Variables

x	y	f_0	f_1	f_2	f_3	f_4	f_5	f_6	f_7	f_8	f_9	f_{10}	f_{11}	f_{12}	f_{13}	f_{14}	f_{15}
0	0	0	1	0	1	0	1	0	1	0	1	0	1	0	1	0	1
0	1	0	0	1	1	0	0	1	1	0	0	1	1	0	0	1	1
1	0	0	0	0	0	1	1	1	1	0	0	0	0	1	1	1	1
1	1	0	0	0	0	0	0	0	0	1	1	1	1	1	1	1	1

16 functions, expressing each in terms of logical addition, multiplication, and complementation:

$$
\begin{aligned}
&f_0 = 0 &&f_4 = xy' &&f_8 = xy &&f_{12} = x \\
&f_1 = x'y' &&f_5 = y' &&f_9 = x'y' + xy &&f_{13} = x + y' \\
&f_2 = x'y &&f_6 = xy' + x'y &&f_{10} = y &&f_{14} = x + y \\
&f_3 = x' &&f_7 = x' + y' &&f_{11} = x' + y &&f_{15} = 1
\end{aligned}
$$

While $f_3, f_5, f_6, f_8, f_{10}, f_{12}$, and f_{14} are probably the most used functions, electronic circuitry exists which physically realizes several of the other functions.   Also, several of the other functions are often used in mathematical papers.   Functions f_1, f_2, f_4, f_7, f_{11}, and f_{13} will therefore be described and their properties noted.

Function f_1 is generally referred to as the Peirce function, in honor of C. S. Peirce; a vertical arrow ($\downarrow$) is used to represent the corresponding operation between variables; thus $x \downarrow y = x'y'$.   The Peirce operation is called a *universal operation*, because it may be used to generate the entire algebra.   Showing that addition and complementation can be derived from the Peirce operation will suffice, as we have shown all other operations can be derived from these.

$$x \downarrow x = x'$$
$$(x \downarrow y) \downarrow (x \downarrow y) = x + y$$

Function f_7 is also a *universal operation* and is often referred to as the Sheffer stroke operation.   This operation between variables is designated by a stroke symbol (/), so that $x/y = x' + y'$.   The universality of the operation may be proved by showing that both complementation and multiplication can be derived from the function as follows:

$$x/x = x'$$
$$(x/y)/(x/y) = xy$$

The Sheffer stroke function, like the Peirce function, is commutative but not associative.

Function f_2 ($x'y$ in terms of complementation and multiplication) is of interest in that an electronic circuit referred to as an *inhibit gate* physically realizes this function. Figure 3-9 shows the block diagram for the inhibit gate; the circled input is the inhibit (or *extinguishing*) input and if a pulse (or a d-c voltage representing a 1) is connected to this input, a pulse or d-c signal connected at the other input will be inhibited. The circuit is fairly common and is often extended to a number of inputs, only one of

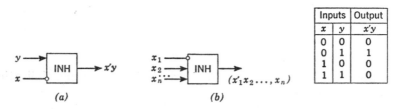

Inputs		Output
x	y	$x'y$
0	0	0
0	1	1
1	0	0
1	1	0

(a)   (b)

FIG. 3-9. Inhibit gates.

which is an inhibit input. Figure 3-9b shows this case. Notice that a multi-input inhibit gate is simply an AND gate with one input signal complemented (the inhibit input).

Function f_4 is essentially the same as f_2 except that a different variable is complemented ($f_2 = x'y$ and $f_4 = xy'$).

Function f_{11}, $x' + y$, is referred to by logicians as the *truth-function conditional*. This book will use the $\Rightarrow$ symbol to indicate the conditional, defining $(x' + y) = (x \Rightarrow y)$. In formal logic the statement $x \Rightarrow y$ is sometimes read *if x then y*. The statement x is referred to as the *antecedent*, and y as the *consequent*.

If $x \Rightarrow y$ and $y \Rightarrow x$ hold, that is, if x then y and if y then x; then $x = y$ holds. This may be seen by stating the relationship symbolically: $[(x \Rightarrow y)(y \Rightarrow x)] \Rightarrow (x = y)$. An expression such as this, which is true for all possible values of the variables, is referred to in logic as a tautology. Simple examples of tautologies are: $x = x$, $x \Rightarrow x$, $[(x \Rightarrow y)(y \Rightarrow z)] \Rightarrow (x \Rightarrow z)$.

Circuits which physically realize the truth-function conditional are not often used; these consist of an OR gate with one input inverted. Function f_{13} is essentially the same relation, but with variables transposed.

3-8. NOT OR Gates and NOT AND Gates. Figure 3-10 shows both circuit diagrams and block diagrams for NOT OR gates and NOT AND gates. These circuits are often used, as they provide gain and are economical, and fairly elaborate combinations of gates are possible. A NOT AND gate has the characteristic that the output signal will represent a 0 only when all the input signals represent 1's. For a NOT AND gate with inputs x, y, and z and an output a, an expression relating the input values to the

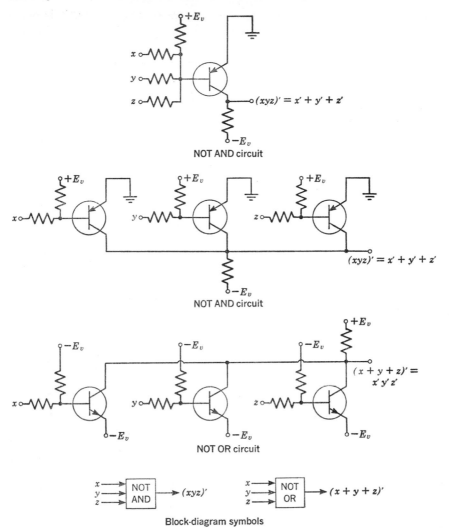

Block-diagram symbols

FIG. 3-10. NOT OR gate and NOT AND gate.

output values is $a = f(x,y,z) = (xyz)'$, so the circuit performance is analogous to that of an AND gate with its output complemented. Since $(xyz)' = x' + y' + z'$, the NOT AND gate also behaves like an OR gate with all the inputs complemented. Similarly, the NOT OR gate has an output representing a 1 only when all of the inputs represent 0's, so the Boolean function for the gate is $f(x,y,z) = (x + y + z)' = x'y'z'$. The NOT OR gate resembles either an OR gate with its output complemented or an AND gate with each input signal complemented.

Since NOT AND gates and NOT OR gates are useful circuits, it is often desirable to synthesize logical networks using them. Consider the expression $xyz + xy'z + x'yz'$. If the AND gates which would normally be used to form the product terms are to be replaced by NOT OR gates, we have only to notice that $(x' + y' + z')' = xyz$, so that by complementing each variable which is uncomplemented in the original expression and uncomplementing complemented variables, the original sum-of-products expression can be changed to the expression for using NOT OR gates. In some cases this is beneficial, since if the NOT OR circuits shown in Fig. 3-10 are used, it is possible to add the output together logically from a set of gates by simply connecting together the leads from the output of each NOT OR gate. In this case there will be no logical elements at the second level of logic and the circuit will be somewhat faster, a delay having been eliminated.

A few observations concerning binary operations and functions should now be made. Logical addition and logical multiplication, as we have defined them, are both *binary operations*. In general, a *binary operation* "o" on a set of elements a, b, c, . . . consists of a rule which assigns a uniquely defined third element $c = a \circ b$ to each pair of elements of the set. (A unary, or singulary, operation assigns an element b to each element a of the set; the singulary operation in previous sections was defined as complementation and indicated by a prime symbol.) The basic binary operations so far defined, logical multiplication and logical addition, are both associative: $a(bc) = (ab)c$ and $(a + b) + c = a + (b + c)$. It might appear that the NOT OR and NOT AND gates could also be mirrored in a symbolic system using binary operations and corresponding symbology. This is not the case, however. The binary operation which corresponds to a NOT OR gate with two inputs is the Peirce operation, which we have assigned a vertical arrow ($\downarrow$) and defined as follows: $x \downarrow y = x'y'$. Now the Peirce operation is not associative, for $(x \downarrow y) \downarrow z \neq x \downarrow (y \downarrow z)$ (consider, for example, $x = 0$, $y = 1$, and $z = 1$), and an n-input NOT OR gate cannot be directly represented, using Peirce operations, while maintaining the isomorphic relation between expressions and circuitry. The operation of a NOT OR gate can be mirrored only by an n-ary operation between the n input variables.

The same reasoning applies to the NOT AND gate, the operation of a two-input gate being a physical realization of the Sheffer stroke operation, $x/y = x' + y'$, which is also not associative $[x/(y/z) \neq (x/y)/z]$.

A reasonable notation can be formed by extending the notation proposed by Lukasiewicz.[6] In order to describe a NOT OR gate with three inputs, we need a ternary operation beween variables; a NOT OR gate with four inputs requires a quaternary operation between variables; and in general an n-input NOT OR gate requires an n-ary operation between variables.   Therefore, define a D operation on n variables as

$$D(x_1, x_2, \ldots, x_n) = (x_1 + x_2 + \cdots + x_n)'$$

(this corresponds to the NOT OR gate), and a T operation on n variables as

$$T(x_1, x_2, \ldots, x_n) = (x_1 x_2 \cdots x_n)'$$

corresponding to the NOT AND gate.

There are eight nondegenerate† two-level logical-network configurations using AND, OR, NOT AND, and NOT OR gates.   These are:

> AND-to-OR
>
> OR-to-AND
>
> NOT AND–to–NOT AND
>
> OR–to–NOT AND
>
> NOT OR–to–NOT OR
>
> AND–to–NOT OR
>
> NOT AND–to–AND
>
> NOT OR–to–OR

The rules for transforming product-of-sums and sum-of-products expressions into expressions for NOT AND and NOT OR gates are straightforward and are listed below.   As examples, the equivalent expressions $x'y + xy'z'$ and $(x' + y')(x + y)(y + z')$ will be used.

AND-to-OR: $x'y + xy'z'$

OR-to-AND: $(x' + y')(x + y)(y + z')$

NOT AND–to–NOT AND: The variables are used as in the sum-of-products expression.   Thus,

$$T[T(x',y), T(x,y'z')] = T[(x + y'), (x' + y + z)] = x'y + xy'z'$$

OR–to–NOT AND: The complements of variables in the sum-of-products expression are used to form the sum terms.   Thus,

$$T[(x + y'), (x' + y + z)] = [(x + y')(x' + y + z)]' = x'y + xy'z'$$

NOT OR–to–NOT OR: The variables are used as in the product-of-sums expression.

$$D[D(x',y'), D(x,y), D(y,z')] = D(xy, x'y', y'z) = (x' + y')(x + y)(y + z')$$

Notice that this requires fewer gates than NOT AND–to–NOT AND.

AND–to–NOT OR: Each variable in the product-of-sums expression is complemented and used to form the expression.

$$D[xy, x'y', y'z] = (xy + x'y' + y'z)' = (x' + y')(x + y)(y + z')$$

---

† A nondegenerate form is one in which any given function of the 2^{2^n} possible functions of n variables may be written.

NOT AND–to–AND: Each variable in the product-of-sums expression is complemented and used.

$$[T(x,y)][T(x',y')][T(y',z)] = (x' + y')(x + y)(y + z')$$

NOT OR–to–OR: Each variable in the sum-of-products expression is complemented and used in the expression for this type of network. $[D(x,y')] + [D(x',y,z)] = x'y + xy'z'$. See Fig. 3-11.

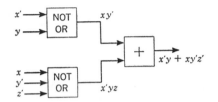

FIG. 3-11. Two-level NOT OR–to–OR network.

Expressions may therefore be derived for a given function in sum-of-products or product-of-sums form, the expressions may be simplified while maintaining the logical multiplication and logical addition operations, and then the expressions may be transformed into those for NOT AND and NOT OR gates by using the above rules.

3-9. The Combinational Network. We have examined all the functions of registers consisting of either one or two storage cells, enumerating each of the functions and showing logical elements constructed of electronic devices which correspond to these functions. It is now possible to

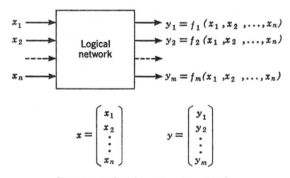

FIG. 3-12. Combinational network.

examine the characteristics of logical networks independent of the devices of which the networks are constructed. Figure 3-12 illustrates a generalized logical network with n input signals designated $x_1, x_2, \ldots, x_n$ and m output signals designated $y_1, y_2, \ldots, y_m$ (n does not necessarily equal m). The input signals are assumed to be from the binary storage cells comprising a register, and the output signals are functions of the input signals. The logical network in Fig. 3-12 is assumed to contain only gating elements and not storage or delay elements, so that the output signals at any given time will be completely determined by the input signals at that time.

If a logical network contains storage or delay elements, the outputs from the network at a time τ will, in general, be determined not only by the inputs at time τ, but also by the previous sequence of inputs to the network (the history of the network). A network of this type is called a *sequential* network. This chapter and Chap. 4 will deal exclusively with the design of combinational networks, and Chaps. 7 and 11 will describe sequential networks.

The set of input values to the combinational network in Fig. 3-12 may be considered to represent an n-dimensional input vector $x = (x_1, x_2, \ldots, x_n)$ and the outputs an m-dimensional vector $y = (y_1, y_2, \ldots, y_m)$, where $y_1 = f_1(x_1, x_2, \ldots, x_n), y_2 = f_2(x_1, x_2, \ldots, x_n), \ldots, y_m = f_m(x_1, x_2, \ldots, x_n)$. Since each component may assume the value 0 or 1, the input vector x can assume any of 2^n distinct values; for each value assumed by x there will be a unique value of y, which will be determined by the transformation performed by the combinational network. The particular transformation performed by the network will be determined by the construction of the network; moreover, a network may be constructed which will perform any desired transformation. The combinational network therefore yields a vector-valued function and corresponds to the dependent register of Chap. 2.

3-10. Derivation of Transmission Functions. The first step in the procedure for designing a combinational network consists in deriving an expression for each output of the desired network in terms of the inputs. In order to synthesize a network of the type illustrated in Fig. 3-12, an expression is derived relating the output y_1 to the set of n input variables; then an expression is derived for y_2, and so on for each of the m output variables. The complete set of m expressions then completely describes the desired network.

An expression describing the input-output relationship between an output line y_i and the inputs $x_1, x_2, \ldots, x_n$ is called the *transmission-function* expression for output line y_i. Transmission functions are generally derived in canonical form; if a sum-of-products expression is derived, the transmission-function expression consists of a set of product terms logically added together, each of which is a product of the n input variables, certain of which may be complemented.

Let us consider the problem of constructing a combinational network with a single output line and inputs from a three-storage-cell register, using a binary integer representation such as that explained in Sec. 2-4 (001 = decimal 1, 011 = decimal 3, etc.). Let the variable x_1 represent the most significant binary digit of the number stored, and x_2 and x_3 the second least and least significant binary digits of the register. The output line from the network is to have the value 1 only when the number stored in the input register is a prime number. Since the register can represent values from 0 to 7 (decimal), the output from the logical network is to be 1 when the register represents a 1, 2, 3, 5, or 7, and a 0 when the register represents a 0, 4, or 6. Table 3-4 is a table of combinations listing the values taken by x_1, x_2, and x_3 and the desired values for the output

Table 3-4. Derivation of Transmission Functions

Decimal integer representation	Inputs x_1 x_2 x_3	Output y	Product terms	Sum terms
0	0 0 0	0	$x_1'x_2'x_3'$	$x_1 + x_2 + x_3$
1	0 0 1	1	$x_1'x_2'x_3$	$x_1 + x_2 + x_3'$
2	0 1 0	1	$x_1'x_2x_3'$	$x_1 + x_2' + x_3$
3	0 1 1	1	$x_1'x_2x_3$	$x_1 + x_2' + x_3'$
4	1 0 0	0	$x_1x_2'x_3'$	$x_1' + x_2 + x_3$
5	1 0 1	1	$x_1x_2'x_3$	$x_1' + x_2 + x_3'$
6	1 1 0	0	$x_1x_2x_3'$	$x_1' + x_2' + x_3$
7	1 1 1	1	$x_1x_2x_3$	$x_1' + x_2' + x_3'$

$y = f(x_1,x_2,x_3)$. Two additional columns are added; one contains a list of all possible product terms and the other a list of all sum terms. The first column contains product terms, the variables of which are complemented or not complemented depending on whether the respective values for the variables in the same row are 0 or 1. If the inputs to the logical network take the values in a given row, the product term in the same row takes the value 1. The column containing the sum terms is formed by complementing the product terms in each row. For instance, in the row containing the product term $x_1x_2'x_3$, the sum term is $x_1' + x_2 + x_3'$.

Let us first derive a transmission-function expression in sum-of-products form. By logically adding together the product terms for the rows for which the desired output is to be a 1, we shall form the expression for the transmission function. The transmission function for the output $y = f(x_1,x_2,x_3)$ in Table 3-4 is

$$y = x_1'x_2'x_3 + x_1'x_2x_3' + x_1'x_2x_3 + x_1x_2'x_3 + x_1x_2x_3$$

This may be seen to be a simple listing of the conditions for which y is to take the value 1. For instance, if $x_1 = 0$, $x_2 = 0$, and $x_3 = 1$, the first product term in the expression will take the value 1, giving y the desired value 1.

The general rule is that the canonical expansion for the transmission function for an output $y_i = f(x_1, x_2, \ldots, x_n)$ is formed by logically summing each product term in n input variables for which y is to be equal to 1. The general expression for the canonical expansion for the transmission function is therefore

$$y_i = f_i(x_1, x_2, \ldots, x_n) = \sum_{j=0}^{2^n-1} y_{i,j}\beta_j \qquad i = 1, 2, \ldots, m$$

where n is equal to the number of input variables to the network and m is the number of output lines. Each β_j represents a product term formed of a combination of complemented or uncomplemented input variables, with no variable repeated in a term, and with each variable appearing in each term. The set of β_j† is the set of all such terms. Each $y_{i,}$ is either 0 or 1, depending on whether the output line y_i is to represent a 0 or a 1 for the associated input term β_j.

The expression $y = x_1'x_2'x_3 + x_1'x_2x_3' + x_1'x_2x_3 + x_1x_2'x_3 + x_1x_2x_3$, which was derived in the previous example, may be simplified to the expression $y = x_1'x_2 + x_3$. This will require four logical elements if the input register is composed of storage cells having both complemented and uncomplemented outputs, and five logical elements if the register has only single-output line cells.

A product-of-sums expression for the transmission function for Table 3-4 can be derived by multiplying together the sum terms from each row for which the output y has the value 0. The logic of this is as follows: Suppose all the product terms are first removed from the rows in which the output is to have the value 0. Summing the terms yields an equation $y' = (x_1'x_2'x_3' + x_1x_2'x_3' + x_1x_2x_3')'$, both sides of which contain complements of the desired expression. Complementing both sides of the expression gives $y = (x_1 + x_2 + x_3)(x_1' + x_2 + x_3)(x_1' + x_2' + x_3)$, the desired expression for the transmission function in product-of-sums form.

The right side of this expression can be formed directly by multiplying together the sum terms in Table 3-4 from each row in which the output y has the value 0.

† It is a common practice to give the β_j term a decimal number subscript which is equal to a binary number formed by giving each uncomplemented variable in the product term the value 1 and each complemented variable the value 0, and letting the resulting set of values represent a binary number. For instance, $x_1x_2'x_3$ is assigned the binary value 101 or decimal 5, so this particular set of variables would be written β_5. The transmission function derived from Table 3-4 would therefore be written $y = \beta_1 + \beta_2 + \beta_3 + \beta_5 + \beta_7$. Sometimes this is further shortened to $y = \Sigma(1,2,3,5,7)$.

The general formula for deriving the product-of-sums expression for a given transmission function can be derived from the expression

$$y'_i = \sum_{j=0}^{2^n-1} y'_{i_j}\beta_j$$

where the y_{i_j} are the desired values for each β_j (i.e., each product term in n variables). Complementing both sides of the equation gives

$$y_i = \prod_{j=0}^{2^n-1} (y_{i_j} + \beta'_j)$$

If the values from Table 3-4 are substituted into this expression,

$$y = [0 + (x'_1x'_2x'_3)'][1 + (x'_1x'_2x_3)'][1 + (x'_1x_2x'_3)'][1 + (x'_1x_2x_3)']$$
$$[0 + (x_1x'_2x'_3)'][1 + (x_1x'_2x_3)'][0 + (x_1x_2x'_3)'][1 + (x_1x_2x_3)']$$

De Morgan's theorem gives

$$y = (0 + x_1 + x_2 + x_3)(1 + x_1 + x_2 + x'_3)$$
$$(1 + x_1 + x'_2 + x_3)(1 + x_1 + x'_2 + x'_3)(0 + x'_1 + x_2 + x_3)$$
$$(1 + x'_1 + x_2 + x'_3)(0 + x'_1 + x'_2 + x_3)(1 + x'_1 + x'_2 + x'_3)$$

This expression can be shortened by the theorems $1 \cdot \beta = \beta$, $1 + \beta = 1$, and $0 + \beta = \beta$; the sum terms containing 1's will always have the value 1; the 0's may be dropped from the other sum terms. Thus

$$y = (x_1 + x_2 + x_3)(x'_1 + x_2 + x_3)(x'_1 + x'_2 + x_3)$$

This expression can be simplified to $(x_2 + x_3)(x'_1 + x_3)$, requiring six logical elements, or to $x'_1x_2 + x_3$, the same expression derived from the transmission function in sum-of-products form, after simplification.

PROBLEMS

3-1. Simplify the following expressions, if possible, and then draw block diagrams of the resulting expressions, using OR gates and AND gates, first assuming that both complemented and uncomplemented input variables are used and then assuming that complements must be formed using inverters.

(a) $abc + ab'c + abd$ (b) $a(b + c) + ab(c' + d'e)$
(c) $abc + abc'd + ab'c'de$ (d) $abc(c'de + cde + cd'e')$

3-2. Convert the following expressions to sum-of-products form, simplifying where possible:

(a) $(a + b)(a + c)(a + c' + d')$
(b) $(a + b' + c')(a' + b + c)(a + d)$
(c) $(a' + b' + c' + d')(a' + b' + c' + d)(a + b + c' + d')(a + b + c)$

3-3. Expand the expressions in Prob. 3-2 to sum-of-products and product-of-sums canonical form.

3-4. Convert the following expressions to product-of-sums form, simplifying where possible:

(a) $abcd + a'b'c'd + a'b'c'd'$ (b) $abc'd'e + ab'c'd + ac'd'e + a'b'c$

(c) $ab + ac + ad + ae$

3-5. Using De Morgan's theorems, form the complements of the following expressions as indicated; then simplify the resulting expressions:

(a) $(ab + cd + e + f)' = y'$
(b) $[a'b'c + cd + bc(a + d')]' = y'$
(c) $[ab(cd + c'd') + ab(c' + de')]' = y'$
(d) $[(a' + bc + d)(a + b'c' + d)(ab + cd)]' = y'$

3-6. Convert the following expressions to sum-of-products form and simplify:

(a) $ab \oplus a'b'$ (b) $abcd \oplus ab'c'd \oplus a'b'c'd'$

(c) $bcd \oplus a'b'c'(a'b' \oplus c'd'e')$

3-7. Form a table of combinations and then derive sum-of-products and product-of-sums expressions for the following:

(a) A combinational network is connected to a four-cell flip-flop register x with cells x_0, x_1, x_2, and x_3. The network is to have a 1 output when the register value

$$\delta_I(x) = \sum_{i=1}^{n} x_i 2^{n-i} \text{ is an odd number.}$$

(b) A combinational network with four inputs and five outputs is connected to a

four-cell register x with value $\delta_I(x) = \sum_{i=1}^{n} x_i 2^{n-i}$. The five outputs y_0, y_1, y_2, y_3, and

y_4 are to represent the input value in the binary-coded decimal system as described in Sec. 2-4. Derive the five Boolean expressions describing the five output lines in terms of the input variables.

3-8. (a) Prove the following theorems:

$x/(y/y') = x'$
$[x/(y/y')]/[y/(x/x')] = x + y$
$[x\downarrow(y\downarrow y')]\downarrow[y\downarrow(x\downarrow x')] = xy$
$D[D(a,b,c)] = T(a',b',c')$
$T[D(a,b,c)] = D[D(a,b,c)]$

(b) Convert the sum-of-products and product-of-sums expressions for Fig. 3-8 into each of the eight nondegenerate forms described in Sec. 3-8.

3-9. Write an expression for the function $f = abc + a'c'$ using the operation $'$ and

(a) The binary operations $\downarrow$ and $\cdot$ (b) The binary operations $/$ and $+$
(c) The binary operation $/$ (d) The binary operation $\downarrow$
(e) The T operator   (f) The D operator

REFERENCES

1. Boole, George: "An Investigation of the Laws of Thought," London, 1854.
2. Whitehead, A. N.: On Mathematical Concepts of the Material World, *Phil. Trans. Roy. Soc. London*, series A, vol. 205, pp. 465–525, 1906.
3. Russell, Bertrand: "The Principles of Mathematics," Cambridge, England, 1903.

 4. Huntington, E. V.: Sets of Independent Postulates for the Algebra of Logic, *Trans. Am. Math. Soc.*, vol. 5, pp. 288–309, 1904.
 5. Shannon, C. E.: A Symbolic Analysis of Relay and Switching Circuits, *Trans. AIEE*, vol. 57, pp. 713–723, 1938.
 6. Lukasiewicz, Jan: O logice trojwartosciowej, *Ruch Fil.*, vol. 5, pp. 169–171, 1920.
 7. Quine, Willard V.: "Mathematical Logic," Harvard University Press, Cambridge, Mass., 1950.
 8. Bartee, T. C.: "Digital Computer Fundamentals," McGraw-Hill Book Company, Inc., New York, 1960.
 9. Caldwell, S. H.: "Switching Circuits and Logical Design," John Wiley & Sons, Inc., New York, 1958.
10. Halmos, P. R.: The Basic Concepts of Algebraic Logic, *Am. Math. Monthly*, vol. 63, pp. 363–387, June-July, 1956.
11. Reed, I. S.: "Some Mathematical Remarks on the Boolean Machine," MIT Lincoln Laboratory, Lexington, Mass., Tech. Rept. 2, Dec. 19, 1951.
12. Reed, I. S.: "Symbolic Design Techniques Applied to a Generalized Computer," MIT Lincoln Laboratory, Lexington, Mass., Tech. Rept. 141, Jan. 3, 1957.
13. Burks, A. W., and J. B. Wright: Theory of Logical Nets, *Proc. IRE*, vol. 41, pp. 1357–1365, October, 1953.
14. Whitehead, A. N., and B. Russell: "Principia Mathematica," Cambridge, England, vol. 1, 1910; vol. 2, 1912; vol. 3, 1913.
15. Huntington, E. V.: New Sets of Independent Postulates for the Algebra of Logic, *Trans. Am. Math. Soc.*, vol. 35, pp. 274–304, 1933.

4

Systematic Minimization Techniques

4-1. General Considerations. The design of a combinational logical network generally involves two steps. First, the desired relations between the input and output signals are stipulated, from which the Boolean expressions which describe the network are derived. These expressions generally contain considerable redundancy; so the second step consists of simplifying the expressions derived in the first step, thereby reducing the number of electrical components necessary to construct the network. Techniques for deriving a canonical expansion expression given a set of input-output relations have been described. Chapter 3 described the procedure for deriving expressions for combinational networks in both product-of-sums and sum-of-products forms, and Chaps. 7 and 11 will describe several sequential circuit-design procedures. The design procedures for both types of networks involve the same problem: the simplification of the expressions describing a given combinational network.

Consider the design procedure for a combinational network with a single output line. After the table of combinations listing the input-output relations has been formed, an expression may be written which describes the desired network symbolically. Calculations may now be performed using this expression, yielding a number of equivalent expressions. For each expression there will be an actual network which physically realizes it. While each of the physical networks will realize the required input-output relations, some networks will require fewer electronic components than others. We may therefore choose between mathematical expressions, on the basis of the total cost of the physical network, by assigning costs to each type of logical element which is used and then calculating the cost of each network and choosing the lowest cost. There may be other requirements on the network, however. For instance, there may be a restriction on the delay from the time the inputs to the network change values to the time the output signal from the network changes. If the maximum delay time is sufficiently short, the use of a network with more than two levels may be precluded, forcing the transmission-function expressions to be in a form which describes a two-level network. There may be other restrictions on the configurations used. In order to maintain digital machines efficiently, networks having very complicated sets of interconnections are often avoided. Also, many machines are of modular construction, so that only a few basic logical elements are designed, facilitating design and maintenance. In a machine of this type only certain logical configurations may be allowed so that all expressions for combinational networks will be in one of several basic forms.

Because of these considerations, a combinational network which will yield a given function of a register is generally synthesized on the basis of an initial assumption as to the general form of the final network. Since the general form of the final network is predetermined, the form for the Boolean expression which describes the network is also predetermined.

The first sections of this chapter will describe systematic techniques for the minimization (simplification) of a given sum-of-products and product-of-sums expression. This will be followed by a description of the procedure for handling "don't care" input and output conditions. Techniques for the synthesis of multiple-output networks will then be presented, with emphasis on minimizing sets of expressions simultaneously so that logical elements in the corresponding networks may be shared.

The techniques described in this chapter will be principally algebraic with logical overtones. Various other approaches have been taken, and in the references are listed a number of papers containing work which is directed toward algebraic topology, matrix methods, and several other areas of mathematics. Algebraic techniques are described in this chapter because the fundamental work in this area has been described in the terminology of algebra and logic and because these languages provide a basis for the work in other areas.

The references at the end of this chapter are arranged as follows: Refs. 1 to 11 refer to papers describing algebraic techniques for minimization of single expressions, Refs. 12 to 15 to papers describing the simultaneous minimization of sets of expressions, or the multiple-output line problem. References 16 to 18 refer to papers describing minimization from a geometric or topological viewpoint, and Refs. 19 to 21 describe chart methods. Reference 22 contains the best treatment of the multilevel-network problem to date.

4-2. Definitions and Criteria for Minimality. Certain definitions logically precede a description of the formal design procedures for two-level combinational networks. Accordingly the terms *literal, subsume,* and *imply* are defined below.

A *literal* is defined as either a complemented variable or an uncomplemented variable. For instance, x, x', and z are all literals, and $x'yz$ is the product of three literals. (The complement of the literal x' is x; the complement of the literal y is y'.) There are only two values which can be substituted into a given literal, and we will continue to designate these values with the symbols 0 and 1.

An expression Φ *implies* another expression Ψ if there is no assignment of truth values which makes Φ (the antecedent) true (take the value 1) and Ψ (the consequent) false (take the value 0). Examples of valid implications are: abc implies ab, xyz implies $xy + xz$.

A product term ϕ *subsumes* another product term ψ if all the literals in ψ are also in ϕ (abc subsumes ab, and $xy'z$ subsumes xy'). If a given term ϕ in a sum-of-products expression subsumes another term ψ, then ϕ (the subsuming term) can be removed without changing the value of the expression. This is an extension of the absorption law, $x + xy = x$. Examples are: $xy'z + xy' = xy'$; $x'y'z' + x'y' + x' = x'$ or, in general, $\phi\alpha + \phi = \phi$.

There are several criteria which may be used to determine which of several equivalent sum-of-products or product-of-sums expressions is minimal. The three most common criteria are:

1. The minimal expression is the expression with the fewest literals.

2. The minimal expression is the expression with the fewest terms, provided another expression does not have the same number of terms and fewer literals.

3. The minimal expression is that expression which requires the fewest logical elements to construct. The number of logical elements is often determined by adding the number of terms in the expression to the number of occurrences of literals.

The third criterion is often used, as the number of terms plus the number of literals in a sum-of-products expression often equals the number of logical circuit elements required to construct the network. For instance, the expression $abc + a'b'$ requires seven diodes if constructed as a two-level diode network.†

Often, given a set of Boolean expressions describing a function ϕ, the same expression(s) will be the minimal expression(s) when evaluated by any of the three criteria. (There are quite often several minimal equivalent expressions.) The procedures which will be described give the

---

† Notice that an adjustment must be made if one or more of the terms of the expression consists of only one literal or if the expression consists only of one term. For instance, $a + bc$ requires only four diodes. An exact determination of the number of diodes required to construct a given expression may be found by adding the number of terms in the expression to the number of literals, and then subtracting 1 for each term containing a single literal and subtracting 1 if the expression contains only one term.

designer the option of developing all minimal sum-of-products or product-of-sums expressions or a single minimal expression. Also, the procedures to be described make possible the calculation of the minimal expression according to whichever of the three criteria may be selected.

The minimization technique which will be described in the following three sections is based on, for the most part, the work of W. V. Quine[1-3] and E. J. McCluskey.[4] Other workers in the field of logic design will be referenced when their particular contributions are described. The concepts of prime implicant and prime implicant table and the basic approach to the problem were first described by Quine. Subsequent authors, for the most part, invented more efficient techniques for calculating the minimal expressions or presented new insight into the problem.

4-3. Prime Implicants. The first step in the procedure for deriving a minimal sum-of-products expression is to derive a set of shortened product terms which are defined as *prime implicants*. The final minimal expression will consist of a subset of these terms; the second step in the minimization procedure consists of selecting the terms of the minimal expression from the prime implicant terms. Two techniques will be described for deriving the prime implicants: first, a direct technique which starts with the expression to be minimized in canonical form, and second, a technique which starts with the expression in sum-of-products (but not necessarily canonical) form. Each of the techniques has its advantages; the first technique is very direct, straightforward, and easy to check; the second technique will often require less steps than the first, but is less direct. The two procedures for deriving prime implicants will be followed by a description of the procedure for selecting the minimal expression from a set of prime implicants.

Let us define the term prime implicant. *A prime implicant of an expression* Ψ *is a product term* α *which implies the expression* Ψ *and subsumes no shorter product term which implies the expression* Ψ. (A prime implicant will generally consist of several literals multiplied together, but may consist of only one literal.) Each prime implicant implies the expression Ψ to be minimized, so there is no assignment of values to the literals of the expression which will make Ψ take the value 0 and a given prime implicant α take the value 1. Further, each prime implicant must consist of a product term in only the literals of the expression Ψ to be minimized, for if a literal, say ζ, which is not in Ψ were in a given prime implicant $\zeta\alpha$, then $\zeta\alpha$ would subsume α, which would be shorter and would also imply Ψ, since the removal of ζ could not affect the implication.

A minimal sum-of-products expression will consist of a set of prime implicants logically added together. Again, there may be only one prime implicant, but in general there will be several. Let us assume an expression Φ is to be minimized and the set of all prime implicants is α_1,

$\alpha_2, \ldots, \alpha_n$. Now a minimal expression must consist of a subset of these prime implicant terms logically added together. To show this, let us assume that a term β of a minimal expression Ψ is not a prime implicant. Now β must imply Ψ and also Φ, but if β is not a prime implicant there must be a product term α_i which is subsumed by β and which also implies Φ. This term may be used to replace β, thus reducing the number of literals in Ψ, so Ψ cannot have been minimal.

A reasonable way to minimize sum-of-products expressions is therefore to derive all the prime implicants for the expression and then to choose from this set of prime implicants a subset of terms which will comprise the minimal expression.

4-4. Derivation of Prime Implicants from Canonical Expansion Expression. In order to use this procedure the expression to be minimized must first be in canonical form. Since the expressions used to design logical networks for digital machines are generally derived from tables of combinations, this is a normal starting point. However, if an expression is in sum-of-products form but not canonical form, it may be expanded to canonical form by repeated applications of the rule $\beta = \beta\chi + \beta\chi'$, where β is one or more literals logically multiplied together and χ is a single literal. If the expression is not in sum-of-products form, it may be put in sum-of-products form by use of the distributive laws.

The complete set of prime implicants, $\alpha_1, \alpha_2, \ldots, \alpha_n$, for an expression Ψ which is in canonical form may be formed by repeated application of the rule $\beta\chi + \beta\chi' = \beta$, where β is a product term and χ is a single variable. This rule must be applied in all possible ways. Let us call application of the rule $\beta\chi + \beta\chi' = \beta$ a *matching* of $\beta\chi$ with $\beta\chi'$ to form the shorter product term β. Now consider the expression $xyz + xyz' + xy'z$. The term xyz can be *matched* with xyz' yielding xy; but xyz can also be matched with $xy'z$ yielding xz. Both matches must be made, yielding the two prime implicants xy and xz. Notice that both xy and xz imply the original expression, for if we let $x = 1$ and $y = 1$, then either xyz or xyz' must take the value 1, so xy implies $xyz + xyz' + xy'z$; the same reasoning will show xz implies $xyz + xyz' + xy'z$. Since xy and xz will not match, they constitute the only two prime implicants for the expression. [Neither of these terms can be eliminated, so the minimal sum-of-products expression is $xy + xz$. A minimal product-of-sums expression is $x(y + z)$.]

Let us examine the derivation of the prime implicants for a longer expression. In this case it will be convenient to use the symbol $-$ to indicate that a literal has been eliminated in the matching process, so if abc is matched with abc', we will write the resulting term $ab-$. With this notation, one term can be matched with another term only if (1) the $-$'s indicating the missing literal(s) are in the same relative position in each term, and (2) all the literals in each of the terms to be matched are identical, save one. Therefore, $-bcd'-$ will match with $-bcd-$ yielding $-bc--$; but $-bc$ will not match with $ab'-$, for their missing literals are in different positions, nor will $-bc$ match with $-b'c'$, for more than one literal differs in the two terms.

It is convenient to array the terms of the original expression in a column and then to write the terms which result from the matching in columns, so that each set of matches results in a new column and each column contains terms with the same number of $-$'s. Also, each time a term is matched with another term, both terms should be marked with a check. Since the checks will then indicate the terms which have matched with other terms, the unchecked terms will comprise the prime implicants. Notice that there is no need to check a term twice, since once a term is checked it cannot be a prime implicant because it subsumes a shorter term which also implies the expression to be simplified.

Assume the canonical expansion expression to be minimized is: $a'bcd + a'b'cd + a'bcd' + a'b'cd' + abcd$.

Arranging the terms of the expression in a column and then matching will yield the following lists:

$$\begin{array}{lll}
\checkmark a'bcd & \checkmark a'-cd & \\
\checkmark a'b'cd & \checkmark a'bc- & a'-c- \\
\checkmark a'bcd' & -bcd & a'-c- \\
\checkmark a'b'cd' & \checkmark a'b'c- & \\
\checkmark abcd & \checkmark a'-cd' & \\
\end{array}$$

Since the column containing terms with only one literal eliminated (that is, terms with one $-$, as written above) contains one term, $-bcd$, which does not match with any of the other terms, this term is a prime implicant. The last column contains two identical terms $a'-c-$, and since $\alpha + \alpha = \alpha$, one of them may be eliminated; there are only two prime implicants, bcd and $a'c$. Since neither of these can be eliminated, the final minimized expression is $a'c + bcd$.

This technique results in the systematic forming of all product terms which imply the expression to be minimized and the subsequent elimination of all terms which subsume other terms, thus yielding the set of prime implicants. Proof that this technique derives all prime implicants is straightforward. The technique systematically forms *all* terms in n or fewer variables which imply the expression Φ, where n is the number of different variables in Φ. Subsuming terms are then systematically removed, leaving only terms which imply Φ and subsume no shorter terms which imply Φ.

It is possible to reduce substantially the number of matches which must be made for large problems by partitioning the terms in each column.

Also, use of binary notation often makes the matching procedure easier to perform. Let each term of the canonical expansion be represented by a set of binary digits, using a 0 to represent a complemented variable and a 1 to represent an uncomplemented variable. For instance, let $abc'd$ be represented by 1101 and $a'b'c$ by 001. Then let eliminated variables be indicated by $-$'s as before, so that $abcd + abcd' = abc-$ will be represented by $1111 + 1110 = 111-$.

Further, assign an *index number* to each term, letting the index number equal the number of uncomplemented variables (this will equal the number of 1's in a given product term if the binary notation is used). Then

Table 4-1. Derivation of Prime Implicants
Prime implicants are $b'd$, $b'c$, ad, ac, ab

√$a'b'c'd$	√$a'b'-d$	$-b'-d$	√0001	√00$-$1	$-0-1$
√$a'b'cd'$	√$-b'c'd$	$-b'c-$	√0010	√$-$001	$-01-$
√$a'b'cd$	√$a'b'c-$	$a--d$	√0011	√001$-$	$1--1$
√$ab'c'd$	√$-b'cd'$	$a-c-$	√1001	√$-$010	$1-1-$
√$ab'cd'$	√$-b'cd$	$ab--$	√1010	√$-$011	$11--$
√$abc'd'$	√$ab'-d$		√1100	√10$-$1	
√$ab'cd$	√$a-c'd$		√1011	√1$-$00	
√$abc'd$	√$ab'c-$		√1101	√101$-$	
√$abcd'$	√$a-cd'$		√1110	√1$-$10	
√$abcd$	√$abc'-$		√1111	√110$-$	
	√$ab-d'$			√11$-$0	
	√$a-cd$			√1$-$11	
	√$ab-d$			√11$-$1	
	√$abc-$			√111$-$	

partition the column listing the terms of the canonical expression, letting each section of the partition contain terms with the same index number; let the section with terms with the lowest index number be placed at the top of the column, the terms with the next lowest index number directly below, and so on, until the terms with the highest index number are listed in a section at the bottom of the column. The various sections of the table are then separated by means of horizontal lines.

When a column is divided into sections with the index numbers arrayed in ascending order, each term in a given section must be matched only with the terms in the sections immediately above or below. (A given term of the canonical expansion with index number m need then be matched only with the terms having an index number which is either $m + 1$ or $m - 1$.) By starting with the terms having the lowest index number and matching these with the terms in the section immediately beneath, and then taking each of the terms with the second least index number and matching with the terms in the sections below, an orderly matching procedure may be performed. Also, by placing the terms which result from the matches in sections and continuing to match only between adjacent sections the matching process may be shortened. Table 4-1 illustrates the derivation of prime implicants for a four-variable problem, using both letter and binary notation; the unchecked terms are the prime implicants.

4-5. Derivation of Prime Implicant Terms by Iterated Consensus. The technique for deriving prime implicant terms described in Sec. 4-4 must be initiated with the expression to be minimized in canonical expansion form. Also, the number of matches which must be made and the number of terms which must be handled may become quite large for some problems. The technique which will be described in this section is called *iterated consensus* and was first described by Samson and Mills,[5] although the term iterated consensus was coined by Quine.[3] This technique is efficient for expressions such as those used in the design of multiple-output networks in Sec. 4-10, and is generally more efficient for deriving prime implicants from expressions in canonical form, although calculations are less straightforward.

There are two basic concepts which are used in this procedure for obtaining prime implicants. One of these involves the use of the subsuming relation, which was defined in Sec. 4-3, and the other is defined as the *consensus operation*. *If two product terms α and β contain one and only one variable which is complemented in one term and not in the other, then the consensus of these terms is the product term formed by deleting the variable which is opposed and forming a product term of the remaining variables.* (Repeated variables may, of course, be eliminated.) As examples, the consensus of abc and $c'de$ is $abde$; the consensus of abc and $b'cd$ is acd; the consensus of a' and abc is bc.

If γ is the consensus of α and β, then γ implies $\alpha + \beta$. Therefore, if a given expression Φ is in sum-of-products form, adding a term which is the consensus of two of the terms in Φ does not change the function described by the expression. For instance, if α and β are two product terms of the expression Φ, and γ is the consensus of α and β, then $\Phi + \gamma = \Phi$, for γ implies $\alpha + \beta$ and hence Φ, and therefore $\Phi + \gamma = \Phi$. The addition of a consensus term to a product-of-sums expression is therefore an equivalence transformation.

A given expression can be reduced to a set of prime implicants by repeatedly forming new consensus terms which are then added to the expression to be minimized, and at the same time removing all subsuming terms from the expression.

Consensus terms which subsume terms already in the expression are not to be added. For instance the term abd is not added to the expression $ab + acd + bc'$, for abd subsumes ab. The process of forming prime implicants is stopped either when no more consensus terms can be formed or when all consensus terms which can be formed subsume terms already present in the expression.

Consider the expression for which prime implicants were found in Sec. 4-4, which was $\Phi = a'bcd + a'b'cd + a'bcd' + a'b'cd' + abcd$. The consensus of $a'bcd$ and $a'b'cd$ is $a'cd$. Since $a'bcd$ and $a'b'cd$ subsume $a'cd$, they may be eliminated. Then $a'cd'$ is the consensus of $a'bcd' + a'b'cd'$ and the latter two terms may be eliminated. This gives

$$\Phi = a'cd + a'cd' + abcd$$

Now $a'c$ is the consensus of $a'cd + a'cd'$, giving $\Phi = a'c + abcd$, and finally bcd is the consensus of $a'c + abcd$, giving $\Phi = a'c + bcd$.

Table 4-2 illustrates a larger problem; subsuming terms have been checked; the remaining terms are prime implicants.

Table 4-2. Consensus Taking Prime Implicant Derivation

$\checkmark a'b'c'd$	$\checkmark a'b'{-}d$
$\checkmark a'b'cd'$	$\checkmark {-}b'cd'$
$\checkmark a'b'cd$	$\checkmark ab'{-}d$
$\checkmark ab'c'd$	$\checkmark abc'{-}$
$\checkmark ab'cd'$	$\checkmark abc{-}$
$\checkmark abc'd'$	$ab{-}{-}$
$\checkmark ab'cd$	${-}b'{-}d$
$\checkmark abc'd$	${-}b'c{-}$
$\checkmark abcd'$	$a{-}{-}d$
$\checkmark abcd$	$a{-}c{-}$

4-6. Elimination of Redundant Terms. Two techniques have been described for deriving the set of prime implicant terms for a given Boolean-algebra expression. Both techniques add new terms and eliminate former terms without changing the basic function described by the algebraic expression; that is, the new terms are formed and the discarded terms eliminated by means of equivalence transformations, so that the set of prime implicants which are derived may be logically added together to form an expression equivalent to the expression being minimized. Also, the prime implicant terms are unique; for a given expression Φ the prime implicants $\alpha_1, \alpha_2, \ldots, \alpha_n$ are uniquely defined.

The set of prime implicants does not necessarily comprise the minimum expression, however, as generally one or more prime implicants may be eliminated without changing the function described by the expression. Consider the expression $abc + c'de + abde$, consisting of three prime implicants logically added together. The term $abde$ may be eliminated, forming a new and shorter expression equivalent to the original expression; that is, $abc + c'de + abde = abc + c'de$. (The expression $abc +$

$c'de$ will be found to be minimal according to any of the three criteria listed in Sec. 4.2.)

Examination of the simple equivalence $abc + c'de = abc + c'de + abde$ will illustrate several of the important rules concerning the elimination of prime implicants from logical expressions. First, *if a given prime implicant implies the logical sum of the remainder of the prime implicants, that prime implicant may be eliminated.* That is, if prime implicant α_j implies $\alpha_1 + \alpha_2 + \cdots + \alpha_{j-1} + \alpha_{j+1} + \cdots + \alpha_n$, where the α's are prime implicant terms and α_j is a selected term, then it is possible to eliminate α_j from the expression and form a new and shorter equivalent expression.

In general, however, it will not only be possible to eliminate several prime implicants from the expression, but there will be a choice as to which prime implicants are eliminated. Therefore, a systematic technique is needed to determine which prime implicants cannot be eliminated, which may always be eliminated, and which further choices exist.

In a given expression Φ composed of a set of prime implicants logically added together, there will generally be one or more prime implicants which cannot be eliminated without altering the function described by the expression. The set of prime implicants which *cannot* be eliminated is referred to as the *core* of the expression, and the individual terms which comprise this core are designated *core prime implicants.* The general rule for finding these is quite simple: *If a given prime implicant does not imply the expression formed by logically adding together the remainder of the prime implicants, then that prime implicant cannot be eliminated and is a member of the core expression.* That is, if α_j does not imply $(\alpha_1 + \alpha_2 + \cdots + \alpha_{j-1} + \alpha_{j+1} + \alpha_n)$ where α_1 through α_n comprise the complete set of prime implicants for a given expression Φ, then α_j is not eliminable and is a member of the core expression.

For our example set of prime implicants $abc + c'de + abde$, the following relations† may be found:

abc does not imply $c'de + abde$.

$c'de$ does not imply $abc + abde$.

$abde$ implies $abc + c'de$.

The terms abc and $c'de$ are therefore members of the core expression and cannot be eliminated; the term $abde$ is eliminable. Since there is only one term which can be eliminated, the minimal expression may be formed by

---

† A simple test to determine whether a given product term Ψ implies an expression Φ may be performed by giving each of the literals in Ψ the value 1 in Φ, and then seeing if Φ has the value 1 for all assignments of values to the other literal in Φ [i.e., determines if $\Phi = 1$ (is made tautologous)]. For instance, let Ψ be the product term $abde$ and $\Phi = abc + c'de$. Substituting 1's into a, b, d, and e in Φ gives $11c + c'11$ or $c + c'$, which has the value 1.

simply eliminating that term.   Another rule can be stated at this point: *if a given term implies the core of a set of prime implicants, that term is absolutely eliminable.*

Since all the terms of the core expression must be used in any final minimal expression, a term which implies the core expression need not be further considered.   Therefore, having found the core expression, each of the remaining terms can be checked against the core expression and if a given term implies the core expression it is *absolutely eliminable* and may be eliminated from further consideration.

We have now defined two classes of prime implicants, core prime implicants and absolutely eliminable prime implicants.   A third class consists of the remaining prime implicants, that is, those prime implicants which are eliminable, as they imply the expression formed by the remaining prime implicants, but do not imply the core.   There will often be a choice as to which of these prime implicants are eliminated, and the following section will describe a systematic technique for determining the core and the scts of prime implicants which may be eliminated.

4-7. The Prime Implicant Table.   In order to describe a systematic technique for deriving a minimal expression from the set of prime implicants, a definition and two theorems will be useful.

Definition.   An *irredundant* sum-of-products expression Ψ describing a function Φ is an expression such that (1) every product term is a prime implicant and (2) no product term may be eliminated from Ψ without changing the function Φ described by Ψ.

There are often a number of irredundant sum-of-products expressions for a function Φ, some of which will contain fewer terms and fewer literals than others.   However, provided with the set of irredundant expressions, we can easily obtain the expression which is minimal according to the criterion selected by simply comparing the expressions.   This is shown by the following theorem.

Theorem.   The minimal sum-of-products expression Ψ describing a function Φ is an irredundant expression.

If an expression $\Lambda = \Phi$, which is minimal according to any one of the three criteria in Sec. 4-2, is not irredundant, then either (1) Λ contains a term which is not a prime implicant, and such a term always subsumes a shorter term which may be substituted into the expression, thereby forming a more minimal expression, or (2) Λ contains a term which can be eliminated, and this would of course form a new expression which was more minimal with respect to Λ.   Both cases involve contradictions, so minimal expressions are irredundant.

Theorem.   If Ψ is an irredundant sum-of-products expression for Φ, then each term of the canonical expansion for Φ subsumes at least one term of Ψ.

Since each term of Ψ is a prime implicant, each term implies Φ. Further, since Φ implies Ψ and Ψ implies Φ (since Φ and Ψ are equivalent), each term of the canonical expansion for Φ must imply at least one term of Ψ. Otherwise, some set of values substituted into Φ would give Φ the value 1 while no term of Ψ would have the value 1, and Ψ would be equal to 0. Now if a single product term χ implies a single product term ϕ, then χ also subsumes ϕ, for χ must contain all the literals of ϕ or the literal not in χ could be given the value 0 when χ was equal to 1. Therefore each term of Φ must subsume at least one term of Ψ. The theorem

Table 4-3. Prime Implicant Table

Core terms: $ab--$, $-b'-d$, $-b'c-$
Absolutely eliminable terms: $a--d$, $a-c-$

	$a'b'c'd$	$a'b'cd'$	$a'b'cd$	$ab'c'd$	$ab'cd'$	$ab'cd$	$abc'd'$	$abc'd$	$abcd'$	$abcd$
$a--d$				×		×		×		×
$a-c-$					×	×			×	×
$ab--$							×	×	×	×
$-b'-d$	×		×	×		×				
$-b'c-$		×	×		×	×				

is stated using the subsuming relation instead of the implication relation, although it is true for either.

A convenient way to display the relations between the prime implicants for a function Φ and the terms of the canonical expansion for Φ is by means of a *prime implicant table*. Table 4-3 is a prime implicant table for the problem initiated in Table 4-2. The prime implicants for the function are listed along the ordinate of the table, the canonical expansion terms along the abscissa. If a given canonical expansion term subsumes a given prime implicant, then a cross is placed at the intersection point in the table listing the two terms; otherwise the intersection is left blank. The resulting table lists all the subsuming relations between the canonical expansion terms and the prime implicants. If a subset Ψ of the prime implicants is chosen so that each canonical term of the expression for Φ subsumes at least one of the prime implicants in Ψ, then the expression found by logically adding together the prime implicants in Ψ will be equivalent to Φ. The selection of the prime implicants proceeds in three steps: (1) The core terms are selected. (2) Any absolutely eliminable terms are eliminated. (3) A subset of the remaining eliminable terms is selected.

The terms of the core expression are quite easily chosen from the prime implicant table, for they consist of those prime implicants which are subsumed by a canonical expansion term which subsumes no other prime implicant. Therefore, if a column of the table contains only a single $\times$, then the prime implicant lying in the same row as this $\times$ is a member of the core expression. Use of this rule indicates that the terms $ab--$, $-b'-d$, and $-b'c-$ comprise the core for Table 4-3. These terms may be listed and removed from the table, as they are necessary. Also all canonical expansion terms which subsume these terms may be lined out, as they no longer need be considered. These are the absolutely eliminable terms.

If the core terms and those terms which subsume the core are lined out, then it may be seen that certain terms, in this case $a--d$ and $a-c-$, no longer are subsumed by any of the remaining canonical terms. These terms imply

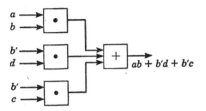

FIG. 4-1. Minimal AND-to-OR network.

the core expression and will not appear in any irredundant expression or in any minimal expression.

The problem in Table 4-3 is completely solved when the core is removed, for all terms of the canonical expansion subsume at least one term of the core expression. Usually this will not be the case, however, and there will be a set of eliminable terms from which a subset must be chosen. Figure 4-1 illustrates a block diagram of the network for the above problem.

Table 4-4. Table of Choices

Core expression $= a'-c'de' + a'b'c-e + ab'-d'- + -b'cd'- + --cd'e$

	$abc'd'e$	$abc'de'$	$abc'de$	$abcd'e'$	$abcde'$	$abcde$
$\alpha_1 = -bc'de'$		$\times$				
$\alpha_2 = a-cd'-$				$\times$		
$\alpha_3 = a--d'e$	$\times$					
$\alpha_4 = abc--$				$\times$	$\times$	$\times$
$\alpha_5 = ab-d-$		$\times$	$\times$		$\times$	$\times$
$\alpha_6 = ab--e$	$\times$		$\times$			$\times$

Table 4-4 is a *table of choices* for a network with five input lines and a single output line. The core expression consists of five terms and has been removed, as have all canonical terms which subsume the core (the five core terms are listed). After the core has been removed, the remaining prime implicants and canonical terms are put in a table of choices. In Table 4-4 there are six prime implicants and six canonical terms. In many cases the choice of the remaining terms will be obvious, but in

Table 4-4 there is some choice about the subset(s) of the prime implicants which are minimal. We describe a technique for selecting a minimal set of terms which was invented by Petrick.[6]

To the left of the prime implicants in Table 4-4 is a column containing the symbols α_1 through α_6, which will be used to represent the prime implicants in the discussion which follows. In order to facilitate the description of this technique we define the term *cover* as follows.

Definition. A product term ϕ covers a product term ψ if ψ contains each literal in ϕ.

Both covers and subsumes are used to denote binary relations; the cover relation is the inverse of the subsume relation, for "ψ subsumes ϕ" implies "ϕ covers ψ," and "μ covers κ" implies "κ subsumes μ." As examples: ab covers abc, xy covers xyz, $xy'z'$ covers $wxy'z'$, and ab covers ab.† Notice also that "ϕ covers and subsumes ψ" implies that ϕ and ψ are identical product terms.

The first column of $\times$'s indicates that either prime implicants α_3 or α_6 may be selected to cover the canonical term $abc'd'e$, and the second column shows that either α_1 or α_5 may be used to cover $abc'de'$. These subsuming relations for the first two columns may be written symbolically thus: $(\alpha_3 + \alpha_6)(\alpha_1 + \alpha_5)$. This indicates that either α_3 or α_6 and α_1 or α_5 must be used to cover $abc'd'e$ and $abc'de'$. Table 4-4 may be written in its entirety in this manner, yielding the expression

$$(\alpha_3 + \alpha_6)(\alpha_1 + \alpha_5)(\alpha_5 + \alpha_6)(\alpha_2 + \alpha_4)(\alpha_4 + \alpha_5)(\alpha_4 + \alpha_5 + \alpha_6)$$

When the sum terms of this expression are multiplied together, the resulting expression will list, in symbolic form, all the sets of α's which may be chosen from Table 4-4 so that each canonical term is covered. It is possible to reduce the product-of-sums expression above, using the theorem $(\gamma + \delta + \epsilon)(\gamma + \delta) = (\gamma + \delta)$, thereby eliminating the sum term $(\alpha_4 + \alpha_5 + \alpha_6)$. The terms of the shortened expression are multiplied together, and the resulting expression simplified, using the theorem $\gamma\delta + \gamma\delta\nu = \gamma\delta$, yielding

$$\alpha_2\alpha_5\alpha_6 + \alpha_3\alpha_4\alpha_5 + \alpha_1\alpha_4\alpha_6 + \alpha_4\alpha_5\alpha_6 + \alpha_2\alpha_3\alpha_5 + \alpha_1\alpha_2\alpha_4\alpha_5$$

Each product term of this expression lists a subset of the prime implicants which may be used to form an irredundant sum-of-products expression for the function in Table 4-4. The six irredundant normal forms are

† The terms subsume and cover are used with about equal frequency in the literature. Generally a given author will use only one of the two terms. Algebraists, logicians, and purists have tended to use the term subsume, which was coined by Quine, while topologists and engineers have tended toward cover. Since these describe inverse relations, and since each becomes rather awkward in certain cases, it is not unreasonable to use both.

$a'c'de' + a'b'ce + ab'd' + b'cd' + cd'e$ plus one of the following sets of terms:

$$acd' + abd + abe \tag{1}$$
$$ad'e + abc + abd \tag{2}$$
$$bc'de' + bc'de' + abe \tag{3}$$
$$abc + abd + abe \tag{4}$$
$$acd' + ad'e + abd \tag{5}$$
$$bc'de' + acd' + abc + abd \tag{6}$$

Expression (6) contains more literals and terms than the first four expressions, and expression (3) more literals, so expressions (1), (2), (4), and (5) are all minimal expressions according to the three criteria in Sec. 4-2.

It is possible to shorten the problem by noticing that if two prime implicants cover the same canonical terms, the prime implicant with the fewer variables is to be preferred. If each prime implicant has the same number of variables, either prime implicant may be picked. Although not all irredundant expressions will result, *one* of the shortest will be derived. Also, if a prime implicant covers all the terms covered by another prime implicant and has the same number of or fewer variables, the prime implicant covering the fewer terms may be eliminated.

Use of these rules will result in the elimination of terms α_1, α_2, and α_3 from the computations on Table 4-4. The expression to be multiplied out and reduced will then be

$$(\alpha_6)(\alpha_5)(\alpha_5 + \alpha_6)(\alpha_4)(\alpha_4 + \alpha_5)$$

The resulting term will then consist of only $\alpha_4\alpha_5\alpha_6$, a correct choice of prime implicants for one of the minimal expressions.

Other rules and examples may be found in Quine[1-3] and McCluskey.[4] Petrick[6] gives further examples of the above technique, Samson and Mills[5] of the consensus-taking operation. Gazale[7] and Mott[8] have shown ways to select the irredundant forms using only the prime implicants, that is, without using the canonical expansion terms. Other references and approaches are listed in the references at the end of this chapter.

4-8. Unspecified Input and Output Conditions. In many cases there are certain states which a given register will never assume. If, for instance, a binary-coded decimal system is used, and each decimal digit is represented in a register by four binary storage cells,† then only 10 of the 16 possible states which the four cells can assume will actually occur. If a combinational network is to realize a function of a four-cell register which assumes only 10 of its 16 possible states, then the question of how most efficiently to design the network arises. Also, if certain of the output

† Refer to the Appendix for a description of binary-coded decimal number systems.

states are of no importance, or if when certain input combinations occur the outputs from a given combinational network are not used, the same problem arises.

The unspecified input states are sometimes referred to as *forbidden* input states, and the unspecified outputs as *indifferent* outputs; also both are referred to as *don't care* conditions. Table 4-5 is a table of combinations for a network which has nine specified output values and seven unspecified output values. The specified values correspond to the

Table 4-5. Incompletely Specified Table of Combinations

| Inputs | | | | Output |
a	b	c	d	Z
0	0	0	0	0
0	0	0	1	1
0	0	1	0	1
0	0	1	1	1
0	1	0	0	0
0	1	0	1	0
0	1	1	0	1
0	1	1	1	1
1	0	0	0	0
1	0	0	1	–
1	0	1	0	–
1	0	1	1	–
1	1	0	0	–
1	1	0	1	–
1	1	1	0	–
1	1	1	1	–

rows containing 0's and 1's in the output columns and the unspecified states to the rows containing –'s in the output column. In order to derive a minimal expression for a problem of this sort, the output values which are unspecified are assumed to be 1's when the canonical expansion expression is formed. When the prime implicants are derived, this allows for maximum shortening of the terms. For Table 4-5, this results in a canonical expression containing 12 terms. The prime implicants derived from this set of product terms consist of the following four terms: $b'd$, ad, ab, and c.

When the prime implicant table is formed, only the canonical terms corresponding to specified 1 outputs are listed along the abscissa. All the prime implicants are listed along the ordinate. For the expression for Table 4-5, two prime implicants are required: c and $b'd$; these cover all the specified 1 states, and the minimal expression is $c + b'd$.

Notice that if the expression $c' + bd$ is evaluated for each of the 16 possible input states, the expression takes both the values 0 and 1 for the unspecified input states (don't cares), but takes the given values for all specified states.

4-9. Minimal Product-of-sums Expressions. Section 3-10 described the procedure for deriving a product-of-sums expression in canonical form, given a table of combinations. The minimization procedure for product-of-sums expressions is the dual of that for sum-of-products expressions. A convenient technique involves forming the complement of the desired expression (which will be in sum-of-products form), minimizing this expression while still in complemented form (using the technique already described), and then complementing the minimized expression.

Table 4-5 may be used to illustrate the technique. The product terms from the rows with specified 0 outputs are logically added to form the complement of the canonical expression for the product-of-sums expression. The terms corresponding to unspecified outputs are then logically added to this expression (these terms are inside the parentheses in the following expression):

$$Z' = a'b'c'd' + a'bc'd' + a'bc'd + ab'c'd'$$
$$+ (ab'c'd + ab'cd' + ab'cd + abc'd' + abc'd + abcd' + abcd)$$

The prime implicants derived from this expression are $c'd' + bc' + a$. If a prime implicant table is formed using only the canonical terms which have specified outputs of 0, the minimal expression is found to be

$$Z' = c'd' + bc'$$

and the complement of this, which is the minimal product-of-sums expression, is $Z = (c + d)(b' + c)$.

4-10. Multiple-output Networks. A block-diagram of a multiple-output logical network is shown in Fig. 3-12. There are n input lines and m output lines, where n does not necessarily equal m. The general technique for synthesizing a network of this type consists in deriving a Boolean expression for each output line in terms of the input variables $x_1, x_2, \ldots, x_n$, so that m canonical expansion expressions are formed. Each of these m expressions is then minimized. The synthesis technique which will be described in this and the following sections will show a method of forming the expressions for the network in minimized form, bypassing the conventional step of forming canonical expansions which then must be simplified. This procedure was first described in Ref. 15.

The synthesis procedure consists of three steps:

1. A set of product terms, which will be referred to as ϵ terms, are derived from the table of combinations describing the desired input-output relations. The algorithm for forming the ϵ terms will be described in Sec. 4-12.

2. The set of ϵ terms derived in the first step are then converted to a set of *multiple-output prime implicants*. Two techniques for doing this will be described in Sec. 4-13.

3. A subset of the multiple-output prime implicants is selected from the terms derived in step 2, and the minimal expressions describing the network are constructed from these terms. The exact procedure for selecting the terms will be dependent on which of the three criteria for minimality is used; the necessary details will be given.

4-11. Criteria for Minimality. Three criteria to determine the respective minimality between equivalent sum-of-products expressions were described in Sec. 4-2. First, the minimal expression is that expression

Table 4-6. Table of Combinations

Inputs			Outputs		
a	b	c	Z_1	Z_2	Z_3
0	0	0	1	1	0
0	0	1	0	0	0
0	1	0	0	0	0
0	1	1	0	1	1
1	0	0	1	1	1
1	0	1	0	0	1
1	1	0	0	0	1
1	1	1	1	0	1

which contains the least number of occurrences of literals; second, the minimal expression is the expression containing the least number of product terms; and third, the minimal expression is the expression which requires the least number of diodes (logical elements) when constructed as an AND-to-OR circuit.

Consider the design of a multiple-output network with three input lines (which will be represented by the variables a, b, and c) and three output lines (which will be represented by the variables Z_1, Z_2, and Z_3). Table 4-6 is a table of combinations listing all possible input states and the corresponding output values for the logical network which is to be designed. By standard techniques the three canonical expansion expressions which describe the network are formed.

$$Z_1 = a'b'c' + ab'c' + abc$$
$$Z_2 = a'b'c' + ab'c' + a'bc$$
$$Z_3 = a'bc + ab'c' + ab'c + abc' + abc$$

If these expressions are now reduced individually, the three sum-of-products expressions will be found to be

$$Z_1 = abc + b'c' \tag{1}$$
$$Z_2 = a'bc + b'c' \tag{2}$$
$$Z_3 = a + bc \tag{3}$$

When these expressions are evaluated according to the three criteria for single-output networks, the results are: (1) there is a total of 13 literals used, (2) there is a total of 6 product terms, and (3) 18 diodes are required to construct the three expressions. If these standards are used, the expressions above will be found to be minimal. Notice, however, that the product term $b'c'$ is repeated in the expressions for Z_1 and Z_2. It is reasonable to consider this in assigning weights to the expressions, since, for instance, only one AND gate need be used to form the logical product of b' and c', and once $b'c'$ has been formed it may be used in several different expressions. The following three criteria take this into consideration and are offered as reasonable criteria for determining the minimality of sets of Boolean expressions.

1. Let Ψ_1, Ψ_2, . . . , Ψ_m be the set of expressions for the m output lines. Then let χ_1, χ_2, . . . , χ_p be the complete set of product terms in the m expressions with *no product term repeated*. (If a given product term, say χ_j, occurs in two or more different expressions, only the first occurrence of χ_j is to be used to form the χ_j's. All other occurrences are to be deleted.) Now let λ_j be the number of literals in χ_j; the total effective number of literals in the set of expressions will be designated W_L, where $W_L = \sum\limits_{j=1}^{p} \lambda_j$.

Using this formula, W_L for expressions (1) to (3) is found to be 11.

2. The integer p in χ_1, χ_2, . . . , χ_p represents the total number of different product terms which occur, and may be used as a measure of relative minimality. For expressions (1) to (3), $p = 5$.

3. A good approximation of the number of diodes used in a given network is W_L plus the total number of product terms in the m expressions, counting duplicates, and this criterion is often used. An exact formula can be formed by again letting Ψ_i's correspond to the expressions to be evaluated, and χ_j's to the different product terms as in criterion 1. Now let $d_i{}^0$ equal 0 if there is one product term in Ψ_i, and otherwise let $d_i{}^0$ equal the number of product terms in Ψ_i. Then let $d_j{}^a$ equal 0 if the number of literals in χ_j is equal to 1, and otherwise let $d_j{}^a$ equal the number of literals in term χ_j. The total number of diodes required is

$$D_t = \sum_{j=1}^{p} d_j{}^a + \sum_{i=1}^{m} d_i{}^0$$

For expressions (1) to (3), $D_t = 16$.

When these new criteria are used, expressions (1) to (3) will no longer be minimal. In particular, the term bc in expression (3) can be formed by logically adding the terms abc and $a'bc$ from (1) and (2) respectively, forming the expressions:

$$Z_1 = abc + b'c' \tag{4}$$
$$Z_2 = a'bc + b'c' \tag{5}$$
$$Z_3 = a + a'bc + abc \tag{6}$$

These new expressions will be found to be minimal according to all three criteria, for p now equals 4, W_L now equals 9, and D_t now equals 15.

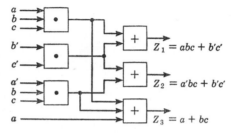

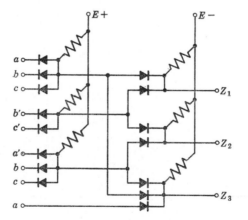

Fig. 4-2. Multiple-output-line logical network.

Figure 4-2 shows a block diagram of the network for expressions (4) to (6) and the corresponding diode logic circuit.

The synthesis technique in the sections which follow systematically yields the minimal expressions describing the logical network to be designed. In some cases a particular set of expressions will be minimal according to all three criteria. More often there will be several sets of expressions, all of which will be minimal, and sometimes a set of expressions which is minimal according to one criterion will not be minimal according to another. The synthesis techniques make possible the derivation of sets of expressions which are minimal using whichever of the three criteria may be selected, by means of slight variations in the basic procedure.

4-12. Forming the ϵ Terms. The first part of the procedure consists of forming a set of product terms, which will be designated ϵ terms. This step is initiated in the customary manner, by listing the input-output relations for the network in a table of combinations. Table 4-7 lists the

Table 4-7. Table of Combinations for Four-input Three-output Network

Inputs				Outputs		
a	b	c	d	Z_1	Z_2	Z_3
0	0	0	0	0	0	0
0	0	0	1	1	0	1
0	0	1	0	0	0	0
0	0	1	1	0	0	0
0	1	0	0	0	1	1
0	1	0	1	0	1	1
0	1	1	0	1	1	0
0	1	1	1	1	1	0
1	0	0	0	0	0	0
1	0	0	1	1	0	1
1	0	1	0	0	1	1
1	0	1	1	1	0	1
1	1	0	0	0	1	1
1	1	0	1	0	1	1
1	1	1	0	1	1	1
1	1	1	1	1	1	1

input and output values for a four-input-line three-output-line problem. Table 4-7 is completely specified; there are no don't care conditions (these will be dealt with in Sec. 4-13).

After the table of combinations has been formed, the next step is to form a set of ϵ terms from the rows of the table. Each ϵ term will consist of two sections, a ν section and a ζ section. The ν section of the ϵ term for the ith row is formed by writing the input variables in the form of a product term, with a given variable complemented or not complemented depending on whether the input value for the variable is 0 or 1, respectively. The ζ section of the ϵ term for the ith row is determined by the values for the outputs Z_i, and is formed from the output variables, with a given variable complemented if its value is 0 in the ith row and omitted if its value is 1. For instance, the fifth row from the top in Table 4-7 is used to form the ϵ term $a'bc'd'Z_1'--$. The set of ϵ terms formed from Table 4-7 (using both letters and binary notation) is listed in Table 4-8.

One other step is taken when the ϵ terms are formed: any row in which all of the output values are 0 may be omitted from further consideration and no ϵ term formed from this row. The reason for this will become apparent: the rule results in the omission of the terms corresponding to rows 1, 3, 4, and 9 in Table 4-7.

Table 4-8. Forming the ϵ Terms

Using letters for variables	*Using binary notation*
$a'b'c'd{-}Z'_2{-}$	$0001{-}0{-}$
$a'bc'd'Z'_1{-}{-}$	$01000{-}{-}$
$a'bc'dZ'_1{-}{-}$	$01010{-}{-}$
$a'bcd'{-}{-}Z'_3$	$0110{-}{-}0$
$a'bcd{-}{-}Z'_3$	$0111{-}{-}0$
$ab'c'd{-}Z'_2{-}$	$1001{-}0{-}$
$ab'cd'Z'_1{-}{-}$	$10100{-}{-}$
$ab'cd{-}Z'_2{-}$	$1011{-}0{-}$
$abc'd'Z'_1{-}{-}$	$11000{-}{-}$
$abc'dZ'_1{-}{-}$	$11010{-}{-}$
$abcd'{-}{-}{-}$	$1110{-}{-}{-}$
$abcd{-}{-}{-}$	$1111{-}{-}{-}$

4-13. Multiple-output Prime Implicants. A set of ϵ terms has been formed in the first step of the procedure. The next step consists of deriving a set of multiple-output prime implicant terms by performing calculations using the ϵ terms. These calculations will form new terms, each having the same format as the original set of ϵ terms (that is, a ν section and a ζ section); the additional terms derived from the original set of ϵ terms will also be called ϵ terms. All the ϵ terms associated with a given problem must have the characteristic that the ν section of each ϵ term implies the expressions for the ouput lines indicated by the missing literals in the ζ section. For instance, the ϵ term $a'bc'd'\,Z'_1{-}{-}$, which was formed from Table 4-7, contains a ν section $a'bc'd'$ which implies the expressions for output lines Z_2 and Z_3. When the ν sections are shortened this rule must still apply; each ν section implies the expressions indicated by the ζ section.

A multiple-output prime implicant is defined as an ϵ term, the ν section of which implies the expressions for the output lines indicated by the ζ section, and the ν section of which subsumes no shorter ν section having the same ζ section or a ζ section containing fewer of the same literals. The multiple-output prime implicants may be derived from the ϵ terms in any of several ways. First, a technique based on the reduction theorem in Sec. 4-4, and second, a variation of the Quine-McCluskey technique will be included.

The first technique requires only two logical operations between ϵ terms. The first consists in determining whether a given ϵ term subsumes or is subsumed by another ϵ term, and the second consists in forming the consensus of two given ϵ terms. (The consensus of two terms ϕ and α was defined in Sec. 4-5.) If ϕ and α contain only one variable which is complemented in one term and not in the others, then the consensus of these terms is the product term formed by deleting the variables which are opposed and forming a product term of the remaining variables. Repeated variables in the consensus term are, of course, eliminated. For example, the consensus of the ϵ terms $abc-- Z_1'-$ and $--c'de\ Z_1'-$ is $ab-de\ Z_1'-$; the consensus of $abc-Z_1'-Z_3'$ and $-b'cd\ Z_1'--$ is $a-cd\ Z_1'-Z_3'$; and the consensus of $a'--Z_1'-$ and $abc\ Z_1'-$ is $-bc\ Z_1'-$.

The multiple-output prime implicants may be formed from the ϵ terms using the following theorem:

Theorem. The list of ϵ terms derived from the table of combinations can be transformed to a set of multiple-output prime implicants by repeatedly forming new consensus terms which are then added to the list of ϵ terms, and at the same time removing all subsuming terms. The ν sections of the multiple-output prime implicants will form the terms of the minimal expressions and the ζ sections will indicate the expressions which are implied by the ν sections.

Consensus terms which subsume terms already in the expression are not to be added, and the process of forming prime implicants is stopped when either no more consensus terms can be formed or when all consensus terms which can be formed subsume terms already in the expression.

The construction of the ϵ terms is such that the ν sections are allowed to form consensus terms in all possible ways, eventually forming all ϵ terms which are candidates for multiple-output prime implicants; at the same time the ζ sections of the ϵ terms perform "bookkeeping," keeping track of which output lines the various ν sections are associated with. For instance, the consensus of $a'bc'd\ Z_1'--$ and $ab-d--\ Z_3'$ is $-bc'd\ Z_1'-Z_3'$. The ν section of the consensus term implies the logical sum of the ν sections of the two original ϵ terms [$bc'd$ implies $(a'bc'd + abd)$]. At the same time the ζ section of the consensus term indicates that the new ν section implies only the expression for output line Z_2.

Notice that the ζ section can never prohibit the forming of a new consensus term, for no uncomplemented variables occur in the ζ sections. This allows the ν sections to form new consensus terms in all possible ways, eventually forming multiple-output prime implicants for all combinations of output lines. Notice also that the ζ sections keep shortened ν sections from eliminating ϵ terms which may be associated with more output lines, while allowing shortened ν sections to eliminate correctly ϵ terms associated with the same or fewer output lines. For instance, $abc'd\ Z_1'--$ will not subsume $-bc'dZ_1'-Z_3'$, while $a'bc'dZ_1'--$ will subsume $a'bc'-Z_1'--$ and be eliminated.

A rule which significantly shortens the process of forming multiple-output prime implicants is: *When the ζ section of the consensus of two ϵ terms contains all the output variables, the new consensus term need not be used.* If a ζ section contains all the output variables, then the ν section cannot be used in any of the output lines and the ϵ term cannot therefore be a multiple-output prime implicant. Further, the ζ section will contain all the output variables in complemented form, and since no ζ section can contain an uncomplemented variable, any consensus term formed with an

Table 4-9a. Deriving Multiple-output Prime Implicants

ϵ terms	Consensus terms
$\checkmark a'b'c'd - Z_2' -$	$\checkmark a'bc' - Z_1' - -$
$\checkmark a'bc'd' Z_1' - -$	$\checkmark a'bc - - - Z_3'$
$\checkmark a'bc'dZ_1' - -$	$-b'c'd - Z_2' -$
$\checkmark a'bcd' - - Z_3'$	$abc - - - -$
$\checkmark a'bcd - - Z_3'$	$\checkmark abc' - Z_1' - -$
$\checkmark ab'c'd - Z_2' -$	$\checkmark ab'c - Z_1' Z_2' -$
$\checkmark ab'cd' Z_1' - -$	$a - cd - Z_2' -$
$\checkmark ab'cd - Z_2' -$	$-bc - - - Z_3'$
$\checkmark abc'd' Z_1' - -$	$-b - - Z_1' - Z_3'$
$\checkmark abc'dZ_1' - - -$	$a - c - Z_1' Z_2' -$
$\checkmark abcd' - - -$	$a - cd' Z_1' - - -$
$\checkmark abcd - - -$	$-bc' - Z_1' - -$
	$ab - - Z_1' - -$
	$- - c'dZ_1' Z_2' -$
	$ab' - d - Z_2' -$
	$a - - dZ_1' Z_2' -$

ϵ term with a ζ section containing all the output variables cannot be a multiple-output prime implicant.

Table 4-9a illustrates the process of deriving multiple-output prime implicants from the ϵ terms. The first column lists the ϵ terms and the next column contains the consensus terms formed. Terms which subsume other terms are checked, the remaining terms are the multiple-output prime implicant terms.

An extension of the Quine-McCluskey technique for forming prime implicants is illustrated in Table 4-9b, again using the ϵ terms listed in Table 4-8.

Several deviations from the standard procedure may be made for the multiple-output problem. There is no need to develop the original set of terms so that each term contains all the output variables. $a'b'c'dZ_2' -$ need not be developed to $a'b'c'dZ_1'Z_2'Z_3' + a'b'c'dZ_1'Z_2'Z_3' + a'b'c'dZ_1Z_2'Z_3' + a'b'c'dZ_1Z_2'Z_3'$; there is no harm in this but it is unnecessary if the matching is replaced by the consensus-forming operation. All possible consensuses must now be taken with each term, however, and the consensus operations now cause the ν sections to be matched while the ζ sections perform bookkeeping. Another rule which lessens the number of calculations is: *If a term having all the output variables in the ζ section is formed as a consensus term, this term need not be included in the list of new ϵ terms.* The subsuming terms in Table 4-9b have been checked; the unchecked terms are the multiple-output prime implicants.

The terms in Table 4-9b have been partitioned according to the number of uncomplemented variables present, so that only the consensus of those

Table 4-9b. Solution by Quine-McCluskey Method

$\checkmark a'b'c'd\text{–}Z'_2\text{–}$	$\checkmark a'\text{–}c'dZ'_1Z'_2\text{–}$	$\text{–}\text{–}c'dZ'_1Z'_2\text{–}$	$\text{–}b\text{–}\text{–}Z'_1\text{–}Z'_3$
$\checkmark a'bc'd'Z'_1\text{–}\text{–}$	$\text{–}b'c'd\text{–}Z'_2\text{–}$	$\checkmark a'b\text{–}\text{–}Z'_1\text{–}Z'_3$	
$\checkmark a'bc'dZ'_1\text{–}\text{–}$	$\checkmark a'bc'\text{–}Z'_1\text{–}\text{–}$	$\text{–}bc'\text{–}Z'_1\text{–}\text{–}$	
$\checkmark a'bcd'\text{–}\text{–}Z'_3$	$\checkmark a'b\text{–}d'Z'_1\text{–}Z'_3$		
$\checkmark ab'c'd\text{–}Z'_2\text{–}$	$\checkmark \text{–}bc'd'Z'_1\text{–}\text{–}$	$\checkmark \text{–}b\text{–}d'Z'_1\text{–}Z'_3$	
$\checkmark ab'cd'Z'_1\text{–}\text{–}$	$\checkmark a'b\text{–}dZ'_1\text{–}Z'_3$	$\checkmark \text{–}b\text{–}dZ'_1\text{–}Z'_3$	
$\checkmark abc'd'Z'_1\text{–}\text{–}$	$\checkmark \text{–}bc'dZ'_1\text{–}\text{–}$	$\text{–}bc\text{–}\text{–}\text{–}Z'_3$	
$\checkmark a'bcd\text{–}\text{–}Z'_3$	$\checkmark a'bc\text{–}\text{–}\text{–}Z'_3$	$a\text{–}\text{–}dZ'_1Z'_2\text{–}$	
$ab'cd\text{–}Z'_2\text{–}$	$\checkmark \text{–}bcd'\text{–}\text{–}Z'_3$	$a\text{–}c\text{–}Z'_1Z'_2\text{–}$	
$\checkmark abc'dZ'_1\text{–}\text{–}$	$ab'\text{–}d\text{–}Z'_2\text{–}$	$ab\text{–}\text{–}Z'_1\text{–}\text{–}$	
$\checkmark abcd'\text{–}\text{–}\text{–}$	$\checkmark a\text{–}c'dZ'_1Z'_2\text{–}$		
$\checkmark abcd\text{–}\text{–}\text{–}$	$\checkmark ab'c\text{–}Z'_1Z'_2\text{–}$		
	$a\text{–}cd'Z'_1\text{–}\text{–}$		
	$\checkmark abc'\text{–}Z'_1\text{–}\text{–}$		
	$\checkmark ab\text{–}d'Z'_1\text{–}\text{–}$		
	$\checkmark \text{–}bcd\text{–}\text{–}Z'_3$		
	$a\text{–}cd\text{–}Z'_2\text{–}$		
	$\checkmark ab\text{–}dZ'_1\text{–}\text{–}$		
	$abc\text{–}\text{–}\text{–}\text{–}$		

terms having one less or one more uncomplemented variable than a given term need be formed. In Table 4-9b terms in a given section need only be tried with the terms in the sections immediately beneath and above.

Proof that using either the Samson and Mills algorithm or the Quine-McCluskey technique on the ϵ terms yields the set of multiple-output prime implicants is relatively straightforward.

Assume that the modified Quine-McCluskey technique is used, except (1) that all ϵ terms, including those from rows of the table of combinations in which all the outputs are 0 (that is, ϵ terms with all the variables in the ζ section), are used at the start of the process, and (2) that ϵ terms containing ζ sections with all variables present are included as consensus terms and used for further consensus taking. The ν sections of the ϵ terms in the table of calculations which results will then contain every product term which may be formed in the n input variables, and each ν

section will have an adjoining ζ section which indicates which, if any, expressions are implied by the ν section. The elimination of subsuming terms and terms with all variables in the ζ section (terms which have ν sections which imply none of the expressions for the output lines) will leave only the multiple-output prime implicants.

The omission of the ϵ terms corresponding to rows of the table of combinations which contain all 0's in the output columns and the omission of consensus terms with ζ sections containing all the output variables will not affect the ultimate forming of all the multiple-output prime implicants during the calculations, for since more of the ζ sections contain uncomplemented variables, these terms can never lead to a term having a ζ section with a variable omitted. At the same time the omission of these terms will significantly reduce the number of calculations required, for only ϵ terms containing ν sections which imply some subset of the output expressions will be formed.

4-14. Selection of Terms for Criteria 1 and 2. A set of multiple-output prime implicant terms has been derived, the ν term sections of which subsume no shorter term which also implies each of the expressions indicated by the ζ sections. The remaining part of the synthesis procedure consists in selecting a subset of the multiple-output prime implicants, and then using the ν sections of these terms to form the correct set of expressions. In order to facilitate the description, multiple-output prime implicants will be referred to as ϵ_m terms.

The selection of the ϵ_m terms will depend on which criterion for minimality is used. The same basic technique may be used for criteria 1 and 2, but a basically more complicated technique is required to ensure minimal expressions according to criterion 3. Accordingly, a procedure for selecting a subset of ϵ_m terms and forming expressions which are minimal by criteria 1 and 2 will be described in this section, followed by the procedure for assuring minimality by criterion 3 in Sec. 4-15.

Table 4-10 is a multiple-output prime implicant table for the problem in Table 4-7. The ϵ_m terms derived in Sec. 4-4 are listed along the ordinate of the table. A set of terms derived from the table of combinations in Table 4-8 is then listed along the abscissa; these terms will be designated c terms, a contraction of *canonical* terms. There are as many c terms along the abscissa as there are 1's in the output section of the table of combinations. To form the c terms the output variables are again adjoined to the input variables, except that instead of eliminating the output variables which correspond to 1 outputs, all output variables are included in each term. Further, the number of c terms formed from the ith row is equal to the number of 1's in that row. Each c term consists of the input variables, complemented or not complemented depending on the input values for the row, plus all the output variables, with only

Table 4-10. Prime Implicant Table for Criteria 1 and 2

	$a'b'c'dZ_1Z_2'Z_3'$	$a'b'c'dZ_1'Z_2'Z_3$	$a'bc'd'Z_1'Z_2Z_3'$	$a'bc'd'Z_1'Z_2'Z_3$	$a'bc'dZ_1'Z_2Z_3'$	$a'bc'dZ_1'Z_2'Z_3$	$a'bcd'Z_1Z_2'Z_3'$	$a'bcd'Z_1'Z_2Z_3'$	$a'bcdZ_1Z_2'Z_3'$	$a'bcdZ_1'Z_2Z_3'$	$ab'c'dZ_1Z_2'Z_3'$	$ab'c'dZ_1'Z_2'Z_3$	$ab'cd'Z_1'Z_2Z_3'$
$-b--Z_1'-Z_3'$			×		×			×		×			
$-bc---Z_3'$							×	×	×	×			
$-bc'-Z_1'--$			×	×	×	×							
$--c'dZ_1'Z_2'-$		×				×						×	
$ab--Z_1'--$													
$a-c-Z_1'Z_2'-$													
$a--dZ_1'Z_2'-$												×	
$ab'-d-Z_2'-$											×	×	
$-b'c'd-Z_2'-$	×	×									×	×	
$abc----$													
$a-cd-Z_2'-$													
$a-cd'Z_1'--$													×

	$ab'cd'Z_1'Z_2'Z_3$	$ab'cdZ_1Z_2'Z_3'$	$ab'cdZ_1'Z_2'Z_3$	$abc'd'Z_1'Z_2Z_3'$	$abc'd'Z_1'Z_2'Z_3$	$abc'dZ_1'Z_2Z_3'$	$abc'dZ_1'Z_2'Z_3$	$abcd'Z_1Z_2'Z_3'$	$abcd'Z_1'Z_2Z_3'$	$abcd'Z_1'Z_2'Z_3$	$abcdZ_1Z_2'Z_3'$	$abcdZ_1'Z_2Z_3'$	$abcdZ_1'Z_2'Z_3$
$-b--Z_1'-Z_3'$				×		×			×			×	
$-bc---Z_3'$								×	×		×	×	
$-bc'-Z_1'--$				×	×	×	×						
$--c'dZ_1'Z_2'-$							×						
$ab--Z_1'--$				×	×	×	×		×	×		×	×
$a-c-Z_1'Z_2'-$	×		×							×			×
$a--dZ_1'Z_2'-$			×				×						×
$ab'-d-Z_2'-$		×	×										
$-b'c'd-Z_2'-$													
$abc----$								×	×	×	×	×	×
$a-cd-Z_2'-$		×	×								×		×
$a-cd'Z_1'--$	×								×	×			

one output variable not complemented and that variable corresponding to a 1 in the table. As an example, row 6 of Table 4-7 lists 0101 for the input values and 011 for the output values, and the two c terms formed are $a'bc'dZ_1'Z_2Z_3'$ and $a'bc'dZ_1'Z_2'Z_3$.

The intersection points of Table 4-10 are marked as follows: If a given c term along the absicissa subsumes an ϵ_m term lying along the ordinate, the intersection point is marked with an $\times$. Since $a'b'c'dZ_1Z_2'Z_3'$ subsumes $-b'c'd-Z_2'-$, an $\times$ is placed at the intersection point of these terms. If a given c term along the abscissa does not subsume a certain ϵ_m term along the ordinate, the intersection point is left blank. The problem is

Table 4-11. Selection of Nonessential Terms

	$ab'cdZ_1'Z_2Z_3'$	$ab'cdZ_1'Z_2'Z_3$	$abcdZ_1'Z_2'Z_3$
$ab--Z_1'--$			$\times$
$a-c-Z_1'Z_2'-$		$\times$	$\times$
$a--dZ_1'Z_2'-$		$\times$	$\times$
$ab'-d-Z_2'-$	$\times$	$\times$	
$abc----$			$\times$
$a-cd-Z_2'-$	$\times$	$\times$	$\times$

now to select a minimal subset of the ϵ_m terms such that each of the c terms subsumes at least one of the selected ϵ_m terms.

The first step consists of selecting from the table every ϵ_m term α for which there exists a c term β such that α is the only ϵ_m term subsumed by β. An examination of the columns of Table 4-10 shows that the first, fourth, seventh, ninth, and thirteenth columns, counting from the left, each contain only one $\times$, indicating that each of these c terms subsumes only a single ϵ_m term; $-bc'-Z_1'--$, $a-cd'Z_1'--$, $-bc---Z_3'$, and $-b'c'd-Z_2'-$ are therefore necessary and will be used to form the final expressions. Further, all terms along the abscissa which contain an $\times$ lying in the same row as these necessary ϵ_m terms may be removed from the table. This results in the removal of all terms along the abscissa except three. After this is done, the multiple-output prime implicants $--c'dZ_1'Z_2'-$ and $-b--Z_1'-Z_3'$ will not be subsumed by any of the remaining canonical terms, and these need no longer be considered. The table may now be redrawn as in Table 4-11.

There will now generally be a choice as to which of the remaining ϵ_m terms are used. In Table 4-11, however, the choice is clear; the term $a-cd-Z_2'-$ is subsumed by all three remaining c terms and is selected. If the table indicates that choices are possible, the technique described in Sec. 4-4 may be used. In addition, if not all minimal sets of expressions are required, ϵ_m terms may be eliminated from the table using the rule: *If a given ϵ_m term α has a ν section containing the same number or fewer literals than another ϵ_m term β, and α is subsumed by every c term which subsumes β, then β may be eliminated from the table.*

The ν sections of the ϵ_m terms which are selected contain the product terms to be used in the final set of expressions, and the missing variables in the ζ section indicate with which output expressions a given product term is associated. For instance, the term $-bc'-Z_1'--$ indicates that bc' may be used in the expressions for Z_2 and Z_3. A first approximation can be obtained by logically adding together the ν sections into the expressions indicated by the ζ sections. Five multiple-output terms have been chosen for the problem started in Table 4-7 and the ζ sections of these terms contain ten dashes indicating missing variables; so the list of expressions will contain ten terms and is as follows:

$$Z_1 = bc + b'c'd + acd$$
$$Z_2 = bc + bc' + acd'$$
$$Z_3 = acd + bc' + acd' + b'c'd$$

The expressions formed in this way will be minimal according to criteria 1 and 2. In this case, $W_L = 13$, and the total number of different terms p is 5.

For this particular problem the three expressions above may not be shortened, as none of the terms are superfluous. In some cases, however, one or more, but not all, of the occurrences of certain terms which are repeated in several expressions may be eliminated from the expressions formed from the ϵ_m terms without changing the functions described; this will be illustrated in Sec. 4-16. Notice that removing a product term which occurs in several different expressions from less than all of the expressions does not change either p or W_L for minimality criteria 1 and 2, although this will generally reduce the complexity of the expressions. After a given problem has been solved and the set of expressions derived for least number of literals or least number of different terms, each of the resulting expressions should be checked by listing the terms of each expression as prime implicants along the ordinate of a prime implicant table and listing the canonical expansion for the expression along the abscissa. The minimal subset of the terms in each expression may be selected using the prime implicant table described previously. It is also possible to test each term of the expression to see if the term is eliminable by checking to see if it implies the expression formed by logically adding the remaining terms in the expression. If several terms are eliminable, however, a prime implicant table should be made.

4-15. Selection of Terms for Criterion 3. In order to synthesize a two-level expression and ensure that the minimum number of diodes is used, additional calculation is required, although quite often the minimal expression will be the same as that obtained for the least number of literals or least number of terms. In forming the prime implicant table when criterion 3, Sec. 4-11, is used, the terms along the abscissa remain the same. Additional terms are added along the ordinate of the table, however. These terms are formed from the ϵ_m terms as follows: the ζ section of each ϵ_m term indicates which expressions the ν section implies and in which it can be used. Notice, however, that if a ζ section indicates that a given term can be used in both expressions Z_1 and Z_2, the term might be used only in Z_1 or only in Z_2. Consequently, a set of *expanded* ϵ_m terms is derived from each ϵ_m term, with each ν section the same, but with the ζ section listing each possible combination of Z_1's in which the ν section might be used. From the ϵ_m term $ab'-d-Z_2'-$ we must therefore form the terms $ab'-d-Z_2'Z_3'$, $ab'-dZ_1'Z_2'-$, and $ab'-d-Z_2'-$, for the ν section of this ϵ_m term will imply and can be used in either the expression for Z_1, or Z_3, or both Z_1 and Z_3. For each multiple-output prime implicant formed there will be $2^q - 1$ prime implicants along the ordinate, where q is the largest number of expressions in which the term may be used. Some of the terms formed in this way may be eliminated by use of the rule: *If a term α has a ν section which subsumes the ν section of another term β, and the ζ sections of both terms are identical, α may be eliminated.*

Table 4-12 is the multiple-output prime implicant table for the problem in Table 4-7 with the ϵ_m terms expanded. First, the necessary terms, or core, will be selected as usual (in Table 4-12 there are no necessary terms). A completely rigorous solution may now be obtained by using the Petrick algorithm explained in Sec. 4-7. To the left of each term along the ordinate in Table 4-12, an α_1 which will be used to represent the expanded ϵ_m terms in the same row is listed. If a product-of-sums expression which lists, symbolically, all the subsuming relations in the table is formed, and then certain equivalence transformations are performed on this expression, an expression in the expanded ϵ_m's, listing all the terms which may be used to form a set of expressions containing no redundant terms, may be derived; one of these sets of expressions will describe the minimal diode network. The expression describing Table 4-12 may be formed by starting with the leftmost c term and forming a product term listing the α_1's which are subsumed by this c term. The first sum term formed in this way is $(\alpha_{12} + \alpha_{16})$ and the term for the second column is $(\alpha_3 + \alpha_{16})$. The complete expression for Table 4-12 is of the form $(\alpha_{12} + \alpha_{16})(\alpha_3 + \alpha_{16}) \cdots (\alpha_4 + \alpha_5 + \alpha_7 + \alpha_{10} + \alpha_{17} + \alpha_{18} + \alpha_{20})$. This expression may be shortened by the theorem $(\beta + \gamma + \delta)(\beta + \gamma) = (\beta + \gamma)$, following which the sum terms of the expression are multiplied together, converting the expression to sum-of

Table 4-12. Prime Implicant Table for Least Diodes

Row definitions:

$\alpha_1 = -b--Z_1'-Z_3'$
$\alpha_2 = -bc--Z_2'Z_3'$
$\alpha_3 = -c'dZ_1'Z_2'-$
$\alpha_4 = a-c-Z_1'Z_2'-$
$\alpha_5 = a--dZ_1'Z_2'Z_3'$
$\alpha_6 = -bc'Z_1'Z_2'-$
$\alpha_7 = ab--Z_1'Z_2'-$
$\alpha_8 = -bc---Z_3'$
$\alpha_9 = -bc'-Z_1'--$
$\alpha_{10} = ab--Z_1'---$
$\alpha_{11} = ab'-d-Z_2'Z_3'$
$\alpha_{12} = -b'c'd-Z_2'Z_3'$
$\alpha_{13} = a-cd-Z_2'Z_3'$
$\alpha_{14} = a-cd'Z_1'-Z_3'$
$\alpha_{15} = ab'-d--Z_3'$
$\alpha_{16} = -b'c'd--Z_3'$
$\alpha_{17} = a-cd--Z_2'-$
$\alpha_{18} = abc---Z_2'-$
$\alpha_{19} = a-cd'Z_1'---$
$\alpha_{20} = abc-----$

products form. If this expression is then shortened using the theorem $\beta\gamma + \beta\gamma\delta = \beta\gamma$, each product term in the resulting expression will represent a solution of the problem, and by evaluating the sets of expressions according to criterion 3, a minimal set may be obtained.

The process may be shortened, however, by assigning a weight to each of the ϵ_m terms, equal to the number of variables missing in the ζ section plus zero if there is only one literal in the ν section, or plus the number of literals in the ν section. Now, if a given term β has a weight equal to or less than another term γ, and β is subsumed by every c term which also subsumes γ, then γ may be omitted. Use of this rule may serve to shorten problems considerably, and the sets of expressions derived in this way will be minimal except for certain cases in which one of the expressions for an output can consist of a single term; these cases are easily detected.

When the problem in Table 4-12 is solved in this way, the set of expanded ϵ_m terms describing the minimal set of expressions and having the lowest total weight will be $-bc'-Z_1'Z_2'-$, $-b--Z_1'-Z_3'$, $a-cd'Z_1'--$, $a-cd-Z_2'-$, $-bc--Z_2'Z_3'$, and $-b'c'd-Z_2'-$. The minimal set of expressions is therefore

$$Z_1 = bc + b'c'd + acd$$
$$Z_2 = b + acd'$$
$$Z_3 = acd + bc' + acd' + b'c'd$$

The network requires 22 diodes to construct, while the expressions which were minimal by criteria 1 and 2 require 23. Notice that W_L for the set of expressions above is 14 and p is 6, so the expressions derived in Sec. 4-14 are still minimal by criteria 1 and 2.

4-16. Derivation of Minimal Product-of-sums Expressions. The procedure which has been described may be used to derive sum-of-products expressions which are minimal according to a criterion selected from the three described. Minimal product-of-sums expressions may also be derived using basically the same procedure.

In order to derive product-of-sums expressions, all the values in the output (Z_i) columns of the original table of combinations are complemented before the ϵ terms are formed. The procedure is then identical to the one described, except that an additional step is required. The expressions formed will be the complements of the desired product-of-sums expressions and will be in sum-of-products form. If each expression is complemented, the desired form will be obtained.

For instance, if the problem in Table 4-12 is approached in this way and a solution is desired which is minimal according to criteria 1 and 2, the ϵ_m terms which are selected according to the procedure in Sec. 4-14 will all be necessary terms and will be

$$-b'c'd'---$$
$$a'-c-Z_1'Z_2'-$$
$$-b'-dZ_1'-Z_3'$$
$$-b'-d'-Z_2'Z_3'$$
$$-bc'--Z_2'Z_3'$$
$$a'b'c----$$

The first set of expressions formed will be

$$Z_1' = b'c'd' + b'd' + bc' + a'b'c \qquad (1)$$
$$Z_2' = b'c'd' + b'd + a'b'c \qquad (2)$$
$$Z_3' = b'c'd' + a'c + a'b'c \qquad (3)$$

This set of expressions offers an example of the rule that certain terms may be eliminated from the expressions formed by the procedure in Sec. 4-14.

The terms $b'c'd'$ and $a'b'c$ in expressions (1) and (3), respectively, are clearly eliminable, as they subsume other terms in the same expression and hence imply the logical sum of the other terms. After these terms have been eliminated, the three expressions are each complemented, forming the desired minimal product-of-sums expressions

$$Z_1 = (b + d)(b' + c)(a + b + c')$$
$$Z_2 = (b + c + d)(b + d')(a + b + c')$$
$$Z_3 = (b + c + d)(a + c')$$

If the technique described in Sec. 4-15 is used, the set of expanded ϵ terms which will be selected is

$$-b'c'd'Z_1'--$$
$$a'-c-Z_1'Z_2'-$$
$$-b'-dZ_1'-Z_3'$$
$$-b'-d'-Z_2'Z_3'$$
$$-bc'--Z_2'Z_3'$$
$$a'b'c---Z_3'$$

These terms detail a set of expressions with no redundant terms, a characteristic of the technique in Sec. 4-15.

4-17. Unspecified Input and Output States. Unspecified input and output values are handled in the following manner: In the original table all possible input states are listed. If certain input conditions will never occur, each output in the corresponding rows is assumed to be a 1 when the table of combinations is written; similarly, 1's are placed in the don't care output values. When the ϵ terms are derived, they will contain $-$'s in the ζ section for the don't care conditions. The don't care conditions are noted, however, and when the abscissa of the prime implicant table is filled in, the c terms, which would ordinarily be made from the table-of-combinations entries in which outputs are unspecified, are omitted; the abscissa of the table lists only the c terms which come from do care conditions. The remainder of the procedure is unaltered.

PROBLEMS

4-1. For an expression in four variables let m_0 represent the product term $x_1'x_2'x_3'x_4'$, m_1 the product term $x_1'x_2'x_3'x_4$, m_2 the product term $x_1'x_2'x_3x_4'$, ..., and finally m_{15} the product term $x_1x_2x_3x_4$. Similarly, let M_0 represent the sum term $x_1' + x_2' + x_3' + x_4'$; M_1 the sum term $x_1' + x_2' + x_3' + x_4$; etc. Find minimal two-level product-of-sums and sum-of-products expressions for the following, using the criterion of least terms to determine minimality:

(a) $f = m_0 + m_3 + m_5 + m_8$: a problem in three variables
(b) $f = m_1 + m_5 + m_{11} + m_{12} + m_{13} + m_{14}$: a problem in four variables
(c) $f = m_0 + m_7$: how many simplest product-of-sums expressions are there?
(d) $f = M_1 \cdot M_3 \cdot M_4 \cdot M_5 \cdot M_6 \cdot M_7$
(e) $f = M_0 \cdot M_1 \cdot M_2 \cdot M_3 \cdot M_4 \cdot M_8 \cdot M_{13} \cdot M_{14}$
(f) $f = (M_0 \cdot M_1 \cdot M_2) + (m_0 + m_2 + m_5)$
(g) $f = m_1 + m_6 + m_7$

4-2. With the notation of Prob. 4-1, simplify the expressions for the following multiple-output networks, using the criterion of least variables:
(a) A network with three outputs y_1, y_2, and y_3, where

$$y_1 = \Sigma m_0, m_1, m_3, m_5$$
$$y_2 = \Sigma m_1, m_3, m_4, m_5$$
$$y_3 = \Sigma m_1, m_5, m_6, m_7$$

(b) A network with four outputs y_1, y_2, y_3, and y_4, where

$$y_1 = M_1 \cdot M_3 \cdot M_5 \cdot M_7$$
$$y_2 = M_3 \cdot M_5 \cdot M_7$$
$$y_3 = M_1 \cdot M_3 \cdot M_4 \cdot M_5$$
$$y_4 = M_1 \cdot M_2 \cdot M_3 \cdot M_4$$

4-3. Simplify the following expressions, allowing three-level logic in the network which realizes the expressions (the notation is that of Prob. 4-1):

(a) $f = m_0 + m_5 + m_6 + m_7$ (b) $f = m_2 + m_4 + m_7$
(c) $f = m_1 + m_6 + m_7$

4-4. Let m_1, m_2, ..., m_n be defined as in Prob. 4-1. Minimize the following expressions for a multiple-output network, using all three criteria:

$y_1 = m_{14} + m_{15}$ $y_2 = m_3 + m_{11} + m_{15}$
$y_3 = m_0 + m_{14} + m_{15}$ $y_4 = m_0 + m_3 + m_{11} + m_{14}$

4-5. Assume that the input combination $x_1 = 0$, $x_2 = 0$, $x_3 = 1$, and $x_4 = 0$ never occurs for the network in Prob. 4-2a (m_2 is a don't care). Minimize the expressions using this fact.

4-6. Let m_2, m_4, m_6, and m_7 be don't cares for the multiple-output network described in Prob. 4-4. Minimize the expressions using this fact.

4-7. Let α, β, and γ be Boolean product terms. Show that α implies $\beta + \gamma$ if and only if $\beta + \alpha\gamma$ is equivalent to $\beta + \alpha$ (Quine).

4-8. Show that if Φ is the sum-of-products canonical expansion for a given function f and Ψ is an irredundant sum-of-products expression equivalent to Φ, then any completion of a product term of Ψ is a term of Φ. [A completion of a product term α which is a term of an expression Ψ is formed by adding to α all possible combinations of complemented and uncomplemented literals not in α but in Ψ (Quine)].

4-9. Show that the only irredundant normal-form expressions which always have the value 1 are of the form $a + a'$, $b + b'$, etc. (Quine).

REFERENCES

1. Quine, W. V.: The Problem of Simplifying Truth Functions, *Am. Math. Monthly*, vol. 59, pp. 521–531, October, 1952.
2. Quine, W. V.: A Way to Simplify Truth Functions, *Am. Math. Monthly*, vol. 62, pp. 627–631, November, 1955.
3. Quine, W. V.: On Cores and Prime Implicants of Truth Functions, *Am. Math. Monthly*, vol. 66, pp. 755–760, November, 1959.
4. McCluskey, E. J., Jr.: Minimization of Boolean Functions, *Bell System Tech. J.*, vol. 35, pp. 1417–1444, November, 1956.
5. Samson, E. W., and B. E. Mills: "Circuit Minimization: Algebra and Algorithm for New Boolean Canonical Expensions," AFCRC-TR-54-21, Cambridge, Mass., 1954.
6. Petrick, S. R.: "A Direct Determination of the Irredundant Forms of a Boolean Function from the Set of Prime Implicants," AFCRC-TR-56-110, Cambridge, Mass., 1956.
7. Gazale, M. J.: Irredundant Disjunctive and Conjunctive Forms of a Boolean Function, *IBM J. Research and Develop.*, vol. 1, pp. 171–176, April, 1957.
8. Mott, T. H.: Determination of the Irredundant Normal Forms of a Truth Function by Iterated Consensus of the Prime Implicants, *IRE Trans. on Electronic Computers*, vol. EC-9, pp. 245–252, June, 1960.
9. Dunham, B., and R. Fridshal: The Problem of Simplifying Logical Expressions, *J. Symbolic Logic*, vol. 24, pp. 17–19, March, 1959.
10. Nelson, R. J.: Simplest Normal Truth Functions, *J. Symbolic Logic*, vol. 20, pp. 105–108, June, 1955.
11. Nelson, R. J.: Weak Simplest Normal Truth Functions, *J. Symbolic Logic*, vol. 20, pp. 232–234, September, 1955.
12. McNaughton, R., and B. Mitchell: The Minimality of Rectifier Nets with Multiple-outputs Incompletely Specified, *J. Franklin Inst.*, vol. 264, pp. 457–480, December, 1957.
13. Muller, D. E.: Application of Boolean Algebra to Switching Circuit Design and to Error Detection, *IRE Trans. on Electronic Computers*, vol. 3, pp. 6–12, September, 1954.
14. Polansky, R. B.: "Further Notes on Simplifying Multiple-output Switching Circuits," MIT Electronics Systems Laboratory, Cambridge, Mass., Mem. 7849-M-330, pp. 1–6, Oct. 26, 1959.
15. Bartee, T. C.: Computer Design of Multiple-output Networks, *IRE Trans. on Electronic Computers*, vol. EC-10, no. 1, March, 1959.
16. Roth, J. P.: Combinational Topological Methods in the Synthesis of Switching

Cirauits, *Proc. Intern. Symposium on Theory of Switching*, Harvard University, Cambridge, Mass., April, 1957.

17. Urbano, R. H., and R. K. Mueller: A Topological Method for the Determination of the Minimal Forms of a Boolean Function, *IRE Trans. on Electronic Computers*, vol. EC-5, pp. 126–132, September, 1956.
18. Harris, B.: An Algorithm for Determining Minimal Representations of a Logic function, *IRE Trans. on Electronic Computers*, vol. EC-6, pp. 103–108, June, 1957.
19. Staff of the Harvard Computation Laboratory: Annals of the Computation Laboratory, in "Synthesis of Electronic Computing and Control Circuits," vol. 27, Harvard University Press, Cambridge, Mass., 1951.
20. Karnaugh, M.: The Map Method for Synthesis of Combinational Logic Circuits, *Communs. and Electronics, Trans. AIEE*, part. I, vol. 72, November, 1953.
21. Veitch, E. W.: Third and Higher Order Minimal Solutions of Logical Equations, presented at Symposium on Switching Algebra, ICIP, Paris, France, June, 1959.
22. Ashenhurst, R. L.: The Decomposition of Switching Functions, Annuals of the Computation Laboratory, Harvard University Press, Cambridge, Mass., 1959.

5

Mathematical Foundations

5-1. The Boolean Ring. Chapter 2 introduced the concept of registers and functions of registers. Chapters 3 and 4 showed how functions of an n-cell register lead naturally to the introduction of Boolean algebra as an algebraic means of expressing such functional relations. In this chapter we shall reinvestigate the properties of Boolean algebra, but this time as an abstract entity, and shall follow this more rigorous treatment of Boolean algebra with an introduction to some of the mathematical concepts underlying the symbolic design techniques in this book.

We shall deviate considerably from the standard approach to Boolean algebra (which can be found elsewhere)[1] and introduce it via the concept of another equivalent entity, the Boolean ring.† Our reasons for this approach to the subject are twofold. Firstly, a minimal set of postulates for a ring sometimes appears more natural to one unacquainted with the subject than a minimal set of postulates for a Boolean algebra. Secondly, it is often useful to be able to think clearly in terms of a Boolean ring as well as in terms of the algebra.

First consider the postulates of a general algebraic ring, which the Boolean ring ultimately will satisfy as well. One should keep in mind, in reading these postulates, that the most familiar example of a ring is the set of positive and negative integers with their associated rules of addition and multiplication. Specifically, a ring may be described as follows:

A ring R is defined as a set of elements $a, b, c, \ldots$ such that for any two elements a and b in R, a sum $a \oplus b$ and a product $a \cdot b$, belonging to R, are uniquely defined.‡ These operations on R satisfy the following minimal set of postulates.

† This approach to Boolean algebra was first taken by M. H. Stone[2] in 1935.

‡ This chapter will refer to the $\oplus$ operation between variables as *addition*, a contraction of *mod 2 addition* as defined in preceding chapters. Later the logical addition operation $+$ defined in preceding chapters will be introduced. Calling $\oplus$ addition is in accordance with common usage in the field of modern algebra as is terming $a \oplus b$ the sum of a and b. Electrical engineers, computer designers, programmers, etc., use the term mod 2 addition, and we shall revert to this terminology after Chap. 5.

POSTULATES FOR A RING R

1. *Laws of addition*
   a. Associative law: $a \oplus (b \oplus c) = (a \oplus b) \oplus c$
   b. Commutative law: $a \oplus b = b \oplus a$
   c. Solvability of the equation $a \oplus x = b$: For all a and b in R there exists a solution x in R of the equation $a \oplus x = b$.
2. *Law of multiplication*
   Associativity: $a(bc) = (ab)c$
3. *Distributive laws*
   a. $a(b \oplus c) = ab \oplus ac$
   b. $(b \oplus c)a = ba \oplus ca$

As we have already mentioned, the reader may readily verify that the set of positive and negative integers satisfies the above set of postulates for a ring. Other examples of rings include the set of all positive and negative even integers, the set of all rational numbers, all complex numbers, and finally the set of all $n \times n$ matrices, all with their usual rules for addition and multiplication.

Each of the above examples of rings has a zero element and the property that every element of the ring has an additive inverse (or negative), i.e., if a is an element of the ring, there is an element $-a$ with the property $a + (-a) = 0$. Consequently, we infer that it is possible to deduce the existence and uniqueness of a zero element and additive inverses from the above postulates for a ring.

Lemma 5-1. There exists one and only one zero element 0 such that $0 \oplus a = a$ for all a in R. For every element a in R there exists a unique additive inverse $-a$ such that $a \oplus (-a) = 0$.

Proof. For an element a in R there exists at least one element, which we denote by 0_a, such that

$$0_a \oplus a = a$$

We now show the equation

$$0_a \oplus b = b$$

is true for all b in R. In other words, the particular element 0_a is a zero element for all elements of the ring. To show this consider a solution x_b in R of the equation

$$a \oplus x_b = b$$

where a is the same element used above and b is an arbitrary element in R. Then

$$0_a \oplus b = 0_a \oplus (a \oplus x_b) = (0_a \oplus a) \oplus x_b = a \oplus x_b = b$$

for all b in R. For the uniqueness of 0_a, suppose 0 is another zero in R, and substitute it for b in the above equation; then

$$0 = 0_a \oplus 0 = 0 \oplus 0_a = 0_a$$

so that a unique zero element, which we denote by 0, exists in R.   Thus the first part of the lemma is proved.

For the second part let $-a$ be an element in R which satisfies the equation

$$a \oplus x = 0$$

To prove its uniqueness, let c be another element satisfying this equation; then

$$a \oplus (-a) = a \oplus c = 0$$

If we add $-a$ to both sides of this equation,

$$(-a) \oplus [a \oplus (-a)] = (-a) \oplus (a \oplus c)$$

or by postulate $1a$

$$[(-a) \oplus a] \oplus (-a) = [(-a) \oplus a] \oplus c$$

so that

$$-a = c$$

Thus the lemma is proved.

In the following corollary the equation of postulate $1c$ is explicitly solved.

Corollary.   The equation $a \oplus x = b$ has the unique solution $x = -a \oplus b$ for all a and b in R.

Proof.   If we add $-b$ to both sides of the above equation, we have the equivalent equation

$$-b \oplus a \oplus x = 0$$

That the above solution satisfies this equation is evident.   The uniqueness of the solution follows from the lemma that the additive inverse of the element $-b \oplus a$ is unique.

The next postulate, combined with the previous postulates for the general ring, defines a Boolean ring B.   This ring will be shown later to be related to a Boolean algebra.

POSTULATES FOR A BOOLEAN RING B.   A Boolean ring B is a ring satisfying the preceding general postulates for a ring plus the following additional postulate:

4. *Idempotence law*
   For all a contained in B, $a \cdot a = a$.

The next lemma is concerned with some of the elementary properties of B which are deducible from the above postulates.

Lemma 5-2.   The Boolean ring B is commutative with respect to multiplication; that is, $ab = ba$ for all a and b in B.   Each element of B

is its own additive inverse, or $a \oplus a = 0$ for all a in B.   Finally, $0 \cdot a = 0$ for all a in B.

Proof.   The idempotence law implies that for all a and b in B,

$$a \oplus b = (a \oplus b)(a \oplus b) = a \cdot a \oplus a \cdot b \oplus b \cdot a \oplus b \cdot b$$
$$= (a \oplus b) \oplus ab \oplus ba$$

If we add $-(a \oplus b)$ to both sides of this equation,

$$ab \oplus ba = 0$$

or in particular, setting $b = a$, $a \oplus a = 0$, and

$$ab = ab \oplus 0 = ab \oplus (ab \oplus ba) = (ab \oplus ab) \oplus ba = ba$$

for ab in B.   Also $a = -a$ for all a in B, and both the first and second parts of the lemma are proved.   For the last part,

$$0 \cdot a = (b \oplus b)a = ba \oplus ba = 0$$

for all a in B, so that the lemma is proved.

It should be noted here that nothing has been said as yet about a unity element† for multiplication.   In a previous example of a general ring R, the set of all positive and negative even integers, there evidently was no need for a unity element for multiplication in order for this set of objects to satisfy the postulates for a ring.   Consequently, it would appear that it is not possible to deduce the existence of a unity element from the postulates for a general ring.   In order to ensure the existence of a unity element for multiplication, it would be necessary either to postulate its existence directly or to impose a postulate or condition from which its existence is deducible.   We shall show later by example that the imposition of the idempotence law on the general ring is not in general sufficient to guarantee the existence of a unity element in the Boolean ring B.   However, if the number of elements in the Boolean ring B is finite we shall be able to show that the law of idempotence alone enables one to deduce the existence of a unity element for multiplication in B.   We shall return to this topic after we discuss some examples of a Boolean ring in the next two sections.

5-2. Elementary Examples and the Boolean Field.   Since the operational rules for a Boolean ring, given in the postulates and lemmas, are closely related to the rules of arithmetic and in fact are less complicated than arithmetic, one suspects that examples of Boolean rings can be formed in arithmetic or the algebra of complex numbers.   The following elementary examples illustrate the plausibility of this inference.

---

† A unity element is an element I such that $aI = a$ for all a.

EXAMPLE 1. Let $B_0 = \{0,1\}$, the integers zero and one. For x and y in B_0, let

$$x \cdot y = xy$$

and
$$x \oplus y = x + y - 2xy$$

where the operations on the right are those of ordinary arithmetic. From this definition we readily obtain the addition and multiplication table shown in Table 5-1. The law of idempotence and the other ring postulates may be verified easily to show that B_0 is an example of a Boolean ring.

Table 5-1. Two Binary Operations in B_0

x	y	$x \cdot y$	$x \oplus y$
0	0	0	0
0	1	0	1
1	0	0	1
1	1	1	0

The next three examples are again examples of two-element Boolean rings; they will differ from the above example only in the meanings of their elements.

EXAMPLE 2. Let

$$\bar{0} = \{0, \pm 2, \pm 4, \pm 6, \ldots, \pm 2m, \ldots\}$$

and
$$\bar{1} = \{\pm 1, \pm 3, \pm 5, \pm 7, \ldots, \pm(2n + 1), \ldots\}$$

that is, $\bar{0}$ is the set of all positive and negative even integers, and $\bar{1}$ is the set of all positive and negative odd integers. These two sets or classes of integers are usually called the residue classes of the even and odd integers, respectively. Let the sum of two such sets of integers be the total set of sums of integers in the respective sets; e.g.,

$$\bar{0} \oplus \bar{1} = \text{set of integers of form } 2m + 2n + 1$$
$$= \text{set of integers of form } 2(m + n) + 1$$
$$= \bar{1}$$

since n and m are defined on all positive and negative integers. Similarly, let the product of two such sets of integers be the total set of products of the integers in the respective sets. It is not difficult to show that the addition and multiplication table for these two elements is the same as that given in Table 5-1 if 0 is replaced by $\bar{0}$ and 1 is replaced by $\bar{1}$.

EXAMPLE 3. Let a and b be two distinct complex numbers and let

$$x \cdot y = \frac{(x - a)(y - a)}{b - a} + a$$

$$x \oplus y = b - \frac{(b - x)(b - y) + (x - a)(y - a)}{b - a}$$

where the operations on the right are those of complex numbers. By a direct calculation it is easy to show that the two elements a and b satisfy Table 5-1 with 0 replaced by a and 1 replaced by b, so that the two elements a and b with the above operations again satisfy the postulates of a Boolean ring.

EXAMPLE 4. Let T and F mean "true" and "false," respectively. Let x and y be statements, and define

$$x \cdot y = x \text{ and } y$$
$$x \oplus y = x \text{ or } y \text{ but not both (exclusive or)}$$

Since $x \cdot y$ is true only when both x and y are true and $x \oplus y$ is true only when x is false and y is true or x is true and y is false, the two va ues T and F satisfy the truth table of Table 5-1, with 0 replaced by F and 1 replaced by T. Thus if B_L consists of the two elements F and T, it is a Boolean ring.

It might have been noticed that the elementary Boolean rings in the above four examples satisfy a constraint that we did not impose in the general definition of a Boolean ring. This constraint has to do with the fact that a multiplicative inverse of the nonzero element exists in these examples; e.g., since $\bar{1} \cdot \bar{1} = \bar{1}$, the element $\bar{1}$ is a unity element for multiplication as well as its own multiplicative inverse. More generally these rings satisfy the division law for multiplication.

5. *Division law*

For all a and b not equal to zero in B, there exists a solution x in B of the equation $ax = b$.

If a commutative ring R with multiplicative unity element satisfies this postulate, it is usually called a field. We see what happens if we impose this postulate on a Boolean ring B in the following theorem.

Theorem 5-1. If the above division law is satisfied by a Boolean ring B, it will be called a *Boolean field* B_F. B_F has precisely two elements, a zero element 0 and a unity I. The addition and multiplication table of B_F is given in Table 5-2.

Table 5-2. Two Binary Operations in B_F

x	y	$x \cdot y$	$x \oplus y$
0	0	0	0
0	I	0	I
I	0	0	I
I	I	I	0

Proof. Let a and b be two nonzero elements in B. By the law of idempotence a is a solution of the equation $ax = a$. Assume the division law; let c be a solution of $ax = b$. Then

$$ab = a(ac) = (a \cdot a)c = ac = b$$

If we replace b by a and a by b and again use the above argument, we obtain also $ba = a$, so that, by commutativity, $a = b$ for all nonzero elements a and b in B. Thus all the nonzero elements of B are the same element, which we denote by I. The addition and multiplication tables for the two elements of B_F follow immediately from the law of idempotence and lemma 5-2. Hence the theorem is proved.

Since the addition and multiplication tables of B_F and the Boolean fields of the preceding four examples are identical except for the meaning and labeling of their corresponding elements, we say that these fields are *isomorphic* to one another. Two rings, A and B, with elements a_1, a_2, a_3, . . . and b_1, b_2, b_3, . . . , are isomorphic if there is a one-to-one function $\alpha(a_i) = b_j$ from A onto B such that $\alpha a_i \oplus \alpha a_j = \alpha(a_i \oplus a_j)$ and $(\alpha a_i) \cdot (\alpha a_j) = \alpha(a_i \cdot a_j)$ for all a_i and a_j in A.

For this reason, unless one is interested in preserving the identity or the symbology of the elements, there would be no loss of information to speak only of B_F when discussing Boolean fields. Since B_F was obtained deductively from a set of operational constraints or postulates imposed on an arbitrary collection of elements, its two elements are arbitrary except for the operational tables which relate them. Consequently, B_F is isomorphic to all Boolean fields which can be constructed with specified elements; it is *the* abstract Boolean field.

Let us now illustrate isomorphism between fields more explicitly. In particular let us establish a one-to-one correspondence between B_0 and B_F. This can be done letting a function $\phi(x)$ of the two values in B_0 be the two corresponding values in B_F. That is,

and
$$\phi(0) = 0$$
$$\phi(1) = I$$

in B_F. By the operational equivalence of Tables 5-1 and 5-2 or by direct verification we can show that

and
$$\phi(x \oplus y) = \phi(x) \oplus \phi(y)$$
$$\phi(xy) = \phi(x)\phi(y)$$

so the fields are isomorphic.

5-3. Vector-space Boolean Rings. The Boolean rings presented in the last section were the most elementary examples, except for the trivial example of a ring consisting only of a zero element. The next three examples of Boolean rings will be more general in that their elements will be vectors with components in the abstract Boolean field B_F. The first example of such a ring will be an n-dimensional vector space B_n over the field B_F.

EXAMPLE 5. Let B_n consist of the set of vectors of the form $(a_1, a_2, \ldots, a_n)$, where a_i for $(i = 1, 2, \ldots, n)$ is either the element 0 or the element I in the Boolean field B_F. Evidently there are 2^n different vectors in B_n. Let the addition of two elements in B_n be the usual addition of corresponding components of the vectors; that is, if

$$a = (a_1, a_2, \ldots, a_n) \quad \text{and} \quad b = (b_1, b_2, \ldots, b_n)$$

then $\qquad a \oplus b = (a_1 \oplus b_1, a_2 \oplus b_2, \ldots, a_n \oplus b_n)$

Similarly, for multiplication let

$$a \cdot b = (a_1 b_1, a_2 b_2, \ldots, a_n b_n)$$

Let $a = (a_1, a_2, \ldots, a_n)$, $b = (b_1, b_2, \ldots, b_n)$, and $c = (c_1, c_2, \ldots, c_n)$. To show that postulate $1a$ of the postulates for a ring R is satisfied by B_n,

$$
\begin{aligned}
a \oplus (b \oplus c) &= (a_1, a_2, \ldots, a_n) \oplus [(b_1, b_2, \ldots, b_n) \oplus (c_1, c_2, \ldots, c_n)] \\
&= (a_1, a_2, \ldots, a_n) \oplus (b_1 \oplus c_1, b_2 \oplus c_2, \ldots, b_n \oplus c_n) \\
&= [a_1 \oplus (b_1 \oplus c_1), a_2 \oplus (b_2 \oplus c_2), \ldots, a_n \oplus (b_n \oplus c_n)] \\
&= [(a_1 \oplus b_1) \oplus c_1, (a_2 \oplus b_2) \oplus c_2, \ldots, (a_n \oplus b_n) \oplus c_n] \\
&= (a_1 \oplus b_1, a_2 \oplus b_2, \ldots, a_n \oplus b_n) \oplus (c_1, c_2, \ldots, c_n) \\
&= [(a_1, a_2, \ldots, a_n) \oplus (b_1, b_2, \ldots, b_n)] \oplus (c_1, c_2, \ldots, c_n) \\
&= (a \oplus b) \oplus c
\end{aligned}
$$

so that the elements of B_n are associative with respect to addition. The remainder of the postulates, except for postulate $1c$, may be verified in a similar manner.

For postulate $1c$ we know from the properties of the Boolean field B_F that, for a_m and b_m in B_F, each equation of the set

$$a_m \oplus x_m = b_m$$

for $m = 1, 2, \ldots, n$ is solvable. That is, for each a_m and b_m in B_F there is a solution $x_m = a_m \oplus b_m$ in B_F for $m = 1, 2, \ldots, n$ and the vector of solutions

$$x = (x_1, x_2, \ldots, x_n) = (a_1 \oplus b_1, a_2 \oplus b_2, \ldots, a_n \oplus b_n)$$

is an element of B_n. Now the solvability of this set of equations is equivalent to the solvability of the equation

$$(a_1 \oplus x_1, a_2 \oplus x_2, \ldots, a_n \oplus x_n) = (b_1, b_2, \ldots, b_n)$$

componentwise for each x_m. But by the definition of addition in B_n and the fact that the above vector of solutions x is an element of B_n, the solvability of the latter equation is further equivalent to the solvability of the vector equation

$$(a_1, a_2, \ldots, a_n) \oplus (x_1, x_2, \ldots, x_n) = (b_1, b_2, \ldots, b_n)$$

or finally equivalent to the solvability of

$$a \oplus x = b$$

for all a and b in B_n. Hence postulate 1c is satisfied.

It should be noted here that B_n always has a unity for multiplication; the element $(I = I, I, \ldots, I)$ is such that for any a in B_n, $aI = a$. In the next section we shall show that any Boolean ring with only a finite number of elements possesses a unity for multiplication and later on we shall show that any such ring always contains exactly 2^n elements for some integer n. In fact, as one might suspect, it is possible to show that any Boolean ring with a finite number of elements is isomorphic to the Boolean ring B_n for some n; B_n is a prototype for all Boolean rings having only a finite number of elements.

In order to motivate the next examples, let us first show that B_n is equivalent to a ring of functions on B_F of a discrete variable m (where $m = 1, 2, 3, \ldots, n$). Consider the set of all functions of the form $a(m)$ where $a(m) = 0$ or 1 in B_F for each m. Define addition and multiplication between two such functions by

$$(a \oplus b)(m) = a(m) \oplus b(m)$$
and
$$(ab)(m) = a(m)b(m)$$

Evidently $a(m)$ for some integer m corresponds to the mth component a_m of some n-dimensional vector a in B_n; moreover, the sum $a \oplus b$ and product ab, as defined above for functions a and b, are equivalent to the way in which addition and multiplication were defined between the corresponding vectors a and b. We will now use the functional formalism to illustrate an example of a Boolean ring of vectors where each vector has a countably infinite number of components.

EXAMPLE 6.　Let B_I consist of the set of all functions (or vectors) of the form $a(m)$, where m runs through the set of positive integers (1,2,3, . . .) and where $a(m)$ for a particular value of m is either the element 0 or the element I in the Boolean field B_F. If a and b are two functions in B_I, the operations of addition and multiplication are defined respectively by

$$(a \oplus b)(m) = a(m) \oplus b(m)$$
and
$$(ab)(m) = a(m)b(m)$$

for $(m = 1, 2, 3, \ldots)$. One can verify that B_I satisfies the postulates of a Boolean ring in the same manner in which they were verified in the previous example. This ring has for unity element the function $I(m) = I$ for $(m = 1, 2, \ldots)$ and for zero element the function $0(m) = 0$ for $(m = 1, 2, \ldots)$.

The elements of the above examples are two-valued functions of the positive integers. Let us henceforth call such two-valued functions of some variable *Boolean functions*. Then we have shown that B_I, the set of all Boolean functions of the discrete variable m, where $(m = 1, 2, 3, \ldots)$, is a Boolean ring.

The next example will be very general in that it will be the ring of Boolean functions of an arbitrary variable λ. The arbitrariness of λ means that the domain S (or the set of objects), the elements of which λ assumes as it varies, can be chosen arbitrarily, though in any particular instance S must remain fixed and well defined when once it has been chosen. For example, our choice for domain S might have been the integers from 1 to n, all positive integers, or perhaps the set of all real numbers from $-\infty$ to $+\infty$. We already have examined two choices for S; these two choices gave rise to the Boolean rings B_n and B_I of Examples 5 and 6, respectively. As a consequence our next example will include, as well as generalize, the preceding examples of this section.

EXAMPLE 7. Let S be some set of objects $(a, b, \ldots)$ and let λ denote an arbitrary member of this set; λ is a variable with domain S. Let B_S be the set of all functions $f(\lambda)$, where $f(\lambda)$ for a particular value of λ in S is either the element 0 or the element I in the Boolean field B_F. If f and g are two functions in B_S with domain S, the rules for the operations between the two functions are

$$(f \oplus g)(\lambda) = f(\lambda) \oplus g(\lambda)$$

and

$$(fg)(\lambda) = f(\lambda)g(\lambda)$$

for each value of λ in S.

Evidently the zero element of B_S is a function $0(\lambda)$ such that $0(\lambda) = 0$ for all λ in S; a unity in B_S is the function $I(\lambda)$ such that $I(\lambda) = I$ for all λ in S. Let us show that the distributive law $3a$ is satisfied. If f, g, and h are Boolean functions in B_S, then

$$[f(g \oplus h)](\lambda) = f(\lambda)[(g \oplus h)(\lambda)] = f(\lambda)[g(\lambda) \oplus h(\lambda)]$$
$$= f(\lambda)g(\lambda) \oplus f(\lambda)h(\lambda) = fg(\lambda) \oplus fh(\lambda)$$
$$= (fg \oplus fh)(\lambda)$$

Thus this postulate is satisfied. The remainder of the postulates, except $1c$, are satisfied in a similar manner. For postulate $1c$, the equation

$$f(\lambda) \oplus x(\lambda) = g(\lambda)$$

for any f and g in B_S, is solvable for any λ in S, and the solution is an element of B_S. But by the definition of addition in B_S, the solvability of this equation is equivalent to the solvability of the equation

$$(f \oplus x)(\lambda) = g(\lambda)$$

for any f and g in B_S, and postulate $1c$ is satisfied. Since the postulates for a Boolean ring are satisfied, B_S is an example of a very general Boolean ring.

Further examples of Boolean rings will be introduced later. It suffices to say at this point that most of the rings introduced later will be specializations of B_S or subrings of these specializations. In the next section we return to the postulates and consider the question of whether or not a unity element for multiplication exists in all abstract Boolean rings.

5-4. The Unity Element for Multiplication. The questions that were raised in the latter part of Sec. 4-2, concerning the existence of a unity element for multiplication in the Boolean ring, will now be examined. We first show that a Boolean ring with a finite number of elements always has a unity element. In order to accomplish this, it will be convenient to introduce another operation in the Boolean ring, which for the present we shall call the *join* operation. As we shall see in the next section, the join of two elements in a Boolean ring will be equivalent to the logical addition operation between two elements in a Boolean algebra defined in Chap. 3.

Definition 5-1. The Join Operation. For any two elements a and b in a Boolean ring B, the *join* $+$ is defined by

$$a + b = a \oplus b \oplus ab$$

The next lemma restates the rules given in Sec. 3-10, for the join operation.

Lemma 5-3. If a, b, and c are arbitrary elements in a Boolean ring B, the join between elements of B satisfies the following relations:

1. *Idempotence law*

$$a + a = a$$

2. *Commutative law*

$$a + b = b + a$$

3. *Associative law*

$$a + (b + c) = (a + b) + c$$

4. *Absorption laws*

$$a(a + b) = a$$
$$a + ab = a$$

5. *Distributive laws*

$$a(b + c) = ab + ac \qquad \text{and} \qquad a + bc = (a + b)(a + c)$$

Deductive Proofs. For the idempotence law:

$$a + a = a \oplus a \oplus a \cdot a = (a \oplus a) \oplus a = 0 \oplus a = a$$

The commutative law follows from an inspection of the definition of the join operation. For the associative law:

$$
\begin{aligned}
a + (b + c) &= a \oplus (b \oplus c \oplus bc) \oplus a(b \oplus c \oplus bc) \\
&= a \oplus b \oplus c \oplus bc \oplus ab \oplus ac \oplus abc \\
&= (a \oplus b \oplus ab) \oplus c \oplus (a \oplus b \oplus ab)c = (a + b) + c
\end{aligned}
$$

Let us prove only the first absorption law; the other can be verified in a similar manner.

$$
\begin{aligned}
a(a + b) &= a(a \oplus b \oplus ab) \\
&= a \cdot a \oplus ab \oplus a \cdot ab \\
&= a \oplus ab \oplus ab \\
&= a \oplus 0 = a
\end{aligned}
$$

Finally, for the same reason we verify only the first distributive law, as follows:

$$
\begin{aligned}
a(b + c) &= a(b \oplus c \oplus bc) \\
&= ab \oplus ac \oplus abc \\
ab \oplus ac \oplus (ab)(ac) &= ab + ac
\end{aligned}
$$

Thus the lemma is established. We are now in the position to prove the following theorem.

Theorem 5-2. A Boolean ring B with only a finite number of elements has a unity element for multiplication; that is, there exists in B an element I such that $aI = a$ for all a in B.

Proof. Consider the join of all elements of B, that is, the expression Σb over all elements b in B. By definition 5-1 and the associative law for the join in the previous lemma, Σb is expressible as a finite sum of products of elements in B. Hence Σb is an element of B. Let a be an element of B; then

$$a(\Sigma b) = a(a + \Sigma^* b)$$

where the star on the summation sign means the join of all elements of B, except a. The theorem clearly follows from the first absorption law, and Σb is the unity I for multiplication.

Let us now give an example of a Boolean ring with an infinite number of elements which does not possess a unity for multiplication. This example will show the necessity in general of postulating the existence of a unity element for multiplication in a Boolean ring if the need exists for a unity element in the ring.

EXAMPLE 8. Let B_H be the set of all sequences of elements from the Boolean field B_F such that no sequence has more than a finite number of

I's.   In other words B_H is the set of elements of the form

$$(a_1, a_2, \ldots, a_n, \ldots)$$

where a_i is either the element 0 or the element I from B_F, and where each sequence is restricted to having only a finite number of I's.   Evidently the set B_H is included or contained in the Boolean ring B_I of Example 6; that is, B_H is a subset of B_I.   That B_H is a Boolean ring can be verified in a manner similar to that used for Example 6, so that B_H is a Boolean *subring* of B_I.   We now make the assumption that B_H has a unity element for multiplication and show that this assumption leads to a contradiction. Let the unity be the sequence

$$I_H = (b_1, b_2, b_3, \ldots)$$

Since I_H is an element of B_H, only a finite number of b_n's of the sequence can be the element I of B_F, so for some k, $b_k = 0$.   Let c be the nonzero element

$$c = (c_1, c_2, c_3, \ldots)$$

such that
$$c_i = \begin{cases} 0 & \text{for } i \neq k \\ I & \text{for } i = k \end{cases}$$

for $(i = 1, 2, 3, \ldots)$.   Certainly c is an element of B_H.   But

$$cI_H = 0 \neq c$$

so that our assumption must have been wrong.   Hence the Boolean ring B_H contains no unity element for multiplication.

If a Boolean ring does contain a unity element for multiplication, there still remains a question as to its uniqueness.   In the next theorem this question is answered.

Theorem 5-3.   If a Boolean ring B has a unity element for multiplication, that is, if there exists an element I in B such that $aI = a$ for all a in B, then I is the only such element in B which has this property.   The unity is unique.

Proof.   Suppose there is another element I^* such that $aI^* = a$ for all a in B.   Then

$$I^* \oplus I = II^* \oplus I = I \oplus I = 0$$

so that $I^* = I$.   Hence the theorem is proved.

In the next section we will define a Boolean algebra and show that it is equivalent to a Boolean ring with unity element.

5-5. The Boolean Ring as a Boolean Algebra.   In Sec. 5-1, the Boolean ring B was defined to be a set of elements $(a, b, c, \ldots)$ such that for any two elements a sum $a \oplus b$ and a product $a \cdot b$, belonging to B, were uniquely defined.   An operational structure was created by imposing the postulates for a Boolean ring B in Sec. 5-1.   These were the postulates for the algebraic ring plus the idempotence law.   In the last section we showed that a Boolean ring did not contain, in general, a unity element for multiplication, but that if it did contain a unity, the unity was unique.

This section will show that a Boolean ring with unity is equivalent algebraically to a Boolean algebra—in particular the algebra created in Secs. 3-2 to 3-4 to describe the functions or transformations of n-cell registers.   We first show that the product operation, the join operation (introducted in the last section), and the complement operation (which we will presently define) give rise to certain relations in a Boolean ring with unity which are equivalent to the Boolean algebra theorems of Sec. 3-4.   It will then be shown that the postulates for a Boolean ring with unity can be deduced from the postulates for a Boolean algebra.   This will establish an equivalence between these two entities which will make it possible for one to use their languages interchangeably in further examples.

Now we define the complement operation in a Boolean ring with unity. This operation will be operationally equivalent to the complement operation defined in Sec. 3-2.

Definition 5-2.   *Complement Operation.*   If a is any element in a Boolean ring with unity, the complement of a is

$$a' = a \oplus I$$

where I is the unity for multiplication.

We can now prove the following lemma, which shows that the rules for a Boolean algebra are true in a Boolean ring with unity.

Lemma 5-4.   If a, b, and c are arbitrary elements in a Boolean ring B with unity I, the following set of relations between the product, join, and complement operations hold (where join and complement are defined, respectively, in definitions 5-1 and 5-2):

1. *Idempotence laws*

$$a \cdot a = a \quad \text{and} \quad a + a = a$$

2. *Commutative laws*

$$ab = ba \quad \text{and} \quad a + b = b + a$$

3. *Associative laws*

$$a(bc) = (ab)c \quad \text{and} \quad a + (b + c) = (a + b) + c$$

4. *Distributive laws*

$$a(b + c) = ab + ac \quad \text{and} \quad a + bc = (a + b)(a + c)$$

5. *The zero and unity laws*

$$0 \cdot a = 0 \quad \text{and} \quad Ia = a$$
$$0 + a = a \quad \text{and} \quad I + a = I$$

6. *The laws of complementation*

   a. Complementarity: $aa' = 0$ and $a + a' = I$
   b. Dualization: $(ab)' = a' + b'$ and $(a + b)' = a'b'$
   c. Involution: $(a')' = a$

Proof.   Rules 1 to 4 and the first part of rule 5 were either properties of the Boolean ring or were proved in lemma 5-3.   To prove the latter part of rule 5,

$$0 + a = 0 \oplus a \oplus 0 \cdot a = a$$
and $$I + a = I \oplus a \oplus Ia = I \oplus a \oplus a = I \oplus 0 = I$$

The first law of complementation obtains as follows:

$$aa' = a(I \oplus a) = a \oplus a \cdot a = a \oplus a = 0$$
$$a + a' = a \oplus a \oplus I \oplus a \oplus a = I$$

Let us show only the first dualization rule; the other follows similarly.

$$a' + b' = a' \oplus b' \oplus a'b' = a \oplus I \oplus b \oplus I \oplus (a \oplus I)(b \oplus I)$$
$$= a \oplus b \oplus (ab \oplus a \oplus b \oplus I) = ab \oplus I = (ab)'$$
Finally, $$(a')' = a' \oplus I = (a \oplus I) \oplus I = a$$

so that the rule for involution holds and the lemma is proved.

Let us now define a Boolean algebra.

Definition 5-3.   A Boolean algebra is the set of objects 0, I, a, b, c, d, . . . , such that for any two elements a and b a product ab, a join $a + b$, and a complement a', belonging to the algebra, are uniquely defined. These operations satisfy rules 1 to 5 of lemma 5-4.

The next lemma establishes the fact that a Boolean algebra, as defined above, is likewise a Boolean ring with unity.

Lemma 5-5.   If B is a Boolean algebra, as defined in definition 5-3, then B is a Boolean ring with unity with respect to the product ab and sum

$$a \oplus b = a'b + ab'$$

where a and b are elements in B.

Proof.   The proof consists of a verification of the postulates for a Boolean ring with unity.   First note that $I' = (I + I')' = I'I = 0$ so that for d in B,

$$d \oplus I = d'I + d0 = d'$$

Thus, for $1a$ of the postulates for a ring

$$
\begin{aligned}
a \oplus (b \oplus c) &= a'(b \oplus c) + a[(b \oplus I) \oplus c] \\
&= a'(b'c + bc') + a(bc + b'c') \\
&= a'b'c + a'bc' + abc + ab'c' \\
&= (ab + a'b')c + (a'b + ab')c' = (a \oplus b) \oplus c
\end{aligned}
$$

Proof of postulate $1b$ is elementary and $1c$ may be demonstrated by showing that $a \oplus x = b$, or its equivalent,

$$a'x + ax' = b$$

has a solution.   Let us try for x the quantity $a'b + ab' = a \oplus b$.   First

$$x' = a' \oplus b = a'b' + ab$$

so that
$$
\begin{aligned}
a'x + ax' &= a'(a'b + ab') + a(a'b' + ab) \\
&= a'b + ab = (a' + a)b = b
\end{aligned}
$$

by the rules of lemma 5-4.   Hence postulate $1c$ is satisfied.   Postulate 2 is equivalent to rule 3 of lemma 5-4.   Let us prove only the first distributive law $3a$, for $3b$ follows similarly.   For this,

$$
\begin{aligned}
ab \oplus ac &= (ab)'ac + (ab)(ac)' \\
&= (a' + b')ac + ab(a' + c') \\
&= ab'c + abc' = a(b \oplus c)
\end{aligned}
$$

The idempotence for multiplication is shown by rule 1 of lemma 5-4, and the rule for multiplication by a unity follows from rule 5; hence the postulates for a Boolean ring with unity are satisfied, and the lemma is proved.

Let us now combine the previous two lemmas into the following, now evident, theorem:

Theorem 5-4.   A Boolean ring B with unity with respect to the operations $\cdot$ and $\oplus$ is likewise a Boolean algebra B with respect to the operations $+$ and $'$, and conversely.   The nonproduct operations between these two algebraic structures are related as follows:

$$
\begin{aligned}
a' &= a \oplus I \\
a + b &= a \oplus b \oplus ab \\
a \oplus b &= a'b + ab'
\end{aligned}
$$

Henceforth it will be convenient to assume that a Boolean ring has a unity unless otherwise specified; consequently all further Boolean rings will be Boolean algebras as defined in definition 5-3.

5-6. Boolean Algebra of Subsets.   Consider, now, Boolean algebras of the subsets of a set.   Although set theory is discussed more fully elsewhere (e.g., see Feller,[3] Halmos[4]), for completeness we will briefly discuss set theory as another example of a Boolean algebra.

Definition 5-4. Sets and Notation. A *set* is an arbitrary collection of objects or elements $(a,b,c, \ . \ . \ .)$. If a is an element of set A, the notation

$$a \in A$$

means that a belongs to A. Set A is called a *subset* of B if every element of A is an element of B; the notation

$$A \subset B \quad \text{or} \quad B \supset A$$

means that A is a subset of B. Two sets A and B are said to be *equal* if every element of A is in B and vice versa; that is, $A = B$ if and only if

$$A \subset B \quad \text{and} \quad B \subset A$$

The set which contains no elements is called the *empty* set and is denoted by the symbol 0. A collection or set of subsets of some entire set Ω is called a *class* C_Ω of subsets of Ω. The entire set Ω is sometimes referred to as the *space* to which the class C_Ω refers, and elements of Ω are sometimes referred to as *points* of space Ω.

Let us now define the operations between the subsets of space Ω. It will be with respect to these operations that a class C_Ω may be a Boolean algebra.

Definition 5-5. Set Operations. The common part of two sets A and B is denoted by $A \cap B$ and is called the *intersection* of the two sets. All the elements of A, together with all the elements of B, comprise a set $A \cup B$, which is called the *union* of the two sets. If Ω is a space and set A is part of Ω, then the set of elements (points) which belong to Ω but not to A is called the complement of A and is denoted by $\bar{A}$. The *symmetric difference* of two sets A and B is the set of all elements of A together with all elements of B except for the part common to both A and B; this operation is denoted by $A \triangle B$.

The set operations just defined obey the following laws. Let A, B, and C be subsets of space Ω.

1. *Idempotence laws*

$$A \cap A = A \quad \text{and} \quad A \cup A = A$$

2. *Commutative laws*

$$A \cap B = B \cap A \quad \text{and} \quad A \cup B = B \cup A$$

3. *Associative laws*

$$A \cap (B \cap C) = (A \cap B) \cap C$$
and
$$A \cup (B \cup C) = (A \cup B) \cup C$$

4. *Distributive laws*

$$A \cap (B \cup C) = (A \cap B) \cup (A \cap C)$$
and
$$A \cup (B \cap C) = (A \cup B) \cap (A \cup C)$$

5. *Zero and unit laws*

$$0 \cap A = 0 \quad \text{and} \quad \Omega \cap A = A$$
$$0 \cup A = A \quad \text{and} \quad \Omega \cup A = \Omega$$

6. *Laws of complementation*
   a. Complementarity:

$$A \cap \bar{A} = 0 \quad \text{and} \quad A \cup \bar{A} = \Omega$$

   b. Dualization:

$$\overline{A \cap B} = \bar{A} \cup \bar{B} \quad \text{and} \quad \overline{A \cup B} = \bar{A} \cap \bar{B}$$

   c. Involution:

$$(\bar{\bar{A}}) = A$$

7. *The laws relating symmetric differences with unions and complements*

$$A \,\Delta\, B = (\bar{A} \cap B) \cup (A \cap \bar{B})$$
and
$$A \cup B = A \,\Delta\, B \,\Delta\, (A \cap B)$$

Let us show the method of proof for the above set of rules by demonstrating the first distributive law of rule 4. The proofs for the remainder of the rules are similar.

To prove the distributive law

$$A \cap (B \cup C) = (A \cap B) \cup (A \cap C)$$

suppose first that a is an element of the left-hand side of the identity; that is, suppose

$$a \in A \cap (B \cup C)$$

Then a is in both A and either of the sets B or C, or both, so that by definition 5-5 a is in set $A \cap B$ or set $A \cap C$ or both. Thus

$$a \in (A \cap B) \cup (A \cap C)$$

and since a was arbitrary,

$$A \cap (B \cup C) \subset (A \cap B) \cup (A \cap C)$$

so that the left-hand side of the identity is contained in the right side of the identity. By the definition of equality in definition 5-4, if we now show that the right side is contained in the left side, the proof will be finished. Suppose

$$b \in (A \cap B) \cup (A \cap C)$$

then b is in the common part of A and B or in the common part of A and C or in both of these common parts. Hence b must be in set A as well as in either of the sets B or C or both. Thus

$$b \in A \cap (B \cup C)$$

and consequently

$$A \cap (B \cup C) \supset (A \cap B) \cup (A \cap C)$$

so that the rule is established.

If we let intersection correspond to product, union correspond to join, and the complement of a set correspond to the complement of an element of a Boolean algebra, and let the empty set 0 correspond to 0 of a Boolean algebra and the space Ω correspond to the unity I, then the above rules 1 to 6 correspond precisely to the set of relationships 1 to 6 of lemma 5-4. Thus, in accordance with definition 5-3, if C_Ω is a class of subsets 0, Ω, A, B, C, . . . of space Ω such that for any two elements A and B of C_Ω, the intersection $A \cap B$, union $A \cup B$, and complement $\bar{A}$ also belong to class C_Ω, then C_Ω is a Boolean algebra. If unions and intersections and complements of elements of a class are likewise elements of the class, the class is said to be *algebraically closed*. We can now state the following theorem:

Theorem 5-5. If C_Ω is a class of subsets of space Ω which includes the empty set 0 and set Ω and if C_Ω is algebraically closed, then C_Ω is a Boolean algebra with respect to the operations of intersection, union, and complement. With respect to the operations of intersection and symmetric difference, C_Ω is likewise a Boolean ring with unity Ω.

The latter statement of the above theorem follows clearly from rule 7 of this section and theorem 5-4. Evidently, if C_Ω is a class of subsets of a space Ω which satisfies the hypotheses of the above theorem, C_Ω is the sixth example of a Boolean algebra in this chapter.

5-7. The Concept of Mapping. In Secs. 2-5 and 5-2 and in Example 5 of section 5-3, special cases of functions which map one set into another set have been discussed. Here we formalize the concept of a mapping and apply it to some examples.

Let A and B be two sets of elements. If there corresponds to each element $a \in A$ a unique element $f(a) = b$, where $b \in B$, then f is called a *single-valued* function of set A into set B, and the element $f(a)$ in B is termed the *image* or *function value* of the element a. Such a correspondence between sets A and B is often termed a *mapping* of set A *into* set B (since the set of all images of elements in A need not exhaust the set B). The set A over which the mapping is defined is called the *domain of definition* or *domain*. The set of all images, usually denoted by $f(A)$, is known as the *range* or *image space* of the mapping. If $f(A) = B$, that is, if for each $b \in B$, $f(a) = b$ for some a in A, then the function f determines a mapping of the set A *onto* set B.

If f is a function which determines a mapping of set A into set B (that is, if f is a single-valued function of elements of set A with values in set B), then an element a which has image $f(a) = b$ is called an *inverse image* of b. Generally there is more than one inverse image in domain A of an element in range B (in Fig. 5-1, a and a' in set A are both inverse images of the

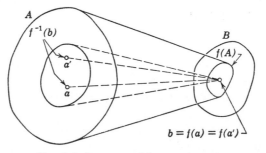

Fig. 5-1. Images and inverse images.

element b). The total set of inverse images of an element b in the image space is called the *complete* inverse image of b and is denoted by $f^{-1}(b)$. Another notation for $f^{-1}(b)$, which is more explicit, is the brace notation

$$f^{-1}(b) = \{x | f(x) = b\}$$

denoting that set of elements (points) x in A for which $f(x) = b$. Evidently (see also Fig. 5-1) $f^{-1}(b)$ is a subset of domain A.

If every complete inverse image contains only *one* element of set A, then the sets A and $f(A)$ are said to be in *one-to-one correspondence* under the mapping f (mapping determined by function f). In this case f^{-1} is a single-valued function of the original image space $f(A)$ onto set A, so that an *inverse mapping* of set $f(A)$ onto A is determined by f^{-1}.

If each complete inverse image of a mapping f contains in general more than one element of the set A, then each complete inverse image of an element in $f(A) \subset B$ is a unique subset of A, and the set of all complete inverse images of elements in $f(A)$ constitutes a subclass (which we denote by $C_{f^{-1}}$) of the class of all subsets of A. Hence, in this case f^{-1} can be regarded as a single-valued function of $f(A)$ with values in the class of all subsets of A, and $C_{f^{-1}}$ is the range of this mapping. Since originally f was a single-valued function of A into B, each element of class $C_{f^{-1}}$ is a subset of A which corresponds to one and only one element of $f(A)$. Hence each complete inverse image of the mapping f^{-1} of $f(A)$ onto $C_{f^{-1}}$ can contain one and only one element of $f(A)$. Thus the set $f(A)$ and the class $C_{f^{-1}}$ of inverse images are in one-to-one correspondence.

In other words the class of all complete inverse images of a mapping is in one-to-one correspondence with the range of the mapping. Evidently $C_{f^{-1}}$ is a *disjoint* class of subsets; that is, a class in which no two sets have elements in common. Such a class is sometimes called a partition of set A. Consider now some examples of mappings.

EXAMPLE 1. For the first example we discuss the mappings of the space of n-cell registers into the space of m-cell registers. As defined in Sec. 2-5, the function $D = f(C)$ of an n-cell register C with values in the m-dimensional vector space V_m is a single-valued function of the set V_n of 2^n points into the set V_m of 2^m points. Hence f determines a mapping of V_n *into* V_m. If $m \leq n$, V_m can be regarded as a subset of V_n, and f may be regarded as a mapping of V_n into itself. Finally, since f and m are arbitrary, the class of all dependent registers of V_n illustrates a particular manner in which the class of all mappings of a set of 2^n objects into itself can be realized.

EXAMPLE 2. As we shall see, this example of a mapping is intimately related to Example 7 of Sec. 5-3. Let C_Ω be a class of subsets of space Ω, such that C_Ω is a Boolean algebra with respect to the set operations intersection and symmetric difference. With each subset $A \in C_\Omega$, define the function $\sigma_A(\lambda)$ for all λ in Ω by the relations

$$\sigma_A(\lambda) = \begin{cases} I & \text{if } \lambda \in A \\ 0 & \text{if } \lambda \in \bar{A} \end{cases}$$

where 0 and I are the elements of the abstract Boolean field B_F (see Sec. 5-2) and $\bar{A}$ is the complement of A. Since the function σ_A is single-valued, it determines a mapping of Ω onto B_F. Evidently,

$$A = \{\lambda | \sigma_A(\lambda) = I\}$$
$$\bar{A} = \{\lambda | \sigma_A(\lambda) = 0\}$$

so that the complete inverse images of the elements I and 0 in B_F are precisely the sets A and $\bar{A}$, respectively, in C_Ω.

Now if the domain S of Example 7 in Sec. 5-3 is the set Ω above, then $\sigma_A(\lambda)$ clearly is a Boolean function of the variable λ and an element of the Boolean ring (algebra) B_Ω of all Boolean functions of the variable λ in set Ω. Thus $\sigma_A(\lambda)$, when regarded as a function of A, where A is an arbitrary element of class C_Ω, is a single-valued function of class C_Ω into the set B_Ω of all Boolean functions of Ω. For each set A in class C_Ω, there is a uniquely defined image $\sigma_A(\lambda)$ in B_Ω, and such a correspondence is a mapping of class C_Ω into B_Ω. The range of this mapping is a subset of all Boolean functions, which we denote by σ_{C_Ω}. Since the complete inverse image of the element $\sigma_A(\lambda)$ is given by

$$\{B | \sigma_B(\lambda) = \sigma_A(\lambda)\} = \{B | \sigma_B(\lambda) = I \text{ if and only if } \lambda \in A\} = A$$

the single element A in class C_Ω, the range σ_{C_Ω} is in one-to-one correspondence with domain C_Ω under the mapping.

We have demonstrated above how the function $\sigma_A(\lambda)$ can be interpreted as a function which determines at least two different mappings. If the set A is held fixed, $\sigma_A(\lambda)$ determines a mapping of space Ω onto B_F; if A is a variable in the class of sets C_Ω, then $\sigma_A(\lambda)$ determines a one-to-one mapping of class C_Ω onto the subset σ_{C_Ω} of all Boolean functions B_Ω. Since $\sigma_A(\lambda)$, we have shown, is a function of two variables (A with domain C_Ω and λ with domain Ω), it is possible in a similar way to obtain three further mappings from $\sigma_A(\lambda)$. If variable λ is held fixed, $\sigma_A(\lambda)$ determines a mapping of class C_Ω onto B_F; if λ is allowed to vary over space Ω, then $\sigma_A(\lambda)$ may be regarded as determining a mapping of space Ω onto a subset of the set B_{C_Ω} of all Boolean functions, defined over the *class* of sets C_Ω. Finally, since $\sigma_A(\lambda)$ takes on either the value 0 or the value I in B_F for each set A in C_Ω and each point λ in Ω, the function $\sigma_A(\lambda)$ determines a mapping of the product space (C_Ω, Ω) onto the Boolean field B_F, where the *product space* (C_Ω, Ω) is the set of all ordered pairs (A, λ) with $A \in C_\Omega$ and λ in Ω. Of the five possible mappings determined by $\sigma_A(\lambda)$, the second mapping is of the most immediate interest.

In the next chapter an important special application is shown for the function $\sigma_A(\lambda)$. There it is shown how to use such functions to select one register from many registers. The selection of a particular register obtains if and only if the "state of the machine" λ is in a set A of possible states of the machine, and this is true if and only if $\sigma_A(\lambda) = I$. For this reason one might call $\sigma_A(\lambda)$ a *selector function*, depending on set A. This function has also been termed the *indicator* or *characteristic* function of set A.

If A and B are two sets in class C_Ω, then the following identities are true for the selector function:

Selector identities

$$\sigma_{A \cap B}(\lambda) = \sigma_A \sigma_B(\lambda)$$
$$\sigma_{A \triangle B}(\lambda) = (\sigma_A \oplus \sigma_B)(\lambda)$$
$$\sigma_{A \cup B}(\lambda) = (\sigma_A + \sigma_B)(\lambda)$$
$$\sigma_{\bar{A}}(\lambda) = \sigma'_A(\lambda)$$

Also for the empty set 0 and space Ω,

$$\sigma_0(\lambda) = 0$$
$$\sigma_\Omega(\lambda) = I \qquad \text{for all } \lambda \text{ in } \Omega$$

Let us prove only the first identity; the others follow in a similar manner. To prove

$$\sigma_{A \cap B}(\lambda) = \sigma_A \sigma_B(\lambda)$$

consider separately the cases $\lambda \in A \cap B$ and $\lambda \in \overline{A \cap B}$.

Case I.　If $\lambda \in A \cap B$, λ is in both set A and set B.　Hence

$$\sigma_{A \cap B}(\lambda) = \sigma_A(\lambda) = \sigma_B(\lambda) = I$$

Thus　　　　　$$\sigma_A(\lambda)\sigma_B(\lambda) = \sigma_A\sigma_B(\lambda) = I$$

and the identity is true for this case.

Case II.　If $\lambda \in \overline{A \cap B} = \bar{A} \cup \bar{B}$, λ is in either the set $\bar{A}$ or the set $\bar{B}$ or both.　Hence either $\sigma_A(\lambda)$ or $\sigma_B(\lambda)$ has value 0.　Thus

$$\sigma_{A \cap B}(\lambda) = \sigma_A\sigma_B(\lambda) = 0$$

and the identity is proved.

The first and last two selector identities show that there is an isomorphism between the class C_Ω and set σ_{C_Ω} of Boolean functions over the set Ω. As a consequence, the set σ_{C_Ω} is a Boolean algebra induced by the one-to-one mapping of class C_Ω onto σ_{C_Ω}.　Evidently σ_{C_Ω} is a sub-Boolean algebra of the Boolean algebra B_Ω of all Boolean functions with domain Ω.

It was pointed out in Sec. 5-2 that B_F is the abstract representative of any Boolean field.　In particular, B_F could be any one of the four examples given in that section.　It is of interest to interpret the meaning of selector $\sigma_A(\lambda)$ if B_F is the Boolean field of Example 4.　For this case it is evident that $\sigma_A(\lambda)$ must be interpreted as the *statement* "λ belongs to A." This statement determines a mapping of the product space (Ω, C_Ω) onto the Boolean field consisting of the two elements T (true) and F (false). The Boolean algebra σ_{C_Ω} corresponds to a Boolean algebra of statements of the form "λ belongs to A," where A is an arbitrary element of class C_Ω. Clearly the Boolean algebra of statements is isomorphic to σ_{C_Ω}, as well as to the Boolean algebra C_Ω.

5-8. Boolean Time Functions and Combinational Switching Functions. In Sec. 2-2 real-valued Boolean functions of time (or simple Boolean time functions) were introduced, and in the next section examples of physical devices were introduced which approximated the postulated properties of the binary storage cell.　Since physical limitations require one to discern the state of a cell only during those intervals of time which do not overlap the switching time intervals, meaningful information can be stored physically in a cell only on a subset of all possible points in time. This fact supplies a motivation for defining Boolean time functions with respect to a subset of the time axis, rather than with respect to the set of all points in time.

Definition 5-6.　*Boolean Time Functions.*　Let $R = \{t|-\infty < t < \infty\}$, the set of all points in time, and let T be a subset of R.　Suppose C_T is a class of subsets of T.　Then the selector function (see Example 2, Sec. 5-7)

$$f(t) = \sigma_F(t)$$

is called a Boolean time function, depending on set F of class C_T.　The real-valued Boolean time function $f(t)$, corresponding to $F \in C_T$, is defined by

$$f(t) = \begin{cases} b & \text{if } t \in F \\ a & \text{if } t \in \bar{F} \end{cases}$$

where a and b are real numbers such that $a < b$.

If class C_T in the above definition is a Boolean algebra, the set of Boolean time functions corresponding to class C_T is likewise a Boolean algebra B_{C_T} isomorphic to C_T (refer to the argument in latter part of Example 2, Sec. 5-7). Let B_{C_T} also designate (for simplicity in terminology) the Boolean algebra of real-valued Boolean time functions, corresponding to C_T.

For the latter case B_{C_T} consists of a set of functions $f(t)$, $g(t)$, etc., where an element $f(t)$ is either the real number a or the real number b for each $t \in T$. Moreover, for all $f(t)$ in B_{C_T}, the collection of all sets of the form

$$F = \{t | f(t) = b\} = f^{-1}(b)$$

constitutes the class (and Boolean algebra) C_T of subsets of T. The functions

$$0(t) = a \qquad \text{for all } t \in T$$
and
$$I(t) = b \qquad \text{for all } t \in T$$

are respectively the zero and unity of B_{C_T}. $0(t)$ and $I(t)$ are the only functions in B_{C_T} which are constant for all $t \in T$. The operations between arbitrary elements of B_{C_T} are defined as follows (see Example 5, Sec. 5-3, and Example 2, Sec. 5-7): For f and g in B_{C_T},

$$(f \oplus g)(t) = f(t) \oplus g(t)$$
$$(fg)(t) = f(t) \cdot g(t)$$
$$(f + g)(t) = f(t) + g(t)$$
and
$$(f')(t) = [f(t)]'$$

for every value of t in T, where the operations on the right for particular values of $t \in T$ are defined in Fig. 5-2 (see also Example 7, Sec. 5-3).

Table 5-3. Four Binary Operations in B_{C_T}

$f(t)$	$g(t)$	$f(t) \oplus g(t)$	$f(t)g(t)$	$f(t) + g(t)$	$f'(t)$
a	a	a	a	a	b
a	b	b	a	b	b
b	a	b	a	b	a
b	b	a	b	b	a

Since $a < b$, we have, by Table 5-3,

$$fg(t) = f(t)g(t) = \min [f(t),g(t)]$$
$$(f + g)(t) = f(t) + g(t) = \max [f(t),g(t)]$$
and
$$f'(t) = [f(t)]' = I \oplus f(t) = a + b - f(t)$$

for all $t \in T$, where min and max denote minimum and maximum, respectively, and where the operations $+$ and $-$ in the last expression for $f'(t)$ are the ordinary addition and subtraction operations of arithmetic of real numbers. Although the physical implementation of the above operations was discussed in Sec. 3-3, it is appropriate here to reexamine realizability as a function of real time t.

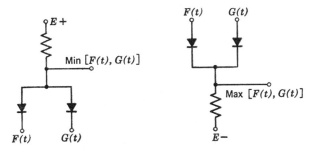

FIG. 5-2. Two gates.

Suppose $F(t)$ and $G(t)$ are two arbitrary voltage signals (real-valued functions of bounded variation) such that

$$E- \; < F(t) < E+ \qquad \text{and} \qquad E- \; < G(t) < E+$$

for all $t \in T$, where $E-$ and $E+$ are two fixed voltages. Then if two ideal unidirectional current elements with zero forward resistance and infinite back resistance are connected† as shown in Fig. 5-2, the output terminals of the two networks will have potentials min $[F(t),G(t)]$ and max $[F(t),G(t)]$, respectively, for all $t \in T$. In particular, if $F(t)$ and $G(t)$ are the real-valued Boolean time functions $f(t)$ and $g(t)$, respectively, where

$$E- \; < a < b < E+$$

the outputs of these two circuits are, respectively, the operations $fg(t)$ and $(f + g)(t)$ for all $t \in T$. Also, $f'(t)$ can be physically realized by an inverter circuit for all $t \in T$ (see Sec. 3-2).

Circuits or combinations of circuits such as shown in Fig. 5-2, which physically realize such operations as complementation and logical multiplication on Boolean time functions, are called combinational circuits. If $f_1(t), f_2(t), \ldots, f_N(t)$ are N Boolean time functions of B_{CT}, a function

$$F[f_1(t),f_2(t), \ldots ,f_N(t)]$$

which can be realized as a combinational switching circuit is termed a

† In Fig. 5-2 we assume current flows from plus to minus and that the diodes have zero forward resistance in the direction in which the arrowhead section of the diode symbol points.

combinational switching function of N *variables.* $F[f_1(t), f_2(t), \ldots, f_N(t)]$
is an element of B_{C_T}.   Our purpose will now be to investigate the fundamental properties of combinational switching functions and in particular to determine that class of mappings of the product algebra

$$\underbrace{B_{C_T} \times B_{C_T} \times \cdots \times B_{C_T}}_{N \text{ times}}$$

into B_{C_T}† which can be realized by combinational switching circuits.

Since the circuit elements of a combinational circuit do not vary with time, a switching function of N variables $F[f_1(t), f_2(t), \ldots, f_N(t)]$ must have the following mathematical property:

Property 1

$$F[f_1(t), f_2(t), \ldots, f_N(t)] = F[f_1(t^*), f_2(t^*), \ldots, f_N(t^*)]$$

if $f_1(t) = f_1(t^*)$, $f_2(t) = f_2(t^*)$, $\ldots$, $f_N(t) = f_N(t^*)$ for all t and t^* in T and $f_1(t), f_2(t), \ldots, f_N(t)$ in B_{C_T}.   In other words, a switching function $F[f_1(t), f_2(t), \ldots, f_N(t)]$ changes its value at only those points in time t for which at least one of the functions $f_1(t), f_2(t), \ldots, f_N(t)$ undergoes a change.

Consider the switching function $G[f(t)]$ of one variable, where $f(t)$ is an element of B_{C_T}.   Let

$$F = \{t | f(t) = b\}$$

be the set of all points of time in T for which $f(t) = b$.   Suppose $t_1 \in F$ and $t_2 \in \bar{F}$; then by property 1, if $t \in F$,

$$G[f(t)] = G[f(t_1)]$$

and if $t \in \bar{F}$,

$$G[f(t)] = G[f(t_2)]$$

Hence in general

$$\begin{aligned} G[f(t)] &= G[f(t_2)]f(t) + G[f(t_1)]f'(t) \\ &= G(a)f(t) + G(b)f'(t) \\ &= G[0(t)]f'(t) + G[I(t)]f(t) \end{aligned}$$

Thus the switching function G determines a mapping of B_{C_T} into B_{C_T}, and $G[f(t)]$ is a Boolean expression of one variable.   Now $0(t)$ and $I(t)$ are functions which are constant for all $t \in T$, so by property 1, $G[0(t)]$ and $G[I(t)]$ are also functions which are constant for all $t \in T$.   Since $0(t)$ and $I(t)$ are the only such functions in B_{C_T} with this property, $G[0(t)]$ must be either $0(t)$ or $I(t)$ for all $t \in T$, and similarly for $G[I(t)]$.
Thus there are exactly *four* possible combinational switching functions of

† Let $A_1, A_2, \ldots, A_n$ be n sets; the *Cartesian product* of these sets is the set of all ordered n-tuples (or vectors) of the form $(a_1, a_2, \ldots, a_n)$, where $a_i \in A_i$, for $i = 1, 2, \ldots, n$.   This product is denoted by $A_1 \times A_2 \times \cdots A_n$.

one variable; they are the functions $0(t)$, $I(t)$, $f(t)$, and $f'(t)$ (compare with Table 3-1). Consequently, there exist only four distinguishable mappings of B_{C_T} into B_{C_T} that can be determined by a combinational switching function of one variable.

Now let B_T be the Boolean field with elements $0(t)$ and $I(t)$. That is,

$$B_T = \{0(t),I(t)\} = \{a,b\}$$

If $x \in B_T$, the Boolean expressions over B_T are determined by the formula

$$F(x) = F[0(t)]x' + F[I(t)]x$$
$$= F(a)x' + F(b)x$$

Substitute $f(t) \in B_{C_T}$ for x; then the formula

$$F[f(t)] = F(a)f'(t) + F(b)f(t)$$

determines all four switching functions of one variable, namely, $0(f)$, $I(f)$, $f(t)$, and $f'(t)$. Let us express this characterization of a switching function as:

Lemma. Let $B_T = \{0(t),I(t)\}$, and let $F(x)$ be any one of the four Boolean expressions of one variable with domain B_T. Then a combinational switching function of one variable with domain B_{C_T} must be of the form $F[f(t)]$, where $f(t) \in B_{C_T}$.

Theorem 5-6. Let $B_T = \{0(t),I(t)\}$, and let $F(x_1,x_2, \ldots ,x_N)$ be any one of the 2^{2^N} Boolean functions of N variables, each with domain B_T. Then any combinational switching function of N variables, each with domain B_{C_T}, must be of the form $F[f_1(t),f_2(t), \ldots ,f_N(t)]$, where $f_1(t)$, $f_2(t), \ldots ,f_N(t)$ are elements in B_{C_T}.

Proof. By the previous lemma the theorem is true for $N = 1$. Now assume, for purposes of mathematical induction, that the theorem is true for all N such that $N \leq M$. Let $G[f_1(t),f_2(t), \ldots ,f_M(t),f_{M+1}(t)]$ be a switching function of $M + 1$ variables, and let

$$F_{M+1} = \{t|f_{M+1}(t) = b\}$$

From property 1, if $t \in F_{M+1}$,

$$G[f_1(t),f_2(t), \ldots ,f_M(t),f_{M+1}(t)] = G[f_1(t),f_2(t), \ldots ,f_M(t),b]$$
$$= G[f_1(t),f_2(t), \ldots ,f_M(t),b] f_{M+1}(t)$$

and if $t \in \bar{F}_{M+1}$,

$$G[f_1(t),f_2(t), \ldots ,f_M(t),f_{M+1}(t)] = G[f_1(t),f_2(t), \ldots ,f_M(t),a]$$
$$= G[f_1(t),f_2(t), \ldots ,f_M(t),a] f'_{M+1}(t)$$

Hence for $t \in T$,

$$G[f_1(t),f_2(t), \ldots ,f_M(t),f_{M+1}(t)] = G[f_1(t),f_2(t), \ldots ,f_M(t),a] f'_{M+1}(t)$$
$$+ G[f_1(t),f_2(t), \ldots ,f_M(t),b] f_{M+1}(t)$$

Since $G[f_1(t),f_2(t), \ldots ,f_M(t),a]$ and $G[f_1(t),f_2(t), \ldots ,f_M(t),b]$ are switching functions of only M variables, there exist by induction hypothesis two Boolean expressions of M variables $H(x_1,x_2, \ldots ,x_M)$ and $K(x_1,x_2, \ldots ,x_M)$, where each variable has domain B_T, with the property

$$G[f_1(t),f_2(t), \ldots ,f_M(t),a]$$
$$= H(x_1,x_2, \ldots ,x_M)|x_1 = f_1(t),\ x_2 = f_2(t),\ \ldots ,\ x_M = f_M(t)$$
and $\quad G[f_1(t),f_2(t), \ldots ,f_M(t),b]$
$$= K(x_1,x_2, \ldots ,x_M)|x_1 = f_1(t),\ x_2 = f_2(t),\ \ldots ,\ x_M = f_M(t)$$

Thus

$$G[f_1(t),f_2(t), \ldots ,f_M(t),f_{M+1}(t)] = H(x_1,x_2, \ldots ,x_M)f'_{M+1}(t)$$
$$+\ K(x_1,x_2, \ldots ,x_M)f_{M+1}(t)|x_1 = f_1(t),\ x_2 = f_2(t),\ \ldots ,\ x_M = f_M(t)$$
$$= H(x_1,x_2, \ldots ,x_M)x'_{M+1}$$
$$+\ K(x_1,x_2, \ldots ,x_M)x_{M+1}|x_1 = f_1(t),\ x_2 = f_2(t),\ \ldots ,\ x_{M+1} = f_{M+1}(t)$$

Since $\quad H(x_1,x_2, \ldots ,x_M)x'_{M+1} + K(x_1,x_2, \ldots ,x_M)x_{M+1}$

is a Boolean expression of $M + 1$ variables, each with domain B_T, the induction is complete and the theorem is proved.

In this chapter we have developed mathematically the concept of a Boolean algebra. In particular we have given attention to the concept of mapping one Boolean algebra into another, for, as we have already seen in Chaps. 2 and 3, this concept plays an important role in the nature and design of digital computers. In the next chapter the transfer (or substitution) of the results of a mapping (from an n-cell register into an m-cell register) into another m-bit register will be defined and illustrated.

PROBLEMS

5-1. Let R be a ring. The single-valued mapping

$$f(x) = a_0 \oplus a_1x \oplus a_2x^2 \oplus \cdots \oplus a_mx^{m-1}$$

where $a_i \in R$ and $x \in R$ for $(i = 0, 1, 2, \ldots , m)$, is called a polynomial mapping of R into R.

(a) Let B be a Boolean ring with unity. Prove that the most general polynomial mapping of B into B is the linear polynomial

$$F(x) = a \oplus bx$$

where x has domain B and a and b are fixed elements of B.

(b) Show that $F(x) = a \oplus bx$ is expressible in the following two forms:

$$F(x) = F(0) \oplus [F(0) \oplus F(I)]x$$
$$F(x) = F(0)x' \oplus F(I)x$$

The latter form is the canonical expansion of the Boolean expression $F(x)$ with coefficients in B.

(*c*) Show that the equation

$$ax = b$$

for *a* and *b* in *B* has *no* solution unless $ba = b$ and that if $ba = b$, the general solution is

$$x = ab \oplus z$$

where $za' = z$.

(*d*) Find the solution of the simultaneous linear equations

$$ax \oplus by = c$$
$$dx \oplus ey = f$$

(See Ref. 7, Chap. 10.)

(*e*) Using part *c*, show that the complete inverse image of $f(x) = ax \oplus b$ is

$$f^{-1}(ax \oplus b) = \{ax \oplus z | za' = z\}$$

that is, $f^{-1}(ax \oplus b)$ is the complete set of elements in *B* for which

$$f(y) = ax \oplus b$$

where $y \in B$.

(*f*) Let $g(x)$ be the single-valued function

$$g(x) = \begin{cases} I & \text{if } cx = c \\ 0 & \text{if } cx \neq c \end{cases}$$

where $c \neq 0$. Show that this function determines a mapping of *B* into *B* which is *not* a polynomial mapping.

(*g*) Prove that *B* is isomorphic to the Boolean field B_F if and only if *every* mapping of *B* into *B* is the polynomial mapping determined by the linear function

$$F(x) = a \oplus bx$$

where $a, b \in B$.

5-2. Let $x_1, x_2, \ldots, x_n$ be *n* variables with domain the Boolean ring *B* with unity. Let $F(x_1, x_2, \ldots, x_n)$ be a single-valued *n*-variable polynomial with coefficients in *B*. Show that $F(x_1, x_2, \ldots, x_n)$ is a Boolean expression of *n* variables with coefficients in *B* by proving that it has the canonical expansion.

(*a*) Show that a Boolean expression of two variables with coefficients in Boolean algebra *B* (see Example 7, Sec. 5-3) has the power series expansion

$$F(x_1, x_2) = F(0,0) \oplus \underset{1}{\Delta} F(0,0)x_1 \oplus \underset{2}{\Delta} F(0,0)x_2 \oplus \underset{1,2}{\Delta^2} F(0,0)x_1x_2$$

where the Δ's are the partial differences:

$$\underset{1}{\Delta} F(x_1, x_2) = F(x_1 \oplus I, x_2) \oplus F(x_1, x_2)$$
$$\underset{2}{\Delta} F(x_1, x_2) = F(x_1, x_2 \oplus I) \oplus F(x_1, x_2)$$
$$\underset{1,2}{\Delta^2} F(x_1, x_2) = \underset{1}{\Delta} F(x_1, x_2 \oplus I) \oplus \underset{1}{\Delta} F(x_1, x_2)$$
$$= \underset{2}{\Delta} F(x_1 \oplus I, x_2) \oplus \underset{2}{\Delta} F(x_1, x_2)$$

(*b*) Define the *m*th multiple partial difference

$$\underset{k_1, k_2, \ldots, k_m}{\Delta^m} F(x_1, x_2, \ldots, x_m)$$

where $k_1, k_2, \ldots, k_m$ are distinct integers from set $(1,2, \ldots, n)$ for $1 \leq m \leq n$, inductively as follows:

$$\underset{k}{\Delta}\, F(x_1, x_2, \ldots, x_n) = F(x_1, \ldots, x_{k-1}, x_k \oplus I, x_{k+1}, \ldots, x_n)$$

$$\oplus F(x_1, \ldots, x_k, \ldots, x_n)$$

$$\underset{k_1, k_2, \ldots, k_m}{\Delta^m}\, F(x_1, x_2, \ldots, x_n) = \underset{k_1, \ldots, k_{p-1}}{\Delta^{m-1}}\, F(x_1, \ldots, x_{k-1}, x_k \oplus I, x_{k+1}, \ldots, x_n)$$

$$\oplus \underset{k_1, \ldots, k_{p-1}}{\Delta^{m-1}}\, F(x_1, \ldots, x_k, \ldots, x_n)$$

Show that the Boolean expression $F(x_1, x_2, \ldots, x_n)$ with coefficients in Boolean algebra B is the power series (interpolation formula)

$$F(x_1, x_2, \ldots, x_n) = F(0, \ldots, 0) \oplus \underset{1}{\Delta} F(0, \ldots, 0)x_1 \oplus \cdots$$

$$\oplus \underset{n}{\Delta} F(0, \ldots, 0)x_n \oplus \underset{1,2}{\Delta^2} F(0, \ldots, 0)x_1 x_2 \oplus \ldots$$

$$\oplus \underset{1,2,\ldots,n}{\Delta^n} F(0, \ldots, 0)x_1 x_2 \ldots x_m$$

5-3. (a) Let $S[f(t), g(t)]$ be a combinational switching function (see Sec. 5-8) of $f(t)$ and $g(t)$, elements of B_{C_T}. Show that there exist particular $f(t)$ and $g(t)$ in B_{C_T} such that more than one Boolean expression obtains the same switching function, i.e., $f(t)$ and $g(t)$ may be such that there are two distinct Boolean expressions $F_1(x, y)$ and $F_2(x, y)$ such that

$$F_1[f(t), g(t)] = F_2[f(t), g(t)] = S[f(t), g(t)]$$

(b) Let the time points $t_1, t_2, t_3,$ and t_4 exhaust all possible configurations (states) of the pair of functions $[f(t), g(t)]$. Show that any desired switching function $S[f(t), g(t)]$ can be obtained by the truth-table (table of combinations) approach of Sec. 3-10, using only the above four configurations.

(c) Generalize Parts a and b to N elements of B_{C_T}.

(d) To obtain a minimal combinational network, prove that only the allowable configurations of the N-tuples $[f_1(t), f_2(t), \ldots, f_N(t)]$, i.e., the specified states, need be considered in deriving the network (see Sec. 4-8).

5-4. Consider a set of elements S. S is called *partially ordered* if there is a binary relation $x \leq y$ [x is less than (or contained in) y] which exists for all x and y in S.

For all x, $x \leq x$ (reflexive property).

If $x \leq y$ and $y \leq x$, then $x = y$ (antisymmetric property).

If $x \leq y$ and $y \leq z$, then $x \leq z$ (transitive property).

A partially ordered set S has an *upper bound* if there exists an element I such that $x \leq I$ for all $x \in S$. It has a *lower bound* if there exists an element 0 such that $0 \leq x$ for all $x \in S$. If a partially ordered set S has both an upper and a lower bound, it is called a *lattice* (for further properties see Ref. 7).

(a) If a partially ordered set has either an upper or a lower bound, show that it is unique.

(b) In a Boolean ring B define $x \leq y$ if and only if $xy = x$. Show that B with respect to this relation is a partially ordered set with a lower bound, the zero element of B.

(c) Show $x \leq y$ (as defined above) if and only if $x + y = y$, where $+$ is the join operation of definition 5-1.

(d) Show that a Boolean ring with unity (Boolean algebra) is a lattice under the containing relation of part b.

(e) The lattice diagrams for B_1 and B_2, where B_n is the Boolean ring of Example 5 Sec. 5-3, are as follows:

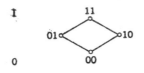

Construct the lattice diagrams for B_3 and B_4.

5-5. Define the majority (or median) operation between any three elements of a Boolean algebra as

$$(x,y,z) = xy + yz + zx$$

(a) Show $(x,y,z) = xy \oplus yz \oplus zx = (x + y)(y + z)(z + x)$.

(b) Show $(x,y,z) = (y,x,z) = (z,x,y)$, $(x + y) = (x,I,y)$, $xy = (x,0,y)$, and $(x,y,z)' = (x',y',z')$.

(c) Show $[(a,b,c),d,e] = [(a,d,e),b,(c,d,e)]$.

(d) Develop a set of postulates, involving the majority operation only, which leads to a Boolean algebra (see Ref. 7).

(e) Show $(x \oplus a,\ y \oplus a,\ z \oplus a) = (x,y,z) \oplus a$, the *translation property* of the majority operation.

(f) Demonstrate that every Boolean expression $F(x_1,x_2, \ldots ,x_n)$ which is *self-dual* (products may be replaced by sums and vice versa), like the majority operation, has the translation property

$$F(x_1 \oplus a_1, x_2 \oplus a, \ldots , x_n \oplus a) = F(x_1,x_2, \ldots ,x_n) \oplus a$$

5-6. An *ideal I* in a commutative ring R is a *subring* of R with the property

$$xy \in I$$

for all $x \in R$ and $y \in I$.

(a) If I is an ideal in a *finite* Boolean ring B, show that I must have the form

$$(a) \equiv \{xa \in B\}$$

where a is some fixed element of B. Element a is the unit element of (a) as a subring of B. [An ideal of the form of (a) is called a *principal ideal*.] *Hint:* Show I is a finite Boolean subring of B. By theorems 5-2 and 5-3 there exists a unique unity element in I, say the element a. For $x \in B$ prove that $xa \in I$. Hence $(a) \subset I$. Next pick an element from I and show it is contained in (a). By the definition of set equality, definition 5-4, the result follows.

(b) Show that $(1) = B$ and $(0) = 0$ where 1 is the unity and 0 is the zero of Boolean ring B. [(0) is called the zero ideal.]

(c) Let (a) be an ideal of a finite Boolean ring B and let I be an ideal of (a); show that I is also an ideal of the total ring B.

(d) Exhibit an ideal in an infinite Boolean ring which is not a principal ideal. *Hint:* See Example 6, Sec. 5-3.

5-7. If an ideal in a ring contains no ideals other than itself and the zero ideal, it is called a *minimal* ideal. Certainly at least one minimal ideal must exist in every finite nonzero ring.

(a) Show that a minimal ideal of a finite Boolean ring B always contains exactly two elements.

(b) By 5-6a there is some finite number n of distinct minimal ideals, say (e_1), (e_2), $\ldots$, (e_n), where (e_j) consists of the two elements 0 and e_j of B for $(j = 1, 2, \ldots , n)$.

Show

$$e_i e_j = 0 \qquad \text{for } j \neq i$$

for $(i,j = 1, 2, 3, \ldots, n)$; that is, $\{e_i\}$ is an orthogonal set of elements in B.

(c) Let x_i be a variable which can assume only the values 0 and 1 for $(i = 1, 2, 3, \ldots, n)$. Show that the 2^n possible sums of the form

$$x_1 e_1 \oplus x_2 e_2 \oplus x_3 e_3 \oplus \cdots x_n e_n$$

form a Boolean ring B' which is isomorphic to B_n, the Boolean ring of Example 5, Sec. 5-3. Show that B' is a subring of B; that is, $B' \subset B$.

(d) Let y be some element of finite Boolean ring B. Let $y e_i = y_i$ for $(i = 1, 2, \ldots, n)$. [Since $y_i \in (e_i)$, y_i is either 0 or e_i.] Consider the sum

$$\bar{y} = y_1 \oplus y_2 \oplus y_3 \oplus \cdots y_n$$

and show that the element $z = y \oplus \bar{y}$ has the property

$$z e_i = 0 \qquad \text{for } (i = 1, 2, 3, \ldots, n)$$

that is, the difference in y and $\bar{y}$ is orthogonal to the nonzero element of every minimal ideal of B.

(e) Consider the ideal (z), generated by $z = y \oplus \bar{y}$. Assume (z) is a nonzero ideal; it must contain by 5-6a a minimal ideal, say (f), of two elements. Show by 5-5c that this assumption contradicts 5-6d, so that $(z) = (0) = 0$ and that $y = \bar{y}$ for all $y \subset B$. Hence $B \subset B'$, and by 5-6c we have finally that $B = B'$. Thus, every finite Boolean ring B has 2^n elements for some integer n and is isomorphic to Boolean ring B_n of Example 5, Sec. 5-3.

5-8. Assume that commutative ring F is a field (by Sec. 5-2 a field satisfies the general ring postulates: commutativity, $xy = yx$ for all x and y in F, and the division law for multiplication). Also suppose F satisfies:

1. F has a finite number of elements.
2. $x \oplus x = 0$ for all $x \in F$.

(a) Show that F must contain exactly 2^n elements, where n is some positive integer (such a finite field is called a Galois Field; see Ref. 6).

(b) Show that every nonzero element $a \in F$ satisfies the equation

$$x^{2^n - 1} = I$$

where I is the identity.

(c) Consider the difference equation

$$a_n = a_{n-2} \oplus a_{n-3}$$

where addition is modulo 2. If initially $a_0 = 0$, $a_1 = 0$, and $a_2 = 1$, show that the periodic sequence

$$\{00101110010111001 \cdots \}$$

of period 7 is the unique solution of the above difference equation (this type of sequence can be generated by a linear sequential network; see Sec. 7-11).

(d) Consider the set of three-bit translates of the above sequence

$$(001), (010), (101), (011), (111), (110), (100)$$

Let $\alpha = (001)$, $\alpha^2 = (010)$, $\alpha^3 = (101)$, $\alpha^4 = (011)$, $\alpha^5 = (111)$, $\alpha^6 = (110)$, and $I = (100)$; and finally let $0 = (000)$. If vector addition of these triples is assumed, show that these eight elements form a Galois field.

REFERENCES

1. Birkhoff, G., and S. MacLane: "A Survey of Modern Algebra," The Macmillan Company, New York, 1941.
2. Stone, M. H.: Postulates for Boolean Algebras and Generalized Boolean Algebras, *Am. J. Math.*, no. 57, pp. 703–732, 1935.
3. Feller, W.: "An Introduction to Probability Theory and Its Applications," John Wiley & Sons, Inc., New York, 1957.
4. Halmos, P. R.: "Measure Theory," D. Van Nostrand Company, Inc., Princeton, N.J., 1950.
5. Sikorski, R.: "Boolean Algebras," Springer-Verlag, Berlin, 1960.
6. Van der Waerden, B. L.: "Modern Algebra," vol. I, 2d ed., Frederick Ungar Publishing Co., New York, 1950.
7. Birkoff, Garrett: "Lattice Theory," American Mathematical Society, 1948.

6

Register Operations

6-1. Introduction. The previous chapters have defined, in a precise way, the concept of the independent register as a vector function of time. We have further defined dependent registers as vector functions of independent registers, and have shown how such functions are described naturally by the formalism of Boolean algebra. In this chapter we begin the discussion of operations between registers with a description of the elementary operation called the transfer. This will lead to a description of digital machines as sets of interacting registers in subsequent chapters.

The transfer operation is essentially a simple operation, but it is the basic operation in the digital machine. The following chapters will show that the function of control primarily consists of the sequencing of the transfers between registers and that the various arithmetic and logical operations performed by digital machines can be implemented by means of transfers of the values of dependent and independent registers from register to register.

This chapter will also introduce much of the symbology which will be used to describe digital machines in the following chapters. At this time it can be safely said that no standard symbology exists; the symbology which will be presented has been developed over the past ten years and has been used in the design of a number of digital machines. Also, most of the symbology now in use can be easily learned once one system has become familiar, for most of the differences between systems consist of variations in detail rather than in over-all structure.

6-2. Interactions of Registers: Transfers. The basic operation within a machine is the transfer of information from one register to another. By a transfer of information we mean, roughly, the constraining of one register to assume a value which had previously been the value of another register. To be more precise, let A and B be independent one-cell registers and let t be a point of time at which a measurement of the state of register A is meaningful, i.e., a time when A is not switching. Let $(t, t + \epsilon)$ be an interval of time. We say that the information in A, or the value of A, is transferred into B (in symbols, $A \rightarrow B$) if $B(t + \epsilon) = A(t)$.

This statement simply says that the function B evaluated at time $t + \epsilon$ is identical to the function A evaluated at time t. It is assumed by this relation that the transfer is initiated at time t and that ϵ seconds later B has been completely switched to assume the value $A(t)$. The function B is undefined during the interval $(t, t + \epsilon)$. In the idealized cell of Sec. 2-2, ϵ is very small, approaching zero.

The statement of the transfer $A \rightarrow B$ requires, circuitwise, an output network associated with register A, and an input network associated with register B, to which the measured value of A can be directed. We shall call these circuits a *transfer path* between A and B.

The foregoing definition of the transfer for one-cell independent registers is extended to n-cell independent registers by applying the one-cell definition to each cell of the n-cell register. Let A and B be n-cell independent registers with cells $A_1, A_2, \ldots, A_n$ and $B_1, B_2, \ldots, B_n$, respectively. We say that the value of A is transferred into B, or in symbols

$$A \rightarrow B$$
if $\qquad\qquad A_i \rightarrow B_i \qquad i = 1, 2, \ldots, n$

By analogy with the previous definition, this statement implies that

$$B(t + \epsilon) = A(t)$$
whenever $\qquad B_i(t + \epsilon) = A_i(t) \qquad i = 1, 2, \ldots, n$

Certain properties of the transfer should be emphasized at this point. The statement of the transfer says nothing about any possible change in the value of register A. In particular, we assume, as in Chap. 2, that the measurement of the state of A implied by the transfer is nondestructive, i.e., the measurement leaves A undisturbed. If the physical implementation of A is such that the measurement is in reality a destructive one, then the change of state of A accompanying the measurement has to be stated explicitly. Also, the state of B before the transfer is lost, unless it is explicitly preserved by a transfer to some other register. Finally, we observe that the transfer is defined only between two registers of the same dimension; this fact allows a relaxation of the labeling of the cells of a register. It will be recalled that the independent register was originally defined as an ordered set of cells. For simplicity the cells have, until now, been labeled from 1 to n consecutively. There is no reason why the cells of an n-cell register A cannot be labeled $A_{i_1}, A_{i_2}, \ldots, A_{i_n}$, where each of the i_k is an integer with

$$i_{k-1} < i_k < i_{k+1}$$

It is often convenient to label the n cells of a register by means of a set of indices other than the integers 1 through n. For example, if the last three cells of A are considered a three-cell subregister, it is most natural to label the cells of this register A_{n-2}, A_{n-1}, and A_n. Accordingly, it is convenient to take this into account in the definition of the transfer in the following way: If $A = (A_{i_1}, A_{i_2}, \ldots, A_{i_n})$ and $B = (B_{j_1}, B_{j_2}, \ldots, B_{j_n})$, then

$$A \to B$$

signifies that

$$A_{i_k} \to B_{j_k} \qquad k = 1, 2, \ldots, n$$

In other words the transfer preserves the ordering without demanding that the cells have identical labels.

The definition of transfer extends simply to subregisters. Let B be an n-cell independent register with cells B_1, B_2, $\ldots$, B_n, and let the first m cells and the last $n - m$ cells define the subregisters $F(B)$ and $G(B)$, respectively. Let A be an m-cell independent register A_1, A_2, $\ldots$, A_m and C an $(n - m)$-cell register C_1, C_2, $\ldots$, C_{n-m}. Then the transfers

$$A \to F(B) \qquad C \to G(B)$$

signify that at time $t + \epsilon$, the first m cells of B assume the value of A at time t and the rest of B assumes the value of C at time t. The transfers

$$A \to F(B) \qquad G(B) \to G(B)$$

state that $F(B)$ is changed as above while $G(B)$ remains unchanged; the transfers

$$F(B) \to F(B) \qquad C \to G(B)$$

assert that $G(B)$ is changed but $F(B)$ remains unchanged, while the transfers

$$F(B) \to F(B) \qquad G(B) \to G(B)$$

or equivalently

$$B \to B$$

assert that B is unchanged during the interval $(t, t + \epsilon)$.

6-3. Transfers between Arbitrary Registers: The Shift Register. All the above transfers represent a very simple extension of the transfer concept to a particular kind of dependent register, namely a subregister. The extension holds for more general kinds of functions of registers. Let A be an arbitrary dependent or independent n-cell register and let B be an n-cell register, either independent or a subregister of an independent register. Then the transfer $A \to B$ is defined exactly as before when A was restricted to independent registers or subregisters. The register B into which the information is transferred must of course be independent in order for the transfer to be meaningful. As an example, consider the function ρ of the n-cell independent register $A = (A_1, A_2, \ldots, A_n)$ defined by

$$\rho(A) = [\rho_1(A), \rho_2(A), \ldots, \rho_n(A)]$$
$$\rho_i(A) = A_{i-1} \qquad i = 2, 3, \ldots, n$$
$$\rho_1(A) = A_n$$

The dependent register $\rho(A)$ thus has the components $(A_n, A_1, A_2, \ldots, A_{n-1})$. It therefore represents the register A with the ordering of the

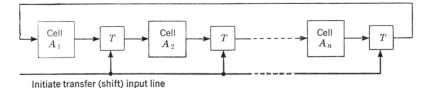

Initiate transfer (shift) input line

FIG. 6-1. Shift or cycle-right operation.   A rectangle with a T inside is used to indicate a *transfer gate*.   The function of the transfer gate connected to cell A_1 and the "initiate transfer line" is to transfer the contents of cell A_1 into cell A_2 only when an initiate or transfer signal occurs on the "initiate transfer line."   Notice that the transfer gate is symbolic as are the transfer lines.   In actual practice a transfer gate might consist of two AND gates or perhaps an AND gate and an inverter, depending on the type of memory cell used.   This points up the fact that the lines in the figure need not be actual wires, nor the blocks single components.

cells changed so as to shift each value one place to the right and insert A_n in the first position.   Thus the transfer

$$\rho(A) \rightarrow A$$

signifies that

$$A_i(t + \epsilon) = \rho_i[A(t)] \qquad i = 1, 2, \ldots, n$$

Therefore, replacing the value of A by the value of $\rho(A)$ has the effect of cycling the register one cell to the right.   This operation can be iterated as follows:

$$\rho^2(A) = \rho[\rho(A)] = (A_{n-1}, A_n, A_1, A_2, \ldots, A_{n-2})$$
$$\rho^k(A) = \rho[\rho^{k-1}(A)] = (A_{n-k+1}, A_{n-k+2}, \ldots, A_{n-k})$$

It evidently follows that

$$\rho^k(A) = A \qquad k = n, 2n, \ldots$$

or that any multiple of n of these cycles restores the original ordering of register A.   The operation of a shift register closed upon itself,† as in Fig. 6-1, is completely described by the transfer $\rho(A) \rightarrow A$ together with a

† The operation performed by a shift register closed upon itself is sometimes referred to as a *cycle operation*.

specification of the time interval $(t, t + \epsilon)$ associated with the transfer. Suppose a time base is established by a clock which produces switching pulses at $t = 0, \tau, 2\tau, \ldots$. These points of time are then the instants at which transfers are initiated. Thus the transfer $\rho(A) \to A$ establishes that

$$A_i(j\tau + \epsilon) = A_{i-1}(j\tau) \qquad i = 2, 3, \ldots, n \qquad j = 0, 1, \ldots$$
$$A_1(j\tau + \epsilon) = A_n(j\tau) \qquad\qquad\qquad\qquad j = 0, 1, \ldots$$

Since the only switching pulses occur at the times $j\tau$, the functions $A_i(t)$ are constant during the intervals $[j\tau + \epsilon, (j + 1)\tau]$, or the functions

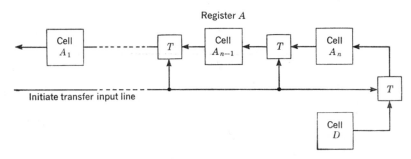

Fig. 6-2. Shift left and transfer operation.

$A_i(t)$ are simply Boolean functions of time defined at all points of time except during the switching intervals $(j\tau, j\tau + \epsilon)$.

An analogous function λ, corresponding to a cycle of one cell to the left, may be defined in a similar fashion:

$$\lambda(A) = (A_2, A_3, \ldots, A_n, A_1)$$
$$\lambda^k(A) = (A_{k+1}, A_{k+2}, \ldots, A_n, A_1, \ldots, A_k)$$

By the same reasoning as above, the statement $\lambda(A) \to A$ describes the cyclic shift register which shifts information one cell to the left at each clock pulse.

A variation of the functions ρ and λ can be used to describe a shift operation which is not cyclic. As before, let A be an n-cell independent register $(A_1, A_2, \ldots, A_n)$. Let $R(A)$ be the $(n - 1)$-cell subregister $(A_2, \ldots, A_n)$, and let $L(A)$ be the $(n - 1)$-cell subregister $(A_1, A_2, \ldots, A_{n-1})$. Let D be a one-cell register. The transfers

$$D \to A_1 \qquad L(A) \to R(A)$$

evidently describe a shift register similar to that of Fig. 6-1 (that is, the information in the register is shifted right by one cell at each clock pulse), but one in which a new bit is inserted in the cell A_1 at each pulse and the bit A_n is "dropped off" the end at each pulse. Similarly, the transfers

$$R(A) \rightarrow L(A) \qquad D \rightarrow A_n$$

describe the analogous "shift left" operation, which inserts new data at the rightmost end of the register with the oldest bits dropping off the left end. This shift register is shown in Fig. 6-2.

6-4. Further Extensions of the Transfer: Arithmetic Operations. The transfer as defined in this chapter is a basic concept for describing the flow of information in a digital machine. Some notion of its usefulness is evident from the previous section, where a concise transfer statement was demonstrated which described precisely the operation of the shift register. The full power of this notation becomes clearer when we apply the transfer concept to more complex interactions involving functions of several registers. In this section we shall demonstrate some of these techniques by considering certain arithmetic functions of two registers.

Let A and B designate two n-cell registers, and let the dependent register $S^H(A,B) = A \oplus B$ be an n-cell function of A and B, where each component $S_i^H = (A \oplus B)_i = A_i \oplus B_i$. The function S^H can be called the *half sum* or *partial sum* of the registers A and B. If A and B are considered to be arbitrary vectors in the n-dimensional vector space V_n, then $S^H = A \oplus B$ defines a vector called simply the mod 2 *sum* of the two vectors in the space V_n. The transfer

$$S^H \rightarrow A$$

represents the computation of the half sum or mod 2 sum of the registers A and B, with the result inserted in A. With this simple transfer statement we have described the operation of the rather complex circuit, shown in Fig. 6-3, consisting of the two registers with n mod 2 adder circuits between them. The mod 2 adder circuit can be called a *half* adder since it computes a sum without carry. (It is often called a quarter adder.)

A full binary addition may be described in a similar way. But before such a description becomes meaningful, a numerical representation for the register must be defined; then an addition operation may be specified consistent with this representation. The simplest example is that of the integer representation defined in Chap. 2. As above, let A and B be n-cell registers with integer representations

$$\delta_I(A) = \sum_{i=1}^{n} A_i 2^{n-i}$$

and
$$\delta_I(B) = \sum_{i=1}^{n} B_i 2^{n-i}$$

respectively. Let $S(A,B)$ be a dependent $(n + 1)$-cell register with integer representation

$$\delta_I(S) = \sum_{i=0}^{n} S_i 2^{n-i}$$

such that $\delta_I(S) = \delta_I(A) + \delta_I(B)$, where the symbol $+$ designates ordinary binary addition. Note that the additional cell S_0 is required, since the

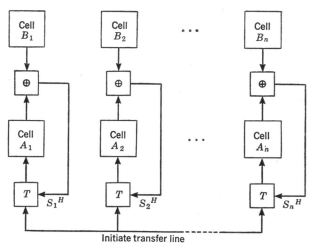

Fig. 6-3. Transfer of half sum.

sum of two n-bit numbers may have $n + 1$ bits. Let $K(A,B)$ be an $(n + 1)$-cell dependent register defined by

$$K_i = A_{i+1}B_{i+1} + K_{i+1}A_{i+1} + K_{i+1}B_{i+1} \qquad i = 0, 1, \ldots, n - 1$$
$$K_n = 0$$

where the plus signs indicate Boolean addition,

$$0 + 0 = 0,\, 0 + 1 = 1 + 0 = 1$$

K_i evidently represents the carry into the ith bit of the sum. The sum function S is then given by†

where
$$S = A + B = S^H \oplus K = A \oplus B \oplus K$$
$$S_i = A_i \oplus B_i \oplus K_i \qquad i = 1, 2, \ldots, n$$
$$S_0 = K_0$$

† The boldface $+$ will be used to indicate conventional (as opposed to logical) addition between *registers*. The boldface $+$ will not be used for such expressions as $n + 1$, $\tau + \epsilon$, etc.

An addition operation consists of the generation of the sum function $S(A,B)$ and its transfer into a register, say A. A little care is necessary in stating the transfer, since S contains $n + 1$ components and A is an n-cell register. To handle this we assume the existence of a one-cell register A_0 which may be considered part of register A. Then the transfer

$$S \rightarrow A$$

which means $\quad\quad S_i \rightarrow A_i \quad\quad i = 1, 2, \ldots, n$

$$S_0 \rightarrow A_0$$

inserts the $(n + 1)$-bit sum into the $(n + 1)$-cell register $A_0, A_1, \ldots, A_n$. If A_0 does not exist, we define the transfer $S \rightarrow A$ to mean $S_i \rightarrow A_i$, $i = 1, 2, \ldots n$, with S_0 being lost. In the latter case, the sum is modulo 2^n; only the n least significant bits of the result are retained. In either case, the entire operation of the adder is specified by the simple statement $S \rightarrow A$. The logical circuitry involved in the adder is no easier to design because of the compressed notation. It is the relationship of this adder to the other operations involved in a computing machine that is clarified by the notation.

6-5. Scalar Functions: Conditional Transfers. A particularly important and useful class of dependent registers is that consisting of one-cell registers, both independent and dependent. It is natural to refer to these registers as *scalars*. Again using the vector space model, we may define a multiplication operation between a scalar function and a vector function. Let α be a scalar register and A an n-cell vector register. The product αA is an n-cell vector

$$\alpha A = (\alpha A_1, \alpha A_2, \ldots, \alpha A_n)$$

where αA_i is the usual Boolean product. Hence, we obtain the simple but useful result

$$\alpha A = A \quad\quad \alpha = 1$$
$$\alpha A = 0 \qu\quad \alpha = 0$$

The scalar multiplication just described provides a convenient method for designating conditional transfers. Let A, B, and C be n-cell registers, and let α and β be scalar registers such that $\alpha\beta = 0$ and $\alpha + \beta = 1$; i.e., one and only one of the functions α and β has the value 1. The function $\alpha A + \beta B$ is then register A if $\alpha = 1$ and is register B if $\beta = 1$. The transfer

$$\alpha A + \beta B \rightarrow C$$

then signifies that

$$A \rightarrow C \quad\quad \alpha = 1$$
$$B \rightarrow C \quad\quad \beta = 1$$

The most common such conditional transfer occurs when $\beta = \alpha'$, or when the decision as to which transfer is to occur is based on the state of a single cell. For example, let A be an n-cell register and let A' be its complement, i.e., the register $(A'_1, A'_2, \ldots, A'_n)$. The transfer

$$A_1 A + A'_1 A' \rightarrow A$$

states that register A is to be complemented if cell A_1 has value 0 and is to remain unchanged if A_1 has value 1. If we let $f(\alpha, \beta, A, B)$ designate $\alpha A + \beta B$, then f is just a dependent register no different from any other dependent register that we have discussed, and the conditional transfer $\alpha A + \beta B \rightarrow C$ is no different conceptually from an ordinary unconditional transfer. However, though nothing is added to the theoretical structure being developed by writing certain transfers explicitly as conditional transfers, the concept is still a useful one in practical design considerations. Whether one chooses to consider $\alpha A + \beta B$ as just another dependent register or as a logical statement describing a binary choice of register A or register B depending upon the functions α and β is determined by the situation at hand. As we shall see in later chapters, rather complex decision situations may be written concisely using this formalism.

In a conditional transfer situation, there is no need to restrict ourselves to binary decisions. The number of alternative functions to be transferred is quite arbitrary. For example, let A, B, C, and D be n-cell registers, and let α, β, and γ be scalar functions, only one of which has the value 1. Then the transfer

$$\alpha A + \beta B + \gamma C \rightarrow D$$

where all but one of the functions α, β, and γ are 0, represents a selection of one of the three registers to be transferred to D, depending upon which of the scalar functions has the value 1. A particularly useful example of this notion occurs when all the decision functions α, β, γ, $\ldots$ are scalar functions of a single register, and the registers A, B, $\ldots$ are all independent registers belonging to a set of storage registers loosely called a *memory*. For convenience let M designate the set of 2^k n-cell registers M^0, M^1, $\ldots$, M^{2^k-1} and let C designate a register containing k cells, labeled C_1, C_2, $\ldots$, C_k. Let c_i, $i = 0, 1, \ldots, 2^k - 1$, define a scalar function of C which has the value 1 if and only if C has integer value i. Then the function

$$c_0 M^0 + c_1 M^1 + \cdots + c_{2^k-1} M^{2^k-1}$$

is a dependent register representing the selection of one of the registers M^i, depending upon the value of register C, and the transfer

$$c_0 M^0 + c_1 M^1 + \cdots + c_{2^k-1} M^{2^k-1} \rightarrow N$$

designates the transfer of the value of the register M^i to N if register C has the value i. It is customary to call C an *address* or *selection* register of the memory M and to call the functions c_i *characteristic* or *selector* functions of register C. For convenience we abbreviate the function $c_0M^0 + c_1M^1 + \cdots + c_{2^k-1}M^{2^k-1}$ by $M\langle C \rangle$, which can be read as the memory register addressed or selected by the value of C. Obviously, a memory need not be restricted to contain 2^k registers. If the memory M contains p registers, then its address register C must have at least k cells, where k is the smallest integer greater than or equal to $\log_2 p$, or

$$\log_2 p \leq k \leq \log_2 2p$$

6-6. Sequences of Transfers: Parallel and Serial Transfers. In any given computing system, not all possible transfer paths are usually available. Transfers between a pair of registers between which there is no transfer path can often be performed by making use of intermediate transfer paths which are available. For example, suppose it is desired to perform the transfer $A \rightarrow B$ when no path exists between registers A and B. The transfer can be accomplished if there is another register or sub-register C of the same dimension as A and B and connected to A and B by transfer paths such that the transfers $A \rightarrow C$ and $C \rightarrow B$ are available. Then the transfer $A \rightarrow B$ is implemented by the sequence of transfers

$$A \rightarrow C$$
$$C \rightarrow B$$

As shown in Sec. 6-2, in order to describe a transfer completely, it is necessary to specify the time interval $(t, t + \epsilon)$ during which the switching occurs. Similarly, to describe a sequence of transfers, the switching intervals for each transfer in the sequence must be specified. If the sequence of transfers shown above is to represent the over-all transfer $A \rightarrow B$, then the second transfer $C \rightarrow B$ must necessarily follow the first transfer. Let $(t_1, t_1 + \epsilon_1)$ represent the switching interval for the transfer $A \rightarrow C$ and $(t_2, t_2 + \epsilon_2)$ represent the interval for the transfer $C \rightarrow B$. Then $t_2 \geq t_1 + \epsilon_1$ assures that the second transfer will not occur too early.

An additional complication occurs if the sequence of transfers under consideration is of the form

$$A \rightarrow C$$
$$F(C) \rightarrow B$$

which realizes the over-all transfer $F(A) \rightarrow B$. $F(A)$ is assumed to be a dependent register. Suppose the first transfer is initiated at time t_1 and is completed by $t_1 + \epsilon_1$. If an additional λ seconds are required for the network realizing the function $F(C)$ to come to equilibrium, then the second transfer may not be initiated until time $t_1 + \epsilon_1 + \lambda$.

At this point we must point out one distinction that, for reasons of accuracy, must be made between so-called *synchronous* and *asynchronous* systems.   In synchronous operation the timing intervals are specified completely by some external clock which generates a sequence of switching pulses.   Transfers occurring in sequence are initiated by these sequentially occurring clock pulses.   In order for a sequence of the form

$$A \to C$$
$$F(C) \to B$$

to be performed by successive clock pulses, the interval τ between these pulses must exceed $\epsilon + \lambda$, where ϵ and λ are, as above, the response times of the cell C and the network $F(C)$, respectively.

In an asynchronous system the sequences of transfers themselves define the timing intervals, thus eliminating the external clock.   We include this in our formalism by requiring that a switching interval always begin at the instant the input function has reached its equilibrium value.   Thus, in an asynchronous system the second transfer in the sequence

$$A \to C$$
$$F(C) \to B$$

is initiated at time $t + \epsilon + \lambda$, where the first transfer is initiated at time t, and ϵ and λ are as defined in the previous paragraph.   In the following sections, when we write sequences of transfers, we shall assume that the timing intervals are consistent with the meanings of the transfers.   The actual design of computer timing is a problem in the design of control units and will be discussed in Chap. 8.

In the example of the transfer $A \to B$ in a system where an intermediate step via another register C is necessary, it obviously follows that if C is the only register in the machine that can provide this function, then writing $A \to B$ uniquely specifies the sequence $A \to C$, $C \to B$.   Thus, to relate the transfer $A \to B$ to other transfers describing some complex operation, it is necessary only to state this transfer, omitting the intermediate step.

Within the above framework we can now discuss an important class of transfers called *serial* transfers.   If A and B are n-cell registers, then, as before, the transfer $A \to B$ designates the transfer of the value of A into B by the end of some time interval $(t, t + \epsilon)$, which, of course, means that the value of each cell A_i is transferred to the corresponding cell B_i by the end of the same interval.   When a transfer path is present between each pair of cells (A_i, B_i), then the transfer between each pair can be performed simultaneously, and the transfer is said to be *parallel*.   When only one path exists between the cells of the registers, and the values of each cell of A must be transferred to B using this one path in sequence, the transfer is called *serial*. We illustrate a typical serial transfer in Fig. 6-4. The only path between the registers is one between A_n and B_1. To preserve the original value of A in A, a path is shown from A_n to A_1. The serial transfer of the value of A into B is performed by a sequence of transfers initiated at times τ, 2τ, 3τ, . . . , $n\tau$. As in Sec. 6-3, let $\rho(A)$ designate the n-cell cycled register $(A_n, A_1, \ldots , A_{n-1})$, let $R(A)$ be the $(n-1)$-cell register $(A_2, \ldots , A_n)$, and let $L(A)$ be the $(n-1)$-cell register $(A_1, \ldots , A_{n-1})$. Then the transfers

$$\rho(A) \to A \qquad A_n \to B_1 \qquad L(B) \to R(B)$$

all occurring during the interval $(\tau, \tau + \epsilon)$, describe the shift of the registers one digit to the right with the value of A_n preserved in A_1 and the

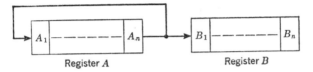

Register A   Register B

Fig. 6-4. Serial transfer.

value of B_n lost. Some slight simplification of notation can be achieved by letting (A,B) designate the $2n$-cell compound register obtained by placing B to the right of A as in Fig. 6-4. The above transfers may then be described by

$$L(A,B) \to R(A,B) \qquad A_n \to A_1$$

Clearly a repetition of the above set of transfers during the succeeding switching interval $(2\tau, 2\tau + \epsilon)$ results in another shift to the right by one digit. If the transfer is performed n times, then the desired transfer of the original value of A to B with the value of A preserved in A has been performed by the time $t = n\tau + \epsilon$, and the same over-all result as in the parallel transfer has been achieved.† For circuits of a given speed a serial transfer will take a longer time than a parallel transfer but will require less equipment.

In the example just described, the transfer was between two independent registers. Suppose now that instead of transferring the value of A to B, the value of some function F of registers A and B is to be transferred to B. If the transfer is to be performed in parallel, then each component of $F(A,B)$ must be generated and transferred to the corresponding cell of B. In the serial transfer of the value of $F(A,B)$ to B,

---

† In a serial machine the interval $(t, t + \tau)$ is generally referred to as a *bit time*, the time required to transfer one binary digit. If the standard word length or register length for the machine is n, so that the basic register of the machine contains n cells, then the interval $(t, t + n\tau)$ is referred to as a *word time*.

it may be necessary to provide only a single function-generating network to be used sequentially during the transfer. As an example, we show in Fig. 6-5 the serial transfer of the half sum of A and B, $S^H(A,B)$, to B. The transfer is described by n performances of the transfers

$$\rho(A) \to A \qquad A_n \oplus B_n \to B_1 \qquad L(B) \to R(B)$$

A considerable saving in equipment may be accomplished here if degradation in speed can be tolerated.

Often some degree of compromise between serial and parallel implementation of transfers is accomplished by breaking the registers into a number

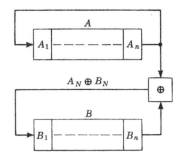

Fig. 6-5. Serial transfer of half sum.

of subregisters and providing simultaneous serial transfers between the subregisters. Thus, for example, if the n-cell registers A and B are each divided into two $n/2$-cell subregisters F and G, then the transfer $A \to B$ can be accomplished by a sequence of $n/2$ repetitions of the serial transfer of $F(A) \to F(B)$ and $G(A) \to G(B)$. This method takes half the time of a completely serial transfer at the cost of another transfer path. Such schemes are called by the obvious name *series-parallel*.

6-7. Transfer Properties of Cells. In all the preceding remarks in this chapter, certain properties of cells and registers were implied. In general, not all the physical representations of cells have all the assumed properties. It obviously follows that before a given transfer can be specified, it must be known whether such a transfer can be performed by the particular equipment at hand. Some limitations of different kinds of cells have already been described in the examples, presented in Chap. 2, of different physical representations of cells.

The fixed cell is the most restricted type of cell since it stores a constant, and hence by its very definition cannot receive information via the transfer operation. Some cells are restricted in the class of functions that can be generated using the cells as the independent variables. The commonly used magnetic-core storage most often consists of a set of 2^k

registers constructed of magnetic cores together with a flip-flop buffer register. The only kind of interaction one of the core registers usually has is to transfer its value to the buffer and to receive a value from the buffer. If a function of a core register is required, a transfer must first be made to the buffer and the function must be generated at the buffer register. For example, suppose it is desired to compute the sum of the values of two registers A and M and store the result in A, where A is a flip-flop register and M is a magnetic-core register. This transfer requires computation of the sum function $S(A,M)$ described in Sec. 6-4. This function cannot be generated with M as independent register; hence a buffer flip-flop register N is used and the desired result is obtained with the sequence

$$M \rightarrow N$$
$$S(A,N) \rightarrow A$$

Most commonly the core register M is even further restricted in its operation in that the measurement of the value of M is destructive. Thus, if it is desired to preserve the original value of M, the above operation must include restoration of the value of M. The entire operation is thus described by

$$M \rightarrow N \qquad 0 \rightarrow M$$

followed by

$$S(A,N) \rightarrow A \qquad N \rightarrow M$$

If both M and A are flip-flop registers, then the transfer $S(A,M) \rightarrow A$ describes the entire operation.

Some varieties of flip-flop registers also have restrictions on their operation. We have emphasized in many places that a measurement of the value of a cell is meaningless if the time of the measurement coincides with the time during which the cell changes state. This problem in circuit design is often called the "race problem." The flip-flop designer must provide adequate circuit safeguards to assure that the measurement of the state of the flip-flop has concluded before the flip-flop is allowed to change state. Otherwise, a transfer of the form $f(A) \rightarrow A$ becomes unreliable. In certain types of circuitry, the race problem is bypassed as a problem in circuit design and imposed as a constraint in logic design. With such flip-flops simultaneous transfers of the form

$$f(A) \rightarrow B \qquad C \rightarrow A$$

are not allowed; the transfers must be performed in sequence. With such flip-flops shift register operation cannot be performed without the aid of an auxiliary register. Thus, to perform the shift operation

$$\rho(A) \rightarrow A$$

it is necessary to perform the sequence

$$A \rightarrow B$$
$$\rho(B) \rightarrow A$$

It is not our purpose to describe all limitations of all kinds of cells. The designer of digital machines must, of course, be aware of the limitations of the cells he has at his disposal. The point of emphasis is that, regardless of the kind of cells, the generalized transfer operations that we have described in this chapter may be used individually or in sequences to provide any desired result in data processing. There is, in general, no unique method of implementation of any operation. It is the function of the designer to specify a method consistent with both the operation to be performed and the class of circuitry available for implementation.

PROBLEMS

6-1. Suppose a register A stores numbers in a 2's-complement fractional representation (see Appendix).

(*a*) Describe transfers to divide the value of A by 2.

(*b*) Describe the transfers to multiply the number in A by 2.

6-2. Write down a sequence of transfers to cycle a six-cell register three places to the left, clear the right half of the register, and finally cycle it three places to the right.

6-3. Describe a transfer to implement the following statement: If register A contains a negative number, clear register B; if A contains a nonnegative number, transfer that number to B.

6-4. Let A and B designate n-cell registers; let $M = (M_1, M_2)$ be a two-cell register. Describe a transfer which implements the following statement: If M has binary value 2, insert the sum of A and B into B; if M has binary value 1, insert the difference between B and A into B; if M_1 and M_2 are alike, leave B unchanged.

6-5. Let S designate a set of 15 registers, and let A be a four-cell register addressing S. Describe a transfer which implements the following statement: Add the selected register in S to register B and store the sum in B; if A is zero, leave B unchanged.

6-6. Let registers A and B contain 12 cells each in a binary-coded decimal representation, with each decimal digit employing four cells. Write a sequence of transfers defining the storage of the sum of A and B in B, where one parallel four-bit adder is employed for the decimal digits in sequence.

7
Introduction to Sequential Networks

7-1. Sequential Network Theory. In the years following the construction of the first successful large-scale digital computers, a significant body of theory was developed concerning what we have termed the sequential network, but which has been referred to by other authors under the titles, sequential machines, finite automata or simply automata, and sequential transducers. The particular set of terms used at some given time is generally determined by the particular emphasis desired or by the generality of the coverage. Authors of papers describing design procedures for actual physical devices are prone to use the term *sequential circuit*, while papers which are more mathematically or philosophically oriented tend to use the term sequential machine or automaton. The present diversity of terminology probably arises from the heterogeneous backgrounds of the designers and research workers in this field; an unlikely mixture of engineers, logicians, mathematicians, and philosophers have contributed to the present-day state of the digital art.

Let us therefore use the term sequential network, without further apology, to refer to a set of memory cells and gates which are interconnected in some rational manner such that the states of the cells may change in time and such that the outputs from the network will be affected by these changes. The outputs are therefore time dependent, and in most cases will be input dependent also.

It is possible to consider the sequential network as a set of dependent and independent registers, and to consider the operation of the network as consisting of the sequencing of the transfers of the values taken by these registers. From this viewpoint an entire machine can be considered to be a sequential network. We can also examine parts of a large machine from this viewpoint; for instance, the shift registers in the preceding chapter comprise examples of sequential networks. In this chapter we first introduce some of the basic concepts of sequential networks and follow them by a discussion of several of the more common types of networks.   The stress in this chapter will be on gating configurations and the characteristics of some of the memory devices commonly used in present-day machines.   Also, we shall primarily be concerned with sequential networks consisting of only a few components, rather than entire machines. These small networks, or parts of machines, can then be interconnected to form large machines, and following chapters will consider machines from this more general, or structural, viewpoint.

7-2. Classes of Sequential Networks.   Figure 7-1 illustrates four different classes of sequential networks.   Sequential networks may be categorized in many ways; other criteria for classifying are often used,

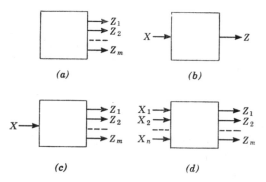

(a)   (b)

(c)   (d)

Fig. 7-1. Sequential networks.   (a) Binary sequence generator; (b) sequential transducer; (c) operation counter; (d) general sequential network.

and some of these will be described in Chap. 11.   The classification of sequential networks in Fig. 7-1 is based on the number of input lines and output lines to each network.   For instance, the binary sequence generator in Fig. 7-1a has output lines but no inputs, while the sequential transducer in Fig. 7-1b has a single input line and a single output line.   The networks have been classified in this way because each of these types of network has certain characteristics which are of interest, and because the design of a given network may be facilitated by considering other networks of the same type.   Also, more complex networks may be constructed by interconnecting several simple networks.   Actually, all the four classes of networks may be included in that of Fig. 7-1d, which shows the generalized network with n input lines and m output lines.   Some preliminary remarks about the networks in Fig. 7-1 follow.   The descriptions assume that the number of elements in a given network is finite and the structure of each network is fixed.

The *binary sequence generator,* or *autonomous network,* has an output which is *periodic* in that after some initial finite time period has passed, the sequence† of outputs generated by the network can always be partitioned into a set of identical subsequences of finite length. That is, the output of a binary sequence generator has the form $\alpha\beta\beta\beta\ \cdot\ \cdot\ \cdot$ where α consists of an ordered set of binary digits which constitute the initial response of the network and the β's consist of identical sets of binary digits. If the number of digits p in each β is minimal, then p is the period of the network. While it is possible to construct networks which generate very complicated binary sequences, a most common type of binary sequence generator is the *clock* for a synchronous machine. The output of a clock often consists of a sequence of pulses, generally with some fixed interval of time between each pulse. This is not necessarily the case, however, for the clock may be designed so that pulses appear in subsequences of some fixed length, with identical intervals between corresponding signals in each subsequence but with different time intervals between members of a subsequence.

The clock for a given machine may also have several output lines on which signals appear at certain intervals and with some fixed relation between the signals on the various lines. Machines have been designed with clocks where the outputs from the clock consist of several such output lines, each carrying sinusoidal signals with a fixed phase relation to the signals on the other lines. It may be seen that the basic function of most binary sequence generators is to provide either timing or reference signals.

The *sequential transducer* has a single input and a single output line. If the input sequence to the sequential transducer is periodic, the output sequence from the network will also be periodic, and the periodicity of the output will be determined by both the input sequence and the structure of the network. Therefore, the output signal from the sequential transducer at time t is determined by not only the structure of the network but also the history of the input signals, and often by the input at time t. Notice that if a clock or sequence generator is connected to the input of a sequential transducer, a sequence generator will be formed.‡ The difference between the sequential transducer and the sequence generator is that given fixed initial states, the output Z at time t from the sequential transducer is a function of the input sequence which occurred prior to t,

---

† A *sequence* $A(0), A(1), \ldots, A(t), \ldots$ is a function defined on the nonnegative integral variable $t = 0, 1, 2, \ldots$. An *infinite sequence* is the ordered set of values of the function when the function is defined for all values of t. A *periodic sequence* is an infinite sequence such that for all t greater than some nonnegative integer m, $A(t) = A(t + p)$. If p is the smallest nonnegative integer for which $A(t) = A(t + p)$ holds, then p is called the *period* of the sequence.

‡ A particular class of sequential transducers and sequence generators has been studied in some detail; these are the *linear sequential transducers*. An introduction to this type of sequential network will appear in Sec. 7-11.

while the output Z at time t of a sequence generator is a function of time only.

Figure 7-1c shows an *operation counter* with a single input line and m output lines. The operation counter is generally used to measure and record the number of occurrences of machine operations (generally transfers) of a certain type, where the performing of an operation is indicated in the system by a specific signal at the input to the operation counter. (In Fig. 7-1c an operation is designated by the occurrence of a pulse or 1 signal.) In order to count, the network must change its internal state each time a specified input signal occurs. Further, the internal states which the counter assumes must be distinguishable in order to be useful, and in Fig. 7-1c the values on the output lines Z_1, Z_2, . . . , Z_m are assumed to be functions of the previous input signals. Since the number of memory cells in a given counter is finite, the counter will have only a finite number of different internal states and will be capable of recording only some fixed number of inputs before the internal and output states become ambiguous. Therefore, a given counter will count modulo some fixed integral value, and we define this number to be the *period* of the counter, or sometimes the *cycle length* of the counter. A given counter generally has a *reset* line by means of which the internal memory cells of the counter can be reset to some fixed initial state.

An operation counter does not necessarily count in some specified number system but simply proceeds through some fixed sequence of internal states. A counter for which the values or states of the internal register generate a sequence which corresponds to the binary integers in a natural way is called a *binary counter*.

The *multi-input multi-output sequential network* is the most general class of network, and all the previous classes are special cases of this particular class of network. The description of the networks which follow will progress from binary sequence generators up to multi-input multi-output networks, but the ordering will not be strict, as certain important networks fall into several of the classes. First, however, the concepts of internal state, input state, and output state will be discussed, and two fundamental network structures, which we shall call class A and class B networks, will be described.

7-3. Class A and Class B Network Structures. Definitions for states or values of memory cells and states or values of registers were given in Sec. 2-2, and these will play important parts in the description which follows. Also, the definitions of synchronous and asynchronous in Secs. 6-6 and 6-7 will be used, as well as the assumptions on timing in those sections.

Consider a sequential network such as that shown in Fig. 7-2a. This configuration consists of a combinational network (dependent register) and a set of m memory cells (an independent register) denoted as register Q and consisting of memory cells Q_1, Q_2, . . . , Q_m. A *feedback loop* leads from the memory cells back into the combinational network. There are n input lines X_1, X_2, . . . , X_n to the sequential network (each of which at any given time t is assumed to carry a signal representing a binary value),

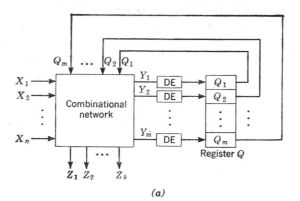

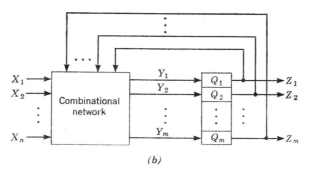

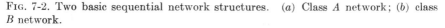

FIG. 7-2. Two basic sequential network structures.   (a) Class A network; (b) class B network.

$n + m$ inputs to the combinational network, and $k + m$ outputs (also assumed to be binary) from the combinational network.

When dealing with a network consisting of actual electronic components, it would be possible to make a set of physical measurements of the voltages, currents, direction of magnetization, etc., at various points in the circuit, and then to use these to specify the *state* of the network at the time of measurement.   Since the circuit elements and signals are assumed to be binary, each measurement would theoretically lie within one of two intervals, and a binary digit could be assigned to the result of each measurement.   If a total of p measurements were made, the set of measurements could be denoted by a p-tuple which would have one of

2^p values at each time of measurement t.   Instead of describing the state of a sequential network by means of a single parameter of this sort, it is convenient to deal with sets of conditions such as the input signals and output signals.   For instance, in analyzing the operation of a sequential circuit such as that in Fig. 7-2a, we will let

$$X(t) = [X_1(t), X_2(t), \; . \; . \; . \; , X_n(t)]$$

denote the input state or input value at time t, where each $X_i(t)$ is a Boolean function of time with value 0 or 1 at a given time t.   Further, let $Y(t) = [Y_1(t), Y_2(t), \; . \; . \; . \; , Y_m(t)]$ denote the set of signals from the combinational network to the m memory cells at time t; let

$$Q(t) = [Q_1(t), Q_2(t), \; . \; . \; . \; , Q_m(t)]$$

denote the state of register Q at time t; and finally, the k output signals at time t will be denoted by $Z(t) = [Z_1(t), Z_2(t), \; . \; . \; . \; , Z_k(t)]$.

The preceding chapters have assumed that signals pass through a combinational network without delay.   For instance, the combinational network in Fig. 7-2b has inputs $X(t)$ and $Q(t)$ and output $Y(t)$.   Therefore we assume that $Y(t) = \nu[Q(t), X(t)]$, with ν a constant function, so that if an input signal to the combinational network changes, any change in the output signals occurs instantaneously.   In actual practice an attempt will generally be made to see that the response time of the network is non-critical, generally by seeing that the maximum time required for signals to propagate through a combinational network is much smaller than the interval between timing signals (refer to Sec. 6-6).

It is important to notice that the memory elements Q_1, Q_2, $\; . \; . \; . \;$, Q_m comprising Q in Fig. 7-2a and b should not change states at the same instant in time as the output Y from the combinational net changes state, for this would cause a condition called the race problem, as described in Sec. 6-4.   If Q changes states at the same time as Y, a change in X will cause a change in Y which will cause a change in Q and hence a change in the inputs to the combinational network, and perhaps another change in Y, etc.   This problem always confronts the designer, and, in synchronous systems, care must be taken to see that memory elements do not change states during clock pulses.   (In asynchronous circuits, a type of circuit action similar to this takes place, greatly complicating design.)   The solutions to the race problem vary with the system and components used. Figure 7-2a shows a set of delay elements, each with a delay of τ seconds, inserted between the combinational network and the memory cells.   Here τ is assumed to be greater than the duration of the clock pulses in the system, so the memory elements do not change states until after the trailing edges of the clock pulses.   Sometimes memory cells are constructed with internal delays, and sometimes memory cells are constructed which change states only during the trailing edges of input pulses.   For the remainder of the chapter, all memory elements will be assumed to have some sort of delay built in, and we will omit this delay from the drawings, unless it is specifically stated otherwise.

Since the memory cells in Fig. 7-2a do not change states until after some time ϵ which is less than the time interval between clock pulses, the output Z from the network at time t will be determined by the input X at time t and the state of register Q at time t. The state of register Q at time t will therefore be a function of the input state X at time $t - 1$ and the state of Q at time $t - 1$. Therefore, given the transformations yielded by the combinational network, the state of register Q at some time $t = \Lambda$, and the input state $X(t)$ for $t = \Lambda, \Lambda + 1, \Lambda + 2, \ldots$, we can calculate the corresponding output values $Z(\Lambda)$, $Z(\Lambda + 1)$, $\ldots$ using recursion.

The operation of the network can be expressed symbolically. Let $Y(t) = \nu[Q(t), X(t)]$ and $Z(t) = \zeta[Q(t), X(t)]$ represent the transformations for the combinational network. Also, assume that the input value $Y_i(t)$ to cell Q_i at time t is taken by the cell by time $t + 1$. Then each clock pulse initiates the transfer $Y \rightarrow Q$, and the formula for the state of register Q at $t + 1$ will then be

$$Q(t + 1) = Y(t) = \nu[Q(t), X(t)]$$

This function ν is called the *next-state* function, for it takes a given pair $X(t)$ and $Q(t)$ and maps them into a unique value which is the next-state value of register Q.

Further, the output Z at time t will be

$$Z(t) = \zeta(\nu[Q(t - 1), X(t - 1)], X(t))$$

The output at a given time t can therefore be calculated using recursion, given Q at some time $t - n$ and the input sequence $X(t - n)$, $X(t - n + 1)$, $\ldots$, $X(t)$.

In many cases the register Q will always be started in some fixed initial state, say q_0. We can then define the operation of the network as follows:

$$Z(0) = \zeta[q_0, X(0)]$$
$$Z(t + 1) = \zeta(\nu[Q(t), X(t)], X(t + 1))$$

The type B network in Fig. 7-2b can be seen to be a special case of the type A network in Fig. 7-1a. Let the structure of the combinational network in Fig. 7-1a be such that the output from cell Q_i is directly connected to output Z_i. The operation of the type A network will now be the same as that of the type B network.† Although the type B net-

---

† The outputs of a class B network might well consist of the outputs from only some subset of the m memory cells. In this case the number of output lines k would be less than the number of memory cells m, and only k connections from the cells to the output lines would be made.

work is a special case, the structure is quite common and there are certain characteristics which make this type of network especially easy to analyze and design.

The output values from the network in Fig. 7-2b can be put in one-to-one correspondence $[Q_i(t) = Z_i(t)]$ with the values of the register Q. The recursion formula for the network is

$$Q(t + 1) = \nu[Q(t),X(t)]$$

or, since we can let $Z(t) = Q(t)$,

$$Z(t + 1) = \nu[Q(t),X(t)]$$

Again assuming the register Q is always started in some fixed state q_0 at time $t = 0$,

$$Q(0) = q_0 = Z(0)$$
$$Q(t + 1) = \nu[Q(t),X(t)] = Z(t + 1)$$

defining the operation of the network recursively.† Notice that the output from a class B network is a function of the state of register Q at time t only, and not of the input state at time t.

7-4. Binary Sequence Generators. The operations performed by a synchronous digital machine are generally sequenced by a "master clock," which is an example of a sequence generator. The output signals from the clock are quite often evenly spaced in time, and the intervals between signals are such that the electronic circuits may make the necessary transitions and stabilize before the next operation is initiated. Since the operating speed of the machine will be to a large part determined by the interval between clock pulses, it is desirable to keep this interval as short as possible. Nevertheless, ample time must be allowed for the transient response of the circuitry, and sufficient time must be allowed for all signals to propagate through the gating circuitry of the machine.

Often the frequency of the clock is determined by some specific component. In most cases the type of internal memory used will be an important factor and will determine the clock frequency. Most machines which use magnetic-core memories have clock rates determined by the operating speeds of the core memory, and if a magnetic drum is used as the basic internal storage device, the drum will not only determine the clock frequency but will probably be used to generate the clock signals.

Figure 7-3a shows two schemes for generating clock signals; the first consists of a delay line and pulse-reshaping amplifier. Once a pulse is started around the loop it will continue circulating indefinitely, producing a sequence of output signals occurring at intervals of Δ seconds. A graph

---

† If the number of output lines k is less than the number of memory cells m, then the formulas above might be written $Z(t) = \zeta[Q(t)]$, $\zeta: z_i = q_i$, $i \leq k$ the number of output lines, and $Z(0) = \zeta[q_0]$; $Q(t + 1) = \nu[Q(t),X(t)]$; and $Z(t + 1) = \zeta(\nu[Q(t)])$.

of the output from this circuit is plotted versus time in Fig. 7-3a.   The time axis is plotted so that the first output signal occurs at 0, the second at 1, the third at 2, etc., thus placing the output pulses in one-to-one correspondence with the nonnegative integers.   The memory cells of the machine are then sensed only when clock pulses occur, and machine operations are initiated by these same signals.   We can therefore assume that time passes in discrete intervals and examine the operation of the machine by examining the states or values of the registers at the unique times $t = 0, 1, 2, \ldots, n, \ldots$.

A magnetic drum is also shown in Fig. 7-3a.   This drum has a timing track around which a set of timing signals has been recorded.   The reading head for the timing track produces a sequence of signals which indicate the positions of the cells around the other tracks which are used as the registers of the machines.   By using the output signals from the timing track as the clock signals, the operations of the machine can be sequenced in a natural way; placing the output signals from the timing track in correspondence with the nonnegative integers will facilitate the analysis and design of the machine.

Figure 7-3b shows a scheme for generating a sequence of signals which occur in groups of four such that the interval between the first and second signal in each group is Δ_1 seconds, the interval between the second and third is Δ_2 seconds, the interval between the third and fourth is Δ_3 seconds, and an interval of Δ_4 seconds then occurs between the fourth signal in a given group and the first signal in the next group.   Sometimes it is convenient to use a clock of this sort.   If speed of operation is very important it may be possible to perform certain operations faster than others, and a clock with unevenly spaced intervals may speed up the over-all operating time.   In this case we can still place the clock signals in one-to-one correspondence with the nonnegative integers in a natural way, thus simplifying our design process, and later take into account the uneven spacing of clock pulses only where actual problems of circuit operation are considered.

Let us now examine two other types of sequence generators.   Figure 7-3c shows a *four-cell shift register with feedback* constructed of four delay lines and an inverter.   Let us assume that 0's are represented in this system by negative pulses and 1's by positive pulses.   If a 0 signal is inserted into each cell of the register at time $t = 0$, the sequence of outputs will be as shown, and the binary sequence generated by the circuit will be 00001111000011 $\cdots$, a sequence of period 8.   If the inverter is capable of restoring and reshaping the pulses in the system, this sequence generator will continue to cycle indefinitely.

The shift register with feedback provides a means for generating any desired sequence.   Figure 7-3d shows a block diagram for a shift register with feedback constructed of magnetic cores, and Fig. 7-3e shows an actual circuit. The sequence produced by this circuit will be the same as that produced by the delay-line shift register in Fig. 7-3c. We will analyze the operation of the circuit using Boolean functions of time. The input line X is assumed to carry a sequence of input pulses occurring at times $t = 0, 1, 2, 3, \ldots$, so the X input could be connected to the output of Fig. 7-3a. The feedback loop contains a single gate which complements the output signal from cell Q_4; the complemented value of Q_4 at time t is read into cell Q_1 by time $t + 1$. In the notation of Sec. 6-6,

$$L(Q) \to R(Q) \qquad Q_4' \to Q_1$$

Output line Z takes the value $Q_4(t)$ at time t. Also, the memory cells are all assumed to be set to the value 0 at time $t = 0$. The formulas for

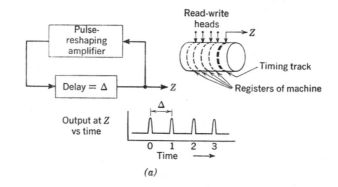

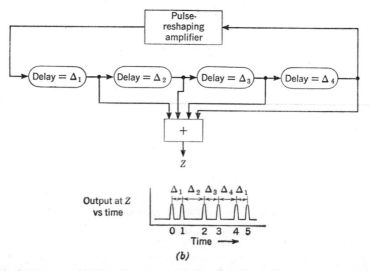

FIG. 7-3. Binary sequence generators. (a) Regular interval sequence generator; (b) sequence generator; (c) to (e) mod 8 operation counter and sequence generator.

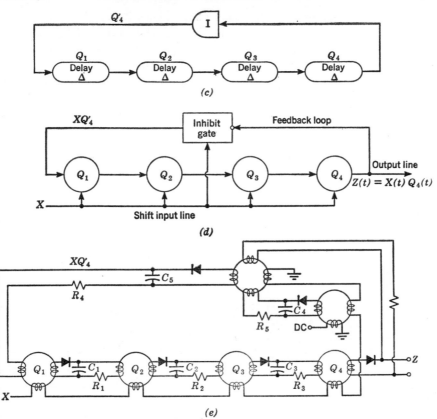

FIG. 7-3. *(Continued)*

this sequence generator are

$$Z(t) = Q_4(t)$$
$$Q(0) = q_0 \qquad Q_1(0) = Q_2(0) = Q_3(0) = Q_4(0) = 0$$
$$Q(t + 1) = \nu[Q(t), X(t)]$$

where $X = 1$ when $t = 0, 1, 2, \ldots$ and 0 at all other times. For this network, ν is defined as

$$Q_1(t + 1) = X(t)Q_4'(t)$$
$$Q_2(t + 1) = X(t)Q_1(t)$$
$$Q_3(t + 1) = X(t)Q_2(t)$$
$$Q_4(t + 1) = X(t)Q_3(t)$$

If $X(t) = 1$ at $t = 0, 1, 2, 3, 4, \ldots$, then

$$Q_1(t + 1) = Q_4'(t) \qquad Q_2(t + 1) = Q_1(t) \qquad \text{etc.}$$

Several remarks should be made about the expressions above and the physical characteristics of the magnetic-core memory cells in Fig. 7-3e. A magnetic-core memory cell stores a binary value which is dependent upon the direction of residual magnetization of the core. Hence we can assume, for instance, that if the core is magnetized clockwise, it represents a 1; a counterclockwise magnetization would then represent a 0. Each core memory cell in Fig. 7-3e will retain its magnetization, and hence its state, unless an electric current inducing a magnetic field which tends to reverse the direction of magnetization of the core is passed through the winding on the core. The windings connected to the input X are arranged so that a positive current pulse on X will reset the cores Q_1, . . . , Q_4 to the 0 state. If a core Q_i contains a 0 and a pulse input is applied at X, the direction of the magnetization of Q_i will not change and no output current will be induced in the output winding of Q_i. If, however, core Q_i contains a 1 when positive current passes through the X winding input, the direction of magnetization of Q_i will be reversed, a current will be induced in the output winding, and a pulse output will result. (The read-out procedure is *destructive* in that the core is always reset to the 0 state when read.) An output signal from core Q_i, $i \leq 3$, will be delayed by either capacitor C_1, C_2, or C_3, and then used to set the next core to the right (Q_{i+1}) to the 1 state. The output from core Q_i at time $t = 1, 2, \ldots$ is therefore equal to the value read into the core at time $t - 1$. Notice also that cores do not have continuous outputs, but only respond when pulsed. The core is therefore a single-input single-output memory cell such as appears in the block diagram, and the capacitor and resistor nets on the output winding of each core correspond to the delay elements which avoid a race situation.

We have assumed that X, the input to the transducer, always has the value 1 when t is an integral value; we therefore have a clock driving a sequential transducer or a binary sequence generator. Let us represent the state of Q at time t as a binary number by simply adjoining the values taken by the Q_i, so that, for instance, $Q(t) = 0001$ when $Q_1(t) = 0$, $Q_2(t) = 0$, $Q_3(t) = 0$, and $Q_4(t) = 1$. If Q is started in initial state $q_0 = 0000$ at time $t = 0$, we can compute the values of Q and Z at time $t = 0, 1, 2, \ldots$ as follows:

$$
\begin{aligned}
Q(0) &= 0000 & Z(0) &= 0 \\
Q(1) &= 1000 & Z(1) &= 0 \\
Q(2) &= 1100 & Z(2) &= 0 \\
Q(3) &= 1110 & Z(3) &= 0 \\
Q(4) &= 1111 & Z(4) &= 1 \\
Q(5) &= 0111 & Z(5) &= 1 \\
Q(6) &= 0011 & Z(6) &= 1 \\
Q(7) &= 0001 & Z(7) &= 1 \\
Q(8) &= 0000 & Z(8) &= 0 \\
Q(9) &= 1000 & Z(9) &= 0
\end{aligned}
$$

The values of Z will form the sequence 0000111100001111 · · · , a periodic sequence with period 8.

7-5. The Flip-flop. The preceding sections have described memory cells with a single input line and a single output line, as defined in Secs. 2-2 and 2-3. A large number of digital machines utilize a memory element known as a flip-flop, a bistable memory cell generally having two

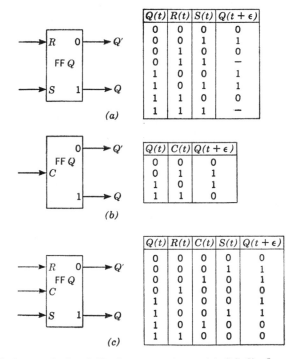

$Q(t)$	$R(t)$	$S(t)$	$Q(t+\epsilon)$
0	0	0	0
0	0	1	1
0	1	0	0
0	1	1	—
1	0	0	1
1	0	1	1
1	1	0	0
1	1	1	—

(a)

$Q(t)$	$C(t)$	$Q(t+\epsilon)$
0	0	0
0	1	1
1	0	1
1	1	0

(b)

$Q(t)$	$R(t)$	$C(t)$	$S(t)$	$Q(t+\epsilon)$
0	0	0	0	0
0	0	0	1	1
0	0	1	0	1
0	1	0	0	0
1	0	0	0	1
1	0	0	1	1
1	0	1	0	0
1	1	0	0	0

(c)

Fig. 7-4. Tabular analysis of flip-flop operation. (a) RS flip-flop; (b) C flip-flop; (c) RCS flip-flop.

output lines, each output line containing a signal which is the complement of the other line. Many different kinds of flip-flops have been designed and used; the operating characteristics of a given flip-flop will present certain specific advantages and disadvantages to the system designer. Instead of a survey of all available types, the characteristics of three of the most common flip-flops will be described in this section, and the following section will analyze several circuits containing flip-flops.

The RS flip-flop in Fig. 7-4a has two input lines and two output lines. The state of the flip-flop is generally defined as the value of the signal on the Q, or 1, output line: if the signal at Q is a 0, the flip-flop is said to be in the 0 state and the Q' output line will contain a 1; if the Q output line contains a 1, the flip-flop is in the 1 state and the Q' line will contain a 0. Figure 7-4a also shows a table of input and output states for this flip-flop. The flip-flop is assumed to have an internal delay of ϵ seconds, so that if an input signal which will change the state of a flip-flop is applied at time t, the flip-flop will not change states until time $t + \epsilon$. Define $Q(t)$ to be the value of the output signal at output Q at time t, let $S(t)$ be the value of the signal at the S input at time t, and let $R(t)$ be the value on the R input at time t.   The general transfer relation for the RS flip-flop is given by

$$S \cdot 1 + R \cdot 0 + R'S' \cdot Q \to Q \qquad SR = 0$$

with the transfer occurring during the time interval $(t, t + \epsilon)$.   The relation states that if the *set* input is 1, a 1 is inserted in Q, if the *reset* input is 1, a 0 is inserted in Q, and if neither input is 1, Q remains undisturbed. The new value of Q has meaning only at least ϵ seconds after the observation of the old values of the functions.   The transfer relation simplifies to

$$S + S'R' \cdot Q = S + R' \cdot Q \to Q$$

which, by the definition of the transfer, implies that

$$Q(t + \epsilon) = S(t) + R'(t)Q(t)$$

This equation may be verified by inspecting the table in Fig. 7-4a which states that the flip-flop will remain in its present state if both inputs are 0's.   Further, if a 1 signal is applied at the S input, the flip-flop will be set to the 1 state, and if a 1 is applied at the R input, the flip-flop will be reset to the 0 state.   Notice that a 1 signal on both the R and S lines is prohibited; the state of the flip-flop after ϵ seconds is then indeterminate. This input state, 1's on both the R and S lines, is called a *forbidden input state*.

The transfer of the value of a function P into an RS flip-flop Q is easily implemented.   Letting

$$S = P \qquad \text{and} \qquad R = P'$$

the transfer relation becomes

$$P + P \cdot Q = P \to Q$$

To establish the time interval $(t, t + \epsilon)$ at which the transfer occurs, we may introduce a clock pulse generator X which has value 1 at time t and value 0 at other times.   Then the functions XP and XP' can be used as inputs to R and S respectively.

The C flip-flop (or *complementing* flip-flop) in Fig. 7-4b has a single input line and the usual two output lines.   Again the flip-flop is assumed to have a delay of ϵ seconds from the time an input which will cause the flip-flop to change values is applied until the state of the flip-flop changes. The operation of the flip-flop is quite simple: the flip-flop always changes states ϵ seconds after a 1 signal is applied.   If $C(t)$ is the value of the input at time t and $Q(t)$ and $Q'(t)$ are as before, the characteristic equation for this flip-flop is

$$Q(t + \epsilon) = Q(t) \oplus C(t)$$

which may be verified from the table.

This is equivalent to the implementation of the transfer

$$C \oplus Q \to Q$$

The RCS flip-flop in Fig. 7-4c is simply a combination of the two flip-flops described above.   There are three inputs and two outputs; the table describing the circuit is in Fig. 7-4c (rows listing forbidden input states are omitted) and the characteristic equation for the circuit is

$$Q(t + \epsilon) = R'(t)Q(t) + [R(t) \oplus Q(t)] + S(t)$$

7-6. Binary Counters and Operation Counters.   Operation counters are used to record the number of occurrences of machine operations of a specific type.   We assume that each time a specified operation occurs, a one-input signal is applied to the counter.

Let the counter consist of an m-cell register Q with cells $Q_1, Q_2, \ldots ,$ Q_m and a combinational network.   Assign the integer representation $\delta_I(Q) = \sum_{i=1}^{m} Q_i 2^{m-i}$ to each value of Q.   Let r be the number of operations which have occurred, and $Q(r)$ the state of Q after r inputs signifying operations.   A *binary counter* is an operation counter for which $\delta_I[Q(r + 1)] = \delta_I[Q(r)] + 1 \pmod{p}$, where $p \leq 2^m$ and p is the cycle length or period of the counter.

There are also binary counters which decrease the number $\delta(Q)$ each time an input is applied, so that $\delta_I[Q(r + 1)] = \delta_I[Q(r)] - 1 \pmod{p}$; these are defined as *reverse binary counters.*

Let us now design a binary counter with 3 output lines using 3 RS flip-flops, and let this counter have a period $p = 8$.   The sequence $\{Q(r)\}$ will be 000, 001, 010, 011, 100, 101, 110, 111, 000, 001, ...   In order to implement a three-cell class B binary counter using RS memory elements, a combinational network with six output lines will be required, one for the R and one for the S input of each flip-flop.   The inputs to the combinational network will be from the outputs of the flip-flops and the X or *stepping* input, so both complemented and uncomplemented values of each input variable except X will be present.   The transitions listed above are to be initiated by means of the X input, which will contain a pulse or 1 signal each time the counter is to be stepped (incremented). Now the RS flip-flop has one characteristic which is often helpful in reducing the expressions for a given circuit: when the flip-flop is in the 1 state and is not required to change to the 0 state, it does not matter whether or not a 1 signal appears at the S input, so d's or don't cares may be entered in the table of combinations in the S column when this situation exists.   An analogous situation occurs at the R input when the flip-flop is in the 0 state and is not required to change to a 1.   A table of combinations for the combinational network for the modulo 8 binary counter follows, with d's entered in the table where output values are don't cares.   The inputs to the network are from Q_1, Q_2, Q_3, and X; the outputs are to the R and S inputs of Q_1, Q_2, and Q_3, and each R and S carries a subscript designating the flip-flop to which it will be connected. Since all outputs from the network are to be 0 whenever X is a 0, we have omitted the rows corresponding to $X = 0$ inputs from the table.

Inputs				Outputs					
X	Q_1	Q_2	Q_3	R_1	S_1	R_2	S_2	R_3	S_3
1	0	0	0	d	0	d	0	0	1
1	0	0	1	d	0	0	1	1	0
1	0	1	0	d	0	0	d	0	1
1	0	1	1	0	1	1	0	1	0
1	1	0	0	0	d	d	0	0	1
1	1	0	1	0	d	0	1	1	0
1	1	1	0	0	d	0	d	0	1
1	1	1	1	1	0	1	0	1	0

A set of reduced expressions for the output lines from the combinational network is

$$R_3 = XQ_3$$
$$S_3 = XQ_3'$$
$$R_2 = XQ_3Q_2$$
$$S_2 = XQ_3Q_2'$$
$$R_1 = XQ_3Q_2Q_1$$
$$S_1 = XQ_3Q_2Q_1'$$

A block diagram for a mod 8 binary counter may be found in Fig. 7-5a.

The binary counter may also be designed using C flip-flops.   In this case a 1 is entered in the table of combinations in column C_i only when flip-flop Q_i is to be complemented, that is, when the flip-flop is to change states.   For instance, when Q_1, Q_2, and Q_3 contain 101 and an $X = 1$ signal occurs, then Q_2 and Q_3 are to be complemented, yielding 110 as the next state.   Therefore, 1's are placed in columns C_2 and C_3 in the row where the input value is $X = 1$, $Q_1 = 1$, $Q_2 = 0$, and $Q_3 = 1$.   The set of expressions derived from the resulting table are sometimes referred to as

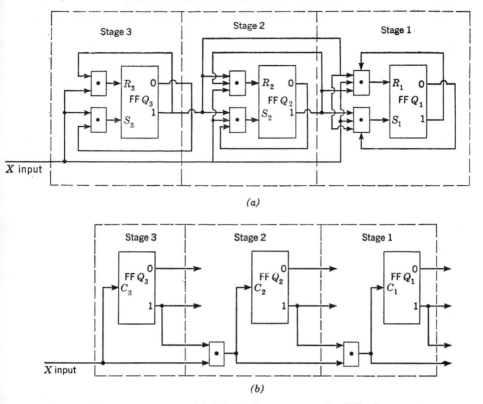

FIG. 7-5. Binary counters. (a) RS flip-flop counter; (b) C flip-flop counter.

change equations, for the expression for C_i represents, in Boolean form, the input states which are to cause Q_i to change value.

	Inputs			Outputs		
X	Q_1	Q_2	Q_3	C_1	C_2	C_3
1	0	0	0	0	0	1
1	0	0	1	0	1	1
1	0	1	0	0	0	1
1	0	1	1	1	1	1
1	1	0	0	0	0	1
1	1	0	1	0	1	1
1	1	1	0	0	0	1
1	1	1	1	1	1	1

The minimal expressions derived from the above table are

$$C_3 = X$$
$$C_2 = XQ_3$$
$$C_1 = XQ_3Q_2$$

A block diagram of a counter using C flip-flops may be found in Fig. 7-5b. Notice that fewer gates are required if C flip-flops are used than if RS flip-flops are used. However, RS flip-flops are liable to be less expensive than C flip-flops, so that the relative costs must be known before a decision can be made.

The combinational networks used to realize a binary counter in Fig. 7-5a and b are iterative networks.† For instance, an m flip-flop counter which will count to $2^m - 1$ can be constructed using m C flip-flops and $m - 1$ AND gates, where flip-flop Q_i, $i \neq m$, is connected as in Fig. 7-5b, stage 1 or stage 2, and flip-flop Q_m is connected as in stage 3. The expression for output C_i of the combinational network is therefore $C_i = XQ_mQ_{m-1} \cdots Q_{i-1}$.

In general there is no way of telling whether a given sequential network can be more inexpensively implemented using RS or C or even RCS flip-flops without designing networks for each. Often only one or two types will be available, however, and sometimes other factors such as speed of operation and output loading will prove the deciding factor.

The most general class of counters we have designated as *operation counters*, and binary counters are members of this class. Many counters generate sequences in which the successive terms do not correspond to any number system. All that is required is that each internal state of the counter be unique within a given period; that is, the sequence of states $\{Q(r)\}$, where $Q(r) = [Q_0(r), Q_1(r), \ldots, Q_m(r)]$ is the m-tuple of binary internal states, should not contain two identical m-tuples within any subsequence of length less than the period of the sequence. Given the sequence of internal states $\{Q(r)\}$, we can design a network which will realize this sequence.

Let us design a counter which generates the output sequence 00, 01, 11, 10 using C flip-flops and a class B network. First a table of combinations is drawn up; there are three inputs to the combinational network (X, Q_1, and Q_2) and two outputs, C_1 to the input of Q_1 and C_2 to the input of Q_2.

---

† An iterative network is a network constructed of identical cells arranged so that the connections between adjacent cells are the same, except possibly for end cells. If cell i of an iterative network has inputs connected to only cell $i - 1$, the network is *unilateral*. If cell i receives inputs from both cell $i - 1$ and $i + 1$, the network is *bilateral*.

Again the rows where $X = 0$ have been omitted, for each contains only 0's in the output section.

Inputs			Outputs	
X	Q_1	Q_2	C_1	C_2
1	0	0	0	1
1	0	1	1	0
1	1	1	0	1
1	1	0	1	0

The network for this counter is shown in Fig. 7-6.   The expressions

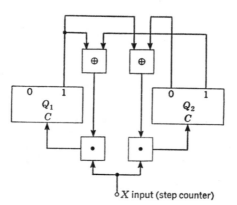

X input (step counter)

FIG. 7-6. Operation counter.

for the combinational network are

$$C_1 = X(Q_1 \oplus Q_2) \qquad C_2 = X(Q_1 \oplus Q_2')$$

or
$$C_1 = X(Q_1 Q_2' + Q_1' Q_2) \qquad C_2 = X(Q_1 Q_2 + Q_1' Q_2')$$

7-7. Shift Registers with Feedback and Constant-weight Counters. The preceding sections have introduced the operation counter and described a systematic procedure for the design of certain counters.   In the quest for simple, inexpensive counters, considerable interest has been directed toward a class of sequential circuits known as *shift registers with feedback*.   In general, a shift register with feedback is of the configuration shown in Fig. 7-7a; the circuits in Fig. 7-3c to e are specific examples.   This type of network comprises a shift register Q, consisting of a set of memory cells $Q_1, Q_2, \ldots, Q_m$, and a combinational network which yields a two-valued function λ of the contents of register Q and the input line X.   Each time an input pulse, signifying an operation has occurred, is applied on line X, the content of cell Q_i is transferred into cell Q_{i+1}, for $i = 1, 2, \ldots, m - 1$, and the value of $\lambda(Q)$ is read into cell Q_1. The counter is assumed to be started in some fixed initial state q_0 (the Q_i not necessarily in the 0 state). Thus, when a binary input sequence

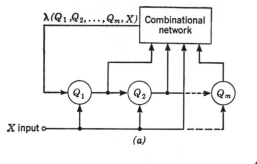

(a)

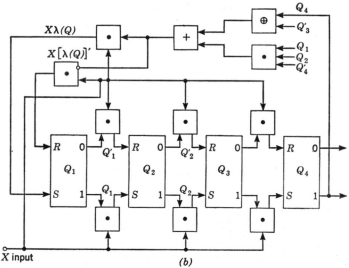

(b)

FIG. 7-7. Shift registers with feedback.

is applied at X, the register Q is stepped through a sequence of values with a cycle length or period p, where p is determined by the number of cells in the register, the initial state q_0, and λ.

Section 7-3 described a circuit of this type, using a four-cell magnetic-core register where $\lambda(Q) = Q_4'$ (refer to Fig. 7-3e). The period of the register was 8 when the initial state q_0 was $Q_i = 0$; a check will indicate it is 8, regardless of the initial state.

A general procedure for designing a shift register counter with a given cycle length p can be based on the use of a *code ring*, or *binary chain*, consisting of a closed sequence of p binary symbols. For instance, the code ring for the register in Fig. 7-3 is 11110000, where, by starting with the rightmost subsequence of length 4 (0000) and moving left one digit each time, the sequence of quadruples 0000, 1000, 1100, 1110, 1111, 0111, 0011, 0001 may be generated; this is the sequence of states taken by register Q when the initial state is 0000. A code ring for a three-cell shift register counter with a p of 5 is 11000; subsequences of length 3 from this code ring contain five different combinations of binary values. Starting with the rightmost subsequence and moving left, the sequence of triples generated is 000, 100, 110, 011, 001. Instead of deriving the Boolean expressions for each cell Q_i of the register, we need derive only the expressions $\lambda(Q)$ for Q_1, since the expressions for the other two cells are known to be $Q_2(t + 1) = Q_1(t)$ and $Q_3(t + 1) = Q_2(t)$. An examination of the sequence of triples above shows that a 1 is read into cell Q_1 at time $t + 1$ when register Q contains 000 or 100 at time t. The expression for $\lambda(Q)$ is therefore

$$\lambda(Q) = Q_1'Q_2'Q_3' + Q_1Q_2'Q_3' + (Q_1'Q_2Q_3' + Q_1Q_2'Q_3 + Q_1Q_2Q_3)\dagger$$

which can be simplified to

$$\lambda(Q) = Q_2'Q_3'$$

We have assumed this circuit is to be used as a counter and always started in an initial state where $Q_i = 0$. Notice that if the register Q is started in state 101, then the sequence generated will be 101, 010, 001, 000, 100, 110, 011, 001, 000, . . . where the first three triples constitute an initial response of length 3 (refer to Sec. 7-2) and the remaining triples form a periodic sequence with period 5.

One more example will suffice. A code ring for $p = 10$ is 0001111001, and the sequence of states generated is 1001, 1100, 1110, 1111, 0111, 0011, 0001, 1000, 0100, 0010. The states of Q for which a 1 is to be read into Q_1 are 1001, 1100, 1110, 0001, and 0010, so the expression for λ is

$$
\begin{aligned}
\lambda(Q) = {} & Q_1Q_2'Q_3'Q_4 + Q_1Q_2Q_3'Q_4' \\
& + Q_1Q_2Q_3Q_4' + Q_1'Q_2'Q_3'Q_4 + Q_1'Q_2'Q_3Q_4' \\
& + (Q_1'Q_2'Q_3'Q_4' + Q_1'Q_2Q_3'Q_4 + Q_1'Q_2Q_3Q_4' \\
& \qquad + Q_1Q_2Q_3'Q_4 + Q_1Q_2'Q_3Q_4' + Q_1Q_2'Q_3Q_4)
\end{aligned}
$$

where the terms in parentheses are don't cares. The simplified expression is

$$\lambda = Q_1Q_2Q_4' + Q_3'Q_4 + Q_3Q_4'$$

The block diagram for a shift register with feedback using RS flip-flops may be found in Fig. 7-7b.

---

† The terms in parentheses are don't cares (see Sec. 4-8 for details).

A procedure for deriving binary chains of any length may be found in Ref. 23. (While the technique described in Ref. 23 will generate some chains of any given length p, at present there is no systematic technique for generating all binary chains of a given length.) Binary chains of length $p = 2^m$ are called *complete* chains, and Ref. 11 contains a proof that there are $2^{2^{m-1}-n}$ complete chains of length 2^m.

Counters are generally used to sequence the operations which occur in digital machines. If we set a binary counter such as that in Fig. 7-5a or b to the $Q_i = 0$ state, the counter will then record unambiguously up to 8 input signals signifying that a specified event has taken place. Suppose that a given machine operation is to take place simultaneously with the occurrence of the fifth operation which the counter is monitoring. We can connect an AND gate which physically realizes the Boolean expression $XQ_1Q_2'\,Q_3'$ to the outputs of the counter memory cells and the X input, and this AND gate will have a 1 output at the time the fifth operation occurs after the counter has been cleared. Notice that the AND gate "senses" the binary number 100 or decimal 4, for the memory cells have an inherent delay of ϵ seconds, which is less than the time between the operations being counted but longer than the duration of the input signal signifying an event, and therefore the fifth input signal signifying an operation will occur when the counter is in the 4 state. If the output line was to contain a 1 *after* the fifth operation, then the AND gate would realize the term $Q_1Q_2'Q_3$.

The counter plus the AND gate comprise a class A network. Suppose that a sequential network with a single input X and with eight output lines, which we designate $Z_0, Z_1, \ldots, Z_7$, is required. Only one of these output lines is to contain a 1 at any given time, and that output line is to correspond to the number of operations which have occurred; that is, line Z_2 is to contain a 1 after two signals signifying an event, line Z_3 after 3 events, etc. This network could consist of the counter in Fig. 7-5a plus 8 AND gates, one for each output line. The network would then be a class A network consisting of a counter and a *many-to-one decoder net*† as shown in Fig. 7-8a and b.

A *complete* many-to-one decoder network is a combinational network with inputs from a register Q consisting of m memory cells $Q_1, Q_2, \ldots,$

† By logically adding together some subset of the output lines Z_i in Fig. 7-8b, one may form a sequential network which will realize any sequence of binary digits with period 8. For instance, the sequence 001100110011 $\cdots$ may be formed by logically adding lines $Z_2, Z_3, Z_6,$ and Z_7. By an obvious extension, any periodic sequence of m-tuples with period 8 can be formed. Further, by designing a counter with period p and connecting a suitable network to the outputs of the memory cells, any periodic sequence of period p may be formed, and since we can always design a counter with period p using $\geq \log_2 p$ memory cells, a class A network with $\geq \log_2 p$ cells can always be designed which will generate any sequence of m-tuples of period p.

Q_m and 2^m output lines $Z_0, Z_1, \ldots, Z_{2^m-1}$, such that (1) for each state of the register Q, one and only one of the output lines Z_j will represent a 1, and (2) for each state of Q a different line Z_j will represent a 1. We may therefore establish a one-to-one correspondence between output lines and the states of Q by assigning the value of the integer representation of Q,

which is $\sum_{i=1}^{m} Q_i 2^{m-i}$, as the index j of Z_j, so that line Z_j will represent a 1

when $\sum_{i=1}^{m} Q_i 2^{m-i} = j$.

The register together with decoder is a special case of the selection register defined in Sec. 6-5, and the functions Z_j are the *selector* or *characteristic* functions of the register Q.

Figure 7-8a shows a complete decoder network connected to a 2-cell binary counter. The decoder network in Fig. 7-8a is called a *diode matrix* and the particular configuration in Fig. 7-8a is often called a *rectangular matrix*, while the decoder network in Fig. 7-8b is often called a *pyramid*, for obvious reasons. Both the types of networks shown require 24 diodes (elements) for a three-cell eight-output network, but for larger networks the pyramid network will require fewer diodes (elements).[†]

The choice of configuration will be quite heavily biased by the type of elements used in the decoder network. For instance the pyramid requires signals to pass through more than a single gating level for m greater than 2, and if many levels are used, gain may be required. The various types of matrices have been studied in some detail, and descriptions of the most common types of decoder networks, physical characteristics, and other considerations may be found in Refs. 5, 22, 34, and 35.

If an m-cell counter is connected to a decoder network and provision is made for AND gating the input pulses with each of the outputs from the decoder matrix, a *time-pulse distributor* or *timing-pulse generator* will be formed. A time-pulse distributor is a network with a single input line and p output lines $Z_1, Z_2, \ldots, Z_p$, where line Z_i will carry a pulse signifying a 1 when the ith (mod p) input pulse occurs. Several types of time-pulse distributors have been described in Refs. 4, 22, 34, and 35.

A particular type of shift register can be used to realize a network of this sort. Figure 7-9a shows a *ring counter* which consists of an m-cell shift register. This counter is always started in an initial state q_0 in which cell Q_0 contains a 1 and all other cells a 0. Notice that the content of cell Q_1 is transferred into Q_{i+1} for $i \leq m - 2$, and the content of cell $m - 1$ is

---

[†] There are several other types of decoder networks, including one called a *balanced decoder matrix* which, for more than three binary input cells, requires fewer diodes than either the pyramid or rectangular matrices, and Burks[5] has shown that balanced decoder matrices require fewer diodes than any other type for $m > 3$.

transferred into cell Q_0 each time an input is applied at X. The shift register therefore circulates a single 1, and the location of the 1 in the memory cells designates the number of events which have taken place; that is, after 6 events only Q_6 will contain a 1. The counter constructed in this way counts modulo m, the number of cells in the register, and if an output line is connected to each cell, a network which externally corresponds to the counter-plus-decoder network can be realized. When a ring counter is used, the network requires more memory cells than the diode counter plus decoder, but fewer gates. The choice will again depend on the components available, etc.

Ring counters are members of a larger class of counters referred to as *constant-weight* counters, so called because the number of 1's (or 0's) in

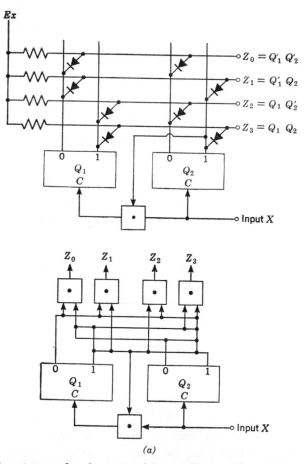

Fig. 7-8. Many-to-one decoder networks. (a) Rectangular matrix; (b) pyramid decoder.

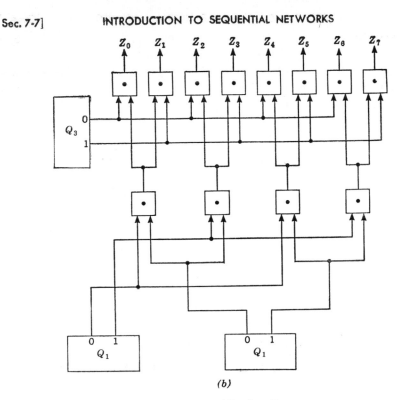

(b)

FIG. 7-8.   (*Continued*)

the memory cells remains constant.   An example[21] of a two-out-of-five counter with decoding net is shown in Fig. 7-9b.   The counter contains five memory cells and counts modulo 10, the sequence generated is $01100 \rightarrow 00110 \rightarrow 00011 \rightarrow 10001 \rightarrow 11000 \rightarrow 01010 \rightarrow 00101 \rightarrow 10010 \rightarrow 01001 \rightarrow 10100$, and the counter is always started in initial state

$$q_0 = 01100$$

The expressions for the counter are

$$Q_1(t + 1) = Q_5(t)$$
$$Q_2(t + 1) = Q_1$$
$$Q_3(t + 1) = Q_1'(t)Q_2(t) + Q_1(t)Q_3(t)$$
$$Q_4(t + 1) = Q_1'(t)Q_3(t) + Q_1(t)Q_2(t)$$
$$Q_5(t + 1) = Q_4(t)$$

The counter in Fig. 7-9b requires fewer (half as many) memory cells than a pure ring counter but requires more gates; it also requires more memory cells than the counter in Fig. 7-8b but fewer gates.   In general, constant-weight counters can always be designed which will count modulo $\binom{m}{w}$, where m is the number of memory cells and w is the number of 1's in the counter, but the choice of w, complexity of the resulting circuitry, and desirability of this type of counter must be determined in each instance. Kautz[21] presents considerable specific information including some comparison of cost for typical configurations. Notice also that

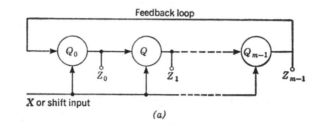

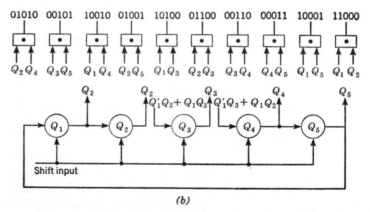

Fig. 7-9. (a) Ring counter; (b) constant-weight counter.

constant-weight counters are very similar to shift registers with feedback, except that some interstage logic is also used.

7-8. The State Diagram. A *state diagram* is a directed linear graph which can be used to represent, in graphical form, the relations between internal states and input and output states. Figure 7-10 shows state diagrams for two networks, each of which has three internal states q_1, q_2, and q_3, two input states x_1 and x_2, and two output states z_1 and z_2. The three *nodes* (vertices) of each graph, labeled q_1, q_2, and q_3, represent the three *internal states* of the network, and the *directed lines* (arcs) show the *transitions* which may be made between the states. Each directed line is labeled with the input state which will cause the state transition q_i to q_j, indicated by the line. In Fig. 7-10a the input states which cause a given transition are written to the left of the /, and the output state corresponding to this input to the right. Since two different input states may cause the same transition, a single transition arrow may carry several different labels; for instance, either x_1 or x_2 will cause a transition from q_3 to q_2 on the two diagrams.

The two types of state diagrams in Fig. 7-10 correspond to the two basic types of network described previously, the class A network being shown in Fig. 7-10a, and the class B network in Fig. 7-10b. The diagrams differ in the locations of the output signals in the diagram. Figure 7-10a has

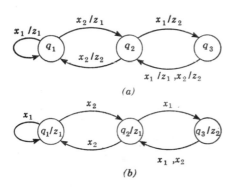

	Next-state section x_1 x_2		Output section x_1 x_2	
q_1	q_1	q_2	z_1	z_1
q_2	q_3	q_1	z_2	z_2
q_3	q_2	q_2	z_1	z_2

(a)

	x_1	x_2
q_1	q_1, z_1	q_2, z_1
q_2	q_3, z_2	q_1, z_2
q_3	q_2, z_1	q_2, z_2

(b)

	Next-states x_1 x_2		Outputs
q_1	q_1	q_2	z_1
q_2	q_3	q_1	z_1
q_3	q_2	q_2	z_2

(c)

FIG. 7-10. The state diagram. (a) Class A machine; (b) class B machine.

FIG. 7-11. State tables.

the output state located on the transition lines and to the right of the / following each input state. The outputs for the type B machines are located in the circles, which also contain the state designations, to the right of the q_i. The diagrams illustrate again that the output of a class A network is a function of its present input as well as its state, and the output of a class B network is a function of the state only.

The state diagram represents the operation of a sequential network in a straightforward manner and is especially useful in preliminary design stages. Assume the network in Fig. 7-10a is started in state q_1 at time $t = 0$. If the input sequence x_2, x_1, x_1, x_2 is applied at times $t = 0, 1, 2, 3$ respectively, the output sequence generated will be z_1, z_2, z_1, z_2, and at time $t = 4$ the network will again be in state q_1.

7-9. The State Table. Another way to represent a sequential network abstractly consists of a *state table*. The state table contains the same information as the state diagram. Figure 7-11 shows three state tables; the tables in Fig. 7-11a and b correspond to the state diagram in Fig. 7-10a, and Fig. 7-11c corresponds to Fig. 7-10b.   The state table in Fig. 7-11a is divided into two basic sections, one containing next-state information and the other the corresponding output data.   These two sections can be combined as in Fig. 7-11b, but we will adhere to the tabular structure of Fig. 7-11a.

The rows of Fig. 7-11a correspond to the states of the network, and the columns to the input states.   For instance, if the network is in state q_2 and an x_1 input state is applied, the *operating point* will be in the first column and second row of both sections of the table, indicating that the next state will be q_3 and the output a z_2.   If the initial state or starting state of the machine is q_2 and an input sequence of x_2, x_1, x_2 is applied, the output sequence generated will be z_2, z_1, z_1, and the *terminal state* will be q_2. The terminal state is, therefore, not the state when the last input is applied, but the state to which the last input takes the network, and we assume the network will stay in that state until another input is applied. Notice that the sequence of outputs and terminal state will be determined by the initial state of the machine as well as the input sequence.

Figure 7-11c shows a state table for the class B network in Fig. 7-10b. Since the output state of a class B network is determined only by the internal state of the network, a single output state is associated with each row of the table, and the output value for each row can be placed either to the right of each state and separated by a comma, or in an additional column as in Fig. 7-11c.

7-10. Analysis and Design of a Serial Binary Adder.   A *serial binary adder* is a sequential circuit with two input lines, which we designate X_1 and X_2, and a single output line S.   The input signals of lines X_1 and X_2 consist of sequences of 0's and 1's represented by electrical signals, where each signal representing a binary digit occupies some finite fixed time interval.   Further, the input signals on X_1 and X_2 represent binary numbers and arrive in sets, each containing some fixed number of digits; the least significant digits in the numbers represented arrive first.   If the machine represents numbers as nonnegative integers, where each number in the basic machine word consists of, let us say, five digits, then integers from 0 to 31 can be represented.   Let 11010 represent the decimal number 26 (reading from right to left, the weights assigned to the digits increase as powers of 2), let this sequence of five digits be presented on line X_1, and let 00011 be presented on line X_2; the output line S is then to carry the sequence 11101, denoting the decimal number 29.   Since the binary digits arrive in sets or words, each set representing some number, it may be necessary, when adding, to reset the memory cell in the serial binary adder after each addition of two machine words.   Also, each set of digits will probably carry a *sign bit*, as explained in Sec. 8-4.   In this section, we shall ignore the problems associated with resetting the adder and with sign digits.   The system for representing negative numbers (whether the sign bit precedes or follows the digits representing a number) and other considerations enter into this problem.   In this section we shall describe only the design of a conventional serial binary adder which adds together two positive binary numbers represented in serial form.   In Chap. 8 a more general description of adders is presented with some discussion of both serial and parallel adders.

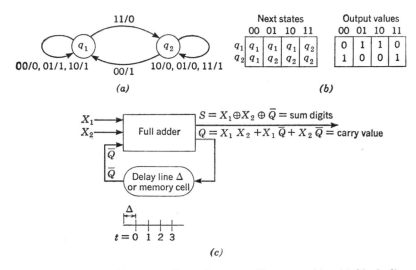

FIG. 7-12. The serial adder.   (a) State diagram; (b) state table; (c) block diagram of serial adder.

The sequence of signals representing the addend and augend are assumed to arrive at regular intervals of time; so let the signals arrive at times $t = 0, 1, 2, \ldots , n$, where $n + 1$ is the number of digits in each number to be added.   The output from the circuit is to be a sequence of signals representing the sum, so that the sum digit on line S at time t is equal to the sum of the two input digits on X_1 and X_2 at time t and any carry from the preceding digits.

The network can be represented by the state diagram in Fig. 7-12a and the state table in Fig. 7-12b, where q_i is the state of the memory cell. State q_1 corresponds to no carry, and state q_2 to a carry of 1.   The network is always started in state q_1 at time $t = 0$.   In Fig. 7-12a and b the input states are represented by actual binary values, with the value on line X_1 to the left of the value on X_2, so that 01 represents a 0 on X_1 and a 1 on X_2.   The S's are also represented by the binary values 0 and 1.   If we let the state q_1 correspond to a 0 in the memory cell, and q_2 to a 1 in the memory cell, and also let Q represent the output of the memory cell and $\bar{Q}$ the input, so that $Q(t + 1) = \bar{Q}(t)$, then the following table of combinations for the circuit may be derived:

Inputs			Outputs	
X_1	X_2	$\bar{Q}$	S	Q
0	0	0	0	0
0	0	1	1	0
0	1	0	1	0
0	1	1	0	1
1	0	0	1	0
1	0	1	0	1
1	1	0	0	1
1	1	1	1	1

The combinational network must therefore realize the relations

$$S = X_1 \oplus X_2 \oplus \bar{Q}$$
$$Q = X_1 X_2 + X_1 \bar{Q} + X_2 \bar{Q}$$

The network can be realized as in Fig. 7-12c.

7-11. Linear Sequential Networks. The networks which have been described have been composed, for the most part, of AND gates, OR gates, memory cells, and inverters. Sequential networks which are constructed of only exclusive-OR gates (or *mod 2 adders*) and memory cells are often referred to as *linear sequential networks*. While linear sequential networks comprise only a small class of networks compared to the more general nonlinear networks which have been described, a number of interesting characteristics of linear networks have been described in the literature. Also, applications of this type of network are of interest in the fields of radar, communication systems, and error-correction devices. Linear networks, when used as binary sequence generators, can be used to generate long strings of quasi-random numbers;[46] linear sequential networks may be used, in many cases, as economical counters. This type of network is of considerable interest in digital communication systems which employ error-correcting codes of certain types, and an excellent description of these codes may be found in Peterson.[29]

The analysis of linear sequential networks is facilitated by the use of linear algebra, and our approach will be algebraic. Figure 7-13 shows a four-cell linear sequential network consisting of a four-cell shift register with feedback. Let us assume that the four cells consist of perfect delay lines, each with a delay of Δ, so that every Δ seconds the contents of the register will be shifted one cell to the right and $Q_3 \oplus Q_4$ read into cell Q_1. The network could, of course, be a clocked network with a clock pulse every Δ seconds, or could be implemented in any of the ways previously mentioned. In this description we will assume ideal delay lines, mod 2 adders, etc., so that if the network cells are started in some state Q^0 at

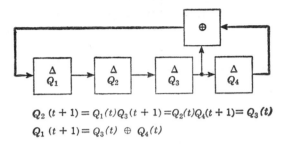

$$Q_2 (t + 1) = Q_1(t)Q_3(t + 1) = Q_2(t)Q_4(t + 1) = Q_3(t)$$
$$Q_1 (t + 1) = Q_3(t) \oplus Q_4(t)$$

FIG. 7-13. Linear sequential network.

time $t = 0$, then the register will continue to shift indefinitely, one shift each Δ seconds. The expressions for the memory cells are

$$Q_1(t + 1) = Q_3(t) \oplus Q_4(t)$$
$$Q_2(t + 1) = Q_1(t)$$
$$Q_3(t + 1) = Q_2(t)$$
$$Q_4(t + 1) = Q_3(t)$$

Let Q_i^t be the state of Q_i at time t, and $Q_i^{t'}$ the state of Q_i at time $t + 1$; then we can represent these relations in matrix† form:

$$\begin{bmatrix} Q_1^{t'} \\ Q_2^{t'} \\ Q_3^{t'} \\ Q_4^{t'} \end{bmatrix} = \begin{bmatrix} 0 & 0 & 1 & 1 \\ 1 & 0 & 0 & 0 \\ 0 & 1 & 0 & 0 \\ 0 & 0 & 1 & 0 \end{bmatrix} \begin{bmatrix} Q_1^t \\ Q_2^t \\ Q_3^t \\ Q_4^t \end{bmatrix}$$

Or, letting the square matrix be denoted as M,

$$Q^{t'} = MQ^t$$

If Q^0 is a column vector representing the initial state of the register Q at time $t = 0$, then the sequence of states the register will generate is

$$Q^0, \, MQ^0, \, M^2Q^0, \, M^3Q^0, \, \ldots, \, M^nQ^0$$

and the state of register Q at time t will be $Q(t) = M^tQ^0$. Also, if the state of Q at time t is Q^t, then the state at time tm will be $Q^{tm} = M^mQ^t$.

If the matrix for a given sequential network M is nonsingular (that is, if its determinant $|M|$ is not equal to 0), then each state will have a unique predecessor which may be calculated using the inverse matrix M^{-1}. The

† This section uses standard matrix notation as defined in Birkhoff and MacLane, for instance (see Ref. 1 in Chap. 5). Addition of the $M_{ij}Q_{ij}$ is mod 2,

matrix for the network in Fig. 7-13 is nonsingular, with an inverse

$$M^{-1} = \begin{bmatrix} 0 & 1 & 0 & 0 \\ 0 & 0 & 1 & 0 \\ 0 & 0 & 0 & 1 \\ 1 & 0 & 0 & 1 \end{bmatrix}$$

which leads to the relation $M^{-1}Q^t = Q^{t-1}$.

A series of calculations using the matrix M will indicate that $M^{15} = I$, so that if register Q at time 0 is in any state except $Q_i = 0, i = 1, \ldots, 4$ (the zero state), the successive states of a given cell Q_i of the network (or the states of register Q) will form an infinite sequence with period 15. This network falls into the category of *maximal-period networks*, which are linear sequential networks which generate sequences with a periodicity of $m = 2^n - 1$, where n is the number of cells in the register. In order to discuss such networks, we first define the *characteristic polynomial*, which can be used to derive certain properties of a given matrix and hence a corresponding linear network, and then mention some of the properties of characteristic polynonials and the corresponding networks.

The characteristic polynomial $\phi(x)$ for a given matrix M is defined as

$$\phi(x) = |M - xI|$$

$\phi(x)$ is therefore formed by subtracting an indeterminate x from the diagonal elements of M and then taking the determinant of the resulting matrix. For our example,†

$$\left| \begin{bmatrix} 0 & 0 & 1 & 1 \\ 1 & 0 & 0 & 0 \\ 0 & 1 & 0 & 0 \\ 0 & 0 & 1 & 0 \end{bmatrix} \oplus \begin{bmatrix} x & 0 & 0 & 0 \\ 0 & x & 0 & 0 \\ 0 & 0 & x & 0 \\ 0 & 0 & 0 & x \end{bmatrix} \right| = x^4 \oplus x^3 \oplus 1$$

One of the theorems concerning matrices is that every matrix satisfies its own characteristic polynomial equation. Therefore

$$\phi(M) = M^4 \oplus M^3 \oplus I = 0$$

Now the period of the matrix and hence the period of the sequence generated by the corresponding linear sequential circuit will be equal to the smallest positive integer p for which $M^p = I$, for then $M^p Q^0 = Q^0$, $M^{p+1}Q^0 = Q^1$, and in general $M^{p+n}Q^0 = Q^n$ for all n, again letting Q^k equal the state of Q at time k.

† Notice that subtraction in a two-element field (GF_2) follows the same rules as sum mod 2 addition.

The following calculations may be performed:

$$\phi(M) = M^4 \oplus M^3 \oplus I = 0$$
$$M^{11}(M^4 \oplus M^3 \oplus I) = 0$$
$$M^{15} \oplus M^{14} \oplus M^{11} = 0$$

which will simplify to

$$M^{15} \oplus I = 0 \qquad \text{or} \qquad M^{15} = I$$

So the period p for the network in Fig. 7-13 is 15. The period p may also be calculated by finding the smallest integer p such that $\phi(x)$ divides $x^p - 1$, for if $\phi(x) \lambda(x) = 0$ and $\phi(x) \lambda(x) = x^p - 1$, then $x^p - 1 = 0$ and $x^p = 1$.

Zierler[47] has shown that a linear sequential network with a characteristic polynomial which is (1) irreducible and (2) does not divide $2^{m-1} - 1$ for any $m < p$ will always generate maximal-period sequences.

Further results on linear sequential networks may be found in Zierler and in the paper of Weiss.[45] Solomon[39] treats the problem in terms of finite difference equations, Elspas[12] presents an excellent summary of much of the known theory, and Friedland[13] contains interesting results. The application of linear sequential networks in the area of error-correcting codes may be found in Peterson,[29] and a technique for analyzing these networks using operator polynomials and explained in terms of transfer ratio may be found in Huffman.[19]

PROBLEMS

7-1. If the circuit in Fig. 7-3c is started in state 0110, describe the output sequence which will be generated. Not counting sequences which are translates (displacements) of other sequences, how many different sequences can be generated by the circuit in Fig. 7-3c? What is the period of each sequence?

7-2. Design a clocked operation counter, using three RS flip-flops, which yields the sequence of internal states 000, 010, 011, 111, 110, 100, 000, 010, ... Derive the Boolean expressions and draw a block diagram for the circuit.

7-3. Consider the RS flip-flop of Sec. 7-5. Let $X(t)$ be a periodic clock pulse train with pulses at $t = n\tau$; that is,

$$X(t) = \begin{cases} 1 & n\tau \le t < n\tau + \delta \\ 0 & n\tau + \delta \le t < (n+1)\tau \end{cases}$$

for $0 < \delta < \tau$ and $(n = 0, \pm 1, \pm 2, \pm 3, \ldots)$. Let the flip-flop have inputs $S(t) = S_1(t)X(t)$ and $R(t) = R_1(t)X(t)$ such that $R_1(t)S_1(t) = 0$ for all t and $R_1(t)$ and $S_1(t)$ are constant in all intervals

$$n\tau + \delta \le t < (n+1)\tau + \delta$$

$R_1(t)$ and $S_1(t)$ can change only at points $n\tau$. Designate the flip-flop as Q and then let the state of Q at time t be $Q(t)$.

(a) Under the above conditions show that flip-flop Q satisfies the time-difference equation $Q(t + \tau) = S_1(t) + R_1(t)Q(t)$ for all t and the discrete time-difference equation $Q(t_{n+1}) = S_1(t_n) + R_1(t_n)Q(t_n)$ for any periodic sequence of times $t_n = n\tau + a$, where $0 \leq a < \tau$ and $(n = 0, \pm1, \pm2, \ldots)$. This is the equation for a *clocked* RS flip-flop.

(b) Draw the state diagram for the clocked RS flip-flop.

(c) We introduce the *up* (positive) *change operator*

$$\underset{+}{\Delta} Q(t) = Q'(t)Q(t + \tau)$$

the *down* (negative) *change operator*

$$\underset{-}{\Delta} Q(t) = Q(t)Q'(t + \tau)$$

the *hold operator*

$$H[Q(t)] = Q(t)Q(t + \tau)$$

and the *change operator*

$$\Delta Q(t) = Q(t) \oplus Q(t + \tau) = \underset{+}{\Delta} Q(t) + \underset{-}{\Delta} Q(t)$$

Now let us agree to write $Q(t)$ as simply Q, $S(t)$ as simply S, etc. Show that for the RS flip-flop,

$$\underset{+}{\Delta} Q = S_1Q'$$
$$\underset{-}{\Delta} Q = S_1'R_1'Q$$

and
if
$$HQ = (S_1 + R_1)Q = (S_1 \oplus R_1)Q$$
$$\Delta Q = S_1Q' + S_1'R_1'Q = S_1Q' \oplus (S_1 \oplus R_1)'Q$$
$$S_1R_1 = 0$$

(d) Derive the following equations from the identities in part c:

$$\Delta Q = (\underset{+}{\Delta} Q)Q' + (\underset{-}{\Delta} Q)'Q = (\underset{+}{\Delta} Q)Q' + (HQ)Q$$
$$Q(\tau) = (\underset{+}{\Delta} Q)Q' + (\underset{-}{\Delta} Q)Q = (\underset{+}{\Delta} Q)Q' + (HQ)'Q$$

where $Q_\tau(t) = Q(\tau + t)$, i.e., Q delayed by time τ.

(e) Suppose q_+, q_-, q_H, and q are *desired* up change, down change, hold, and change functions of a clocked RS flip-flop. If $S_1 = q_+$, $R_1 = q_-$ and $R_1 = q_H'$. Show that the delay equations for Q are $Q_\tau = q_+Q' + q_-Q = q_+Q' + q_H'Q$ and that the change equations for Q are $\Delta Q = q_+Q' + q_-'Q = q_+Q' + q_HQ$. If $S_1 = Q'q$ and $R_1 = Qq$, show that the changed equation for Q is $\Delta Q = q$.

7-4. On the left bank of a river there is a traveler with his wolf, goat, and cabbage. The former wishes to reach the right bank with all his possessions in a boat which holds at most himself and one of his charges. His task is complicated by the fact that if left alone the goat will eat the cabbage and the wolf will eat the goat. However, the wolf has no interest in the cabbage. (Due to Alcuin, a friend of Charlemagne.)

(a) Show that the possible sequences of states of the objects on the left and right banks of the river that the traveler may allow without loss, until he finally reaches the right bank with all his possessions, are equivalent to the solutions of the three simultaneous equations

$$T'(G' + W'C') + T(G + WC) = 1$$
$$\Delta T = W + G + C$$
$$(\Delta W)(\Delta G) + (\Delta W)(\Delta C) + (\Delta G)(\Delta C) = 0$$

where $T = 1(T = 0)$ = traveler on left (right) bank
  $W = 1(W = 0)$ = wolf on left (right) bank
  $G = 1(G = 0)$ = goat on left (right) bank
  $C = 1(C = 0)$ = cabbage on left (right) bank
  τ = time between river crossings

It is assumed without loss of generality that τ is constant and that the crossing time is instantaneous.

(*b*) Deduce all solutions of the above three simultaneous equations starting with initial condition $(TWGC) = (1111)$ (the solution should not be unique).

7-5. Design a shift register with feedback using the code ring 01001100010011000 . . . . Connect an AND-to-OR-gate network to the shift register so that if the register is started in state 0100, there will be an output in the interval between every third and fifth input pulse. Can the network be simplified by starting the counter in another state?

7-6. Draw a block diagram of a complete pyramid many-to-one decoder for a four-cell register. How many diodes are required to construct the network? Can you derive a general rule to determine the number of diodes required for a complete pyramid decoder given the number n of cells to be decoded? Try to design a four-cell decoder using fewer diodes than required by the pyramid decoder.

7-7. Draw a state table for the following machine:

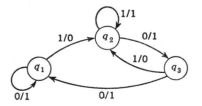

7-8. Design a three-cell shift register with feedback which will generate the following sequence: 0101110101110 . . . . This is a linear recurring sequence of maximal period, so the circuit can be realized using only exclusive-OR gates in the feedback loop. Derive the matrix M for the circuit and show that $M^8 = M$.

REFERENCES

1. Aufenkamp, D. D.: Analysis of Sequential Machines II, *IRE Trans. on Electronic Computers*, vol. EC-7, pp. 299–306, December, 1958.
2. Aufenkamp, D. D., and F. E. Hohn: Analysis of Sequential Machines, *IRE Trans. on Electronic Computers*, vol. EC-6, pp. 276–295, December, 1957.
3. Aufenkamp, D. D., S. Seshu, and F. E. Hohn: The Theory of Nets, *IRE Trans. on Electronic Computers*, vol. EC-6, pp. 154–164, September, 1957.
4. Bartee, T. C.: "Digital Computer Fundamentals," McGraw-Hill Book Company, Inc., New York, 1960.
5. Burks, A. W., R. McNaughton, C. H. Pollmar, et al.: Complete Decoding Nets: General Theory and Minimality, *J. Soc. Indust. Appl. Math.*, vol. 2, pp. 201–243, 1954.
6. Burks, A. W., and H. Wang: The Logic of Automata, *J. Assoc. for Computing Machinery*, vol. 4, pp. 193–218, April, 1957; pp. 279–297, July, 1957.
7. Burks, A. W., and J. B. Wright: Theory of Logical Nets, *Proc. IRE*, vol. 41, pp. 1357–1365, October, 1953.

8. Cadden, W.: "Sequential Circuit Theory," Ph.D. Dissertation, Princeton University, Princeton, N.J., 1956.
9. Caldwell, S. H.: "Switching Circuits and Logical Design," John Wiley & Sons, Inc., New York, 1958.
10. Church, A.: "Introduction to Mathematical Logic," vol. 1, Princeton University Press, Princeton, N.J., 1956.
11. De Bruijn, N. G.: A Combinatorial Problem, *Proc Ned. Akad. Wetenschop.*, vol. 49, pp. 758–764, 1946.
12. Elspas, B.: The Theory of Autonomous Linear Sequential Networks, *IRE Trans. on Circuit Theory*, vol. CT-6, pp. 45–60, March, 1959.
13. Friedland, B.: Linear Modular Sequential Circuits, *IRE Trans. on Circuit Theory*, vol. CT-6, pp. 61–68, March, 1959.
14. Hartmanis, J.: Linear Multivalued Sequential Coding Networks, *IRE Trans. on Circuit Theory*, vol. CT-6, pp. 69–74, March, 1959.
15. Ginsburg, S.: On the Length of the Smallest Uniform Experiment which Distinguishes the Terminal States of a Machine, The National Cash Register Company, Hawthorne, Calif., 1957.   Also in *J. Assoc. for Computing Machinery*, vol. 5, pp. 266–280, July, 1958.
16. Ginsburg, S.: A Technique for the Reduction of a Given Machine to a Minimal State Machine, *IRE Trans. on Electronic Computers*, vol. EC-8, pp. 346–356, September, 1959.
17. Ginsburg, S.: Synthesis of Minimal State Machines, *IRE Trans. on Electronic Computers*, vol. EC-8, pp. 441–449, December, 1959.
18. Huffman, D. A.: The Design and Use of Hazard-free Switching Networks, *J. Assoc. for Computing Machinery*, vol. 4, pp. 47–62, January, 1957.
19. Huffman, D. A.: The Synthesis of Linear Sequential Coding Networks, *Proc. Third London Symposium on Inform. Theory*, pp. 77–95, Academic Press, Inc., New York, 1956.
20. Huffman, D. A.: The Synthesis of Sequential Switching Circuits, *J. Franklin Inst.*, vol. 257, pp. 161–190, March, 1954; pp. 275–303, April, 1954.
21. Kautz, W. H.: Constant-weight Counters and Decoding Trees, *IRE Trans. on Electronic Computers*, vol. EC-9, no. 2, pp. 231–245, June, 1960.
22. Ledley, R. S.: "Digital Computer and Control Engineering," McGraw-Hill Book Company, Inc., New York, 1960.
23. Lippel, B., and I. J. Epstein: "Methods for Obtaining Complete Digital Chains," U.S. Army Signal Engineering Laboratories, Fort Monmouth, N.J., Tech. Mem. NR M-1850 (ASTIA no. 143540).
24. McClusky, E. J., and T. C. Bartee: "A Survey of Modern Switching Theory," McGraw-Hill Book Company, Inc., New York, 1962.
25. Mealy, G. H.: A Method for Synthesizing Sequential Circuits, *Bell System Tech. J.*, vol. 34, pp. 1045–1079, September, 1955.   Also Bell Telephone System Monograph 2458.
26. Metze, G., R. E. Miller, and S. Seshu: Transition Matrices of Sequential Machines, *IRE Trans. on Electronic Computers*, vol. CT-6, pp. 5–11, March, 1959.
27. Moore, E. F.: Gedanken-experiments on Sequential Machines, "Automata Studies," Annals of Mathematics Studies, no. 34, pp. 129–153, Princeton University Press, Princeton, N.J., 1955.
28. Netherwood, D. B.: Minimal Sequential Machines, *IRE Trans. on Electronic Computers*, vol. EC-8, pp. 367–380, September, 1959.
29. Peterson, W.: "Error-correcting Codes," John Wiley & Sons, Inc., New York, 1961.

30. Phister, M., Jr.: "Logical Design of Digital Computers," John Wiley & Sons, Inc., New York, 1958.

31. Radchenko, A. N., and V. I. Filippov: Shift Registers with Logical Feedback and Their Use as Counting and Coding Devices, *Automation and Remote Control*, vol. 20, no. 11, November, 1959.

32. Reed, I. S.: "The Symbolic Design of Digital Computers," MIT Lincoln Laboratory, Lexington, Mass., Tech. Mem. 23, 1953.

33. Reed, I. S.: "Symbolic Design Techniques Applied to a Generalized Computer," MIT Lincoln Laboratory, Lexington, Mass., Tech. Rept. 141, Jan. 3, 1957.

34. Richards, R.K.: "Digital Computer Components and Circuits," D. Van Nostrand Company, Inc., Princeton, N.J., 1957.

35. Scott, N. R.: "Analog and Digital Computer Technology," McGraw-Hill Book Company, Inc., New York, 1960.

36. Shannon, C. E., and J. McCarthy (eds.): "Automata Studies," Annals of Mathematics Studies, no. 34, Princeton University Press, Princeton, N.J., 1956.

37. Simon, J. M.: Some Aspects of the Network Analysis of Sequence Transducers, *J. Franklin Inst.*, vol. 65, pp. 439–450, June, 1958.

38. Simon, J. M.: A Note on the Memory Aspects of Sequence Transducers, *IRE Trans. on Circuit Theory*, vol. CT-6, pp. 26–30, March, 1959.

39. Solomon, G.: "Linear Recursive Sequences as Finite Difference Equations," MIT Lincoln Laboratory, Lexington, Mass., Group Rept. 47.37, March, 1960.

40. Turing, A.: On Computable Numbers with Applications to the Entscheidungs-problem, *Proc. London Math. Soc.*, series 2, vol. 42, pp. 230–265, 1936.

41. Turing, A.: Computing Machinery and Intelligence, *Mind*, vol. 59, pp. 433–460, 1950.

42. Turing, A.: Solvable and Unsolvable Problems, *Science News*, no. 31, pp. 7–23, Penguin Books, Inc., Baltimore, 1954.

43. Unger, S. H.: Hazards and Delays in Asynchronous Sequential Switching Circuits, *IRE Trans. on Circuit Theory*, vol. CT-6, pp. 12–26, March, 1959.

44. Unger, S. H., and M. C. Paull: Minimizing the Number of States in Incompletely Specified Sequential Switching Functions, *IRE Trans. on Electronic Computers*, vol. EC-8, pp. 356–366, September, 1956.

45. Weiss, E.: "Informal Lectures," MIT Lincoln Laboratory, Lexington, Mass., Group Rept. 55.22, 1960.

46. Zierler, N.: "Several Binary-sequence Generators," MIT Lincoln Laboratory, Lexington, Mass., Tech. Rept. No. 95, September 12, 1955.

47. Zierler, N.: Linear Recurring Sequences, *J. Soc. Indust. Appl. Math.*, vol. 7, no. 1, pp. 31–48, March, 1959.

8

Elementary Machines

8-1. Introduction. This chapter begins the discussion of structural design. The level of description will be that of the register and transfer. That is, a machine will be regarded as a set of independent and dependent registers communicating with one another by means of the transfer operation.

The term digital machine has been used throughout this book in quite a general way. A simple binary counter, for example, is a digital machine. A large, high-speed general-purpose computer is also a digital machine. The term structural design, as we shall employ it in this and succeeding chapters, will designate a description of digital machines at the register and transfer level. The operation of a machine is therefore described by a set of transfers. A binary counter may require one or two transfer relations to define its operation; a large general-purpose machine will, in general, require an extensive set of transfers.

The principal advantage of this type of description is one of notation. The transfer is fundamentally a shorthand way of designating a potentially complex set of Boolean relations which may specify a complex electronic operation. Indeed, the Boolean relations defining the transfer need not even be unique. For example, the sum $S(A,B)$ of two registers A and B may be computed in any of a number of different ways. The particular implementation used may be quite irrelevant to the problem of combining the transfer $S(A,B) \rightarrow A$ with other transfers to achieve a desired result. The transfer description then supplies just enough detail for the specification of the over-all machine. Once this description is available, the individual transfers may be implemented by translating the transfer relations into Boolean equations and then by realizing the equations in electronic components.

The design process is then divided into three phases: (1) the system design, which sketches in the general configuration of the machine and specifies the general class of hardware to be used, (2) the structural design, which describes the system in terms of transfer relations, and (3) the logic design, which realizes the transfer relations by means of Boolean equations. Practically speaking, of course, the three levels of design interact with one another. The Boolean equations make sense only if they are compatible with the electronic components. Hence a structural design which yields Boolean equations, which in turn indicate an awkward electronic configuration, must be modified as much as possible to produce as an end result a design consistent with both the available circuitry and the requirements of the system design.

In this chapter and in the two chapters which follow we develop the techniques of machine design in the sense of the previous paragraph. We begin with simple configurations of small numbers of independent and dependent registers and develop the notion of the control unit which specifies the transfers connecting the registers. Some fairly commonplace arithmetic units are used as examples. Beginning with Sec. 8-8 a rather general description of more realistic machines is presented in preparation for Chaps. 9 and 10. In Chap. 9 the techniques of Chap. 8 are expanded into a design of a simple general-purpose computer, and in Chap. 10 special-purpose computers are discussed.

8-2. Elementary Systems: Control. Let us begin with the simplest kind of machine, one with a single repetitive operation, and then generalize to more complex systems. Suppose, then, that the system contains two registers: an independent register A and a dependent register $f(A)$. The operation of the system is simply to replace the value of A with the value of $f(A)$ periodically. Symbolically, the machine is to perform the transfer $f(A) \rightarrow A$ at certain instants of time. The function $f(A)$ may be, for example, the cycled register $\rho(A)$, as in Secs. 6-3 and 7-6. In this case, the machine performs the transfer $\rho(A) \rightarrow A$, which cycles the register A one digit to the right whenever a switching impulse occurs. We assume that the switching pulses are provided by a clock pulse generator at the times $t_j = j\tau$, $j = 1, 2, \ldots$ If ε is the switching interval of the cells in register A, the transfer $f(A) \rightarrow A$ initiated at time t_j implies that

$$A(t_j + \varepsilon) = f[A(t_j)]$$

The interval τ between clock pulses must, of course, exceed $\varepsilon + \lambda$, where λ is the time required for the dependent register $f(A)$ to reach its equilibrium value once A has changed states.

In Fig. 8-1 we show the instants of time $t_j = j\tau$, $j = 1, 2, \ldots$, at which the clock pulses begin, each followed by the interval ε required for the register to switch. Also shown in the figure are the intervals T_1, T_2, . . . of length τ, defined by

$$T_j = (t_{j-1} + \varepsilon, t_j + \varepsilon] \qquad j = 1, 2, \ldots \tag{1}$$

The τ-second interval T_j therefore includes the point t_j, the beginning of the jth clock pulse, or, loosely speaking, T_j contains the clock pulse beginning at t_j.   Since each interval T_j contains within it the start of one and only one clock pulse, we can say that the transfer initiated at time t_j occurs during interval T_j.

Let $\sigma_j(t)$ define a Boolean time function which has the value 1 during T_j and 0 at all other times.   (In Chap. 5, the function σ_j is referred to as

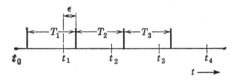

FIG. 8-1. Clock pulses and timing intervals.

the characteristic function of the interval T_j.)   We can therefore say that a transfer initiated at time t_j is initiated during the interval in which σ_j has the value 1.   From now on, when indicating transfers, instead of referring to the clock pulse that initiates the transfer, we shall specify the characteristic function σ_j of the interval T_j during which the pulse occurs. We represent this symbolically by

$$\sigma_j|\quad f(A) \rightarrow A$$

If the transfer is to be performed repetitively by each clock pulse, we specify

$$\sigma_1|\quad f(A) \rightarrow A$$
$$\sigma_2|\quad f(A) \rightarrow A$$
$$\cdot\ \cdot\ \cdot\ \cdot\ \cdot\ \cdot\ \cdot\ \cdot\ \cdot$$

If, on the other hand, the transfer is to be performed by, say, pulses 1, 2, and 4, then we specify

$$\sigma_1|\quad f(A) \rightarrow A$$
$$\sigma_2|\quad f(A) \rightarrow A$$
$$\sigma_3|$$
$$\sigma_4|\quad f(A) \rightarrow A$$

The absence of any transfer during interval T_3 is to be interpreted as the transfer $A \rightarrow A$, which leaves A unchanged.   Another way of describing the above sequence of transfers makes use of the notation of the conditional transfer introduced in Sec. 6-7:

$$(\sigma_1 + \sigma_2 + \sigma_4) \cdot f(A) + \sigma_3 \cdot A \rightarrow A$$

This conditional transfer, which applies to the first four clock pulses, states explicitly that when the function σ_1 or σ_2 or σ_4 has value 1, $f(A)$ is transferred into A, while when σ_3 has value 1, A is left unchanged.

We return once again to the example of the cyclic shift register; the continuous cycling of the register is specified by the sequence of transfers

$$\sigma_1|\quad \rho(A) \to A$$
$$\sigma_2|\quad \rho(A) \to A$$
$$\cdots \cdots \cdots$$

Another elementary computing system of this type is the binary counter described in detail in Chap. 7. As above, let A be an n-cell independent register and let δ_I be the binary integer representation of A. We specify the operation of the counter by the transfer

$$A + 1 \to A$$

initiated at time t, which designates that

$$\delta_I[A(t + \varepsilon)] = \delta_I[A(t)] + 1$$

or that the value of A is augmented by the integer 1 as a result of the transfer. The symbol $+$, used in the transfer statement (but not in the equation designating the representation), denotes addition modulo 2^n unless otherwise specified. The function $A + 1$ is just another function $f(A)$ of the type we have been considering in this section, and a machine consisting only of a counter operating continuously is completely described by the transfer sequence

$$\sigma_j|\quad A + 1 \to A \qquad j = 1, 2, \cdots$$

All the foregoing examples were of machines with the capability of executing a single transfer in a repetitive fashion. The next step in complexity is to consider a machine with two allowable transfers performed in sequence. Suppose we desire to design a binary counter as in the previous example, but one in which the components do not permit the operation $f(A) \to A$. Instead, the counting is performed by the use of two registers A and B and the sequence of transfers

$$A + 1 \to B$$
$$B \to A$$

to be executed on successive clock pulses. The desired operation of the system is therefore given by the sequence

$$\sigma_1|\quad A + 1 \to B$$
$$\sigma_2|\quad \quad\quad B \to A$$
$$\sigma_3|\quad A + 1 \to B \qquad\qquad (2)$$
$$\sigma_4|\quad \quad\quad B \to A$$
$$\cdots \cdots \cdots \cdots$$

The designer must therefore see to it that clock pulses occurring during odd-numbered time intervals initiate the transfer $A + 1 \to B$ while clock pulses occurring during the even-numbered intervals initiate the transfer $B \to A$. The two transfers occurring on alternate clock pulses may be distinguished by associating them with the two states of a cell P. Suppose that initially P has the value 1 and that it operates according to the sequence

$$\begin{array}{ll} \sigma_1| & P' \to P \\ \sigma_2| & P' \to P \\ \multicolumn{2}{c}{\cdot\ \cdot\ \cdot\ \cdot\ \cdot\ \cdot\ \cdot} \end{array} \qquad (3)$$

Then P is the time function shown in Fig. 8-2 which has the value 1 when odd-numbered clock pulses occur and 0 when even-numbered clock pulses originate. If the intervals T_j are defined as in Fig. 8-1, then, except for

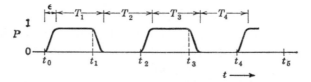

FIG. 8-2. Operation of control flip-flop P for machine of Fig. 8-3.

the ε-second switching intervals at the end of each T interval, the function P has value 1 during T_j for odd j and value 0 during T_j for even j. If follows that

$$\sigma_1 + \sigma_3 + \cdots + \sigma_{2k-1} + \cdots = P$$

and

$$\sigma_2 + \sigma_4 + \cdots + \sigma_{2k} + \cdots = P'$$

if we ignore the ε-second switching interval. From this it is evident that the sequence of transfers (2) may be rewritten as

$$\begin{array}{ll} P| & A + 1 \to B \\ P'| & B \to A \end{array}$$

which is equivalent to the conditional transfer

$$P \cdot (A + 1) + P' \cdot B \to B$$
$$P \cdot A + P' \cdot B \to A$$

These relations state that whenever a clock pulse occurs when P has value 1, B assumes the value $A + 1$; every clock pulse occurring when P has value 0 initiates the transfer $B \to A$. Since $P + P' = 1$, the behavior of the machine for every clock pulse is uniquely defined.

What this formalism has accomplished is the substitution of the functions P and P' for the characteristic functions of time σ_j. The register composed of the cell P together with the clock pulse generator is called the *control unit* of this computing machine. The function of P, once again, is to associate its two values with the two transfers the machine is to implement by creating, in effect, the two sets of time characteristic functions which initiate the two transfers. The control unit must, of course, control itself. How it does this is clear from the sequence (3), which demands that P be complemented during each interval. The system as described is shown in Fig. 8-3.

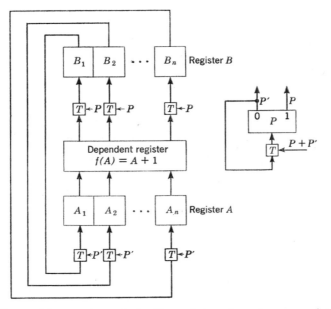

FIG. 8-3. Two-register counter. As in Chap. 6, transfer gates are used to indicate conditional transfer paths rather than physical gates, and transfer gates may consist of several actual gates. See Fig. 6-1 for additional details.

Although the control unit necessary for the specification of this simple machine is elementary, it points the way to the description of control units for more complex systems. In general, a control unit is a set of independent and dependent registers (i.e., a sequential circuit), the states or values of which are placed in correspondence with transfers or sets of transfers in the machine. Then, in order to generate a sequence of transfers, the control unit is constrained to take on a time sequence of values in correspondence with the desired sequence of transfers. Thus, in order to specify that the transfer $A \rightarrow B$ is to occur during interval T_j, we say that the transfer $A \rightarrow B$ is to occur when control has value q_k, and then we demand that control assume the value q_k during interval T_j.

In the discussion of the control for the simple machine of Fig. 8-3, certain simplifications were made for the sake of clarity. First, it is clearly not essential that the instants $t_1, t_2, \ldots$, at which the clock pulses begin, be uniformly spaced. Second, the time intervals T_1, $T_2, \ldots$ need not be defined precisely as in (1), but rather by

$$T_j = (t_{j-1} + \mu, t_j + \mu] \qquad 0 \leq \mu < t_j - t_{j-1}$$

In other words, T_j need be required to contain only the point t_j. Since T_j is essentially defined for the computer by the control unit, this requirement reduces to the practical engineering statement that a control function, say P, must assume its proper value at the time t_j at which the transfer is initiated. The behavior of P at other times within the interval T_j is unimportant. By analogy with the simple machines of this section, every computing machine, however complex, will be described by two sets of transfers, the first set to perform the computation or operation that the machine is expected to do, and the second set to implement the control. It is certainly true that one need not divide the operation into two parts in this way; i.e., one need not distinguish between the operation and its control. We shall see in later sections, however, that machines are conceptually easier to understand, hence easier to design, if this division is made.

8-3. The Shift Register: Start-Stop Control. In the previous section we developed a method of describing systems and applied it to a few very simple examples. We continue this discussion by considering somewhat more complex examples. The approach that is taken is the same as in the examples of Sec. 8-2. Any operation, however complex, is described by a set of transfers time-ordered in a certain way. The transfers are initiated in the proper sequence by the control of the machine, which is itself implemented by another set of transfers.

Consider a machine consisting of a single independent n-cell register A and the dependent register $\rho(A)$, the register cycled right by one digit. This machine is to cycle the digits of A to the right exactly n times and then stop. The control of this computer must evidently contain a counter, which can count at least as high as n, and a start-stop flip-flop. To make things a little more concrete without decreasing the generality of the example, we specify the number of cells in register A as 6. Let P be a three-cell counter which counts from 0 to 5 and then returns to 0. Let Q be a start-stop flip-flop with the value 0 indicating operation and 1 indicating no operation. This configuration is shown in Fig. 8-4. Let p_j be the characteristic functions of register P, i.e., $p_j = 1$ if and only if

$\delta_I(P) = j.$ The functions p_j will play the role of the functions P and P' of the last example of Sec. 8-2.   Hence, p_j represents the characteristic function of the time interval during which P has the binary value j. Furthermore, the function p_jQ' includes the additional condition that $Q = 0$.

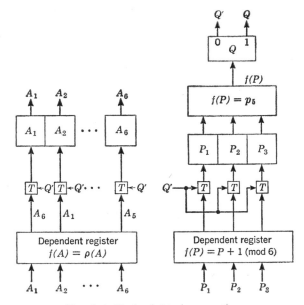

FIG. 8-4. Cycle right six operation.

The operation of the machine may then be defined by the following sequence of transfers [we assume that initially $\delta_I(P) = 0$ and $Q = 0$]:

$$
\begin{array}{lll}
Q'p_0| & \rho(A) \to A & P + 1 \to P \\
Q'p_1| & \rho(A) \to A & P + 1 \to P \\
Q'p_2| & \rho(A) \to A & P + 1 \to P \\
Q'p_3| & \rho(A) \to A & P + 1 \to P \\
Q'p_4| & \rho(A) \to A & P + 1 \to P \\
Q'p_5| & \rho(A) \to A & 0 \to P \quad 1 \to Q
\end{array}
$$

We again have two sequences of transfers, the left sequence specifying six shifts of register A, the right sequence the incrementing of the control counter by 1 until the count of 5, at which point P is reset to 0 and the machine is halted by setting Q to 1.

The initial specification of the system operation demanded that A be shifted six times, but said nothing about P.   The sequence of transfers above stops the operation of P as well as of A. Suppose that for some reason it is desired to continue to count P at the end of the operation. We then specify the sequences

$$Q'p_0| \quad \rho(A) \to A \qquad p_0| \quad P + 1 \to P$$
$$Q'p_1| \quad \rho(A) \to A \qquad p_1| \quad P + 1 \to P$$
$$\cdots \cdots \cdots \cdots \cdots \cdots \cdots \cdots \cdots \cdots$$
$$Q'p_4| \quad \rho(A) \to A \qquad p_4| \quad P + 1 \to P$$
$$Q'p_5| \quad \rho(A) \to A \qquad p_5| \quad 0 \to P \qquad 1 \to Q$$

Here we have explicitly stated that the operation of the machine proper (i.e., the register A) is controlled by P and Q while the operation of P is independent of Q.

This example is hardly more complicated than those of Sec. 8-2. The only real difference is the fact that the control is a little more involved. In the first examples, the control consisted only of a clock. At the next level of complication, the control contained, in addition to the clock pulse generator, a single flip-flop. In the example we have just considered, the control was expanded still more to include a counter and some rudimentary start-stop circuitry. As we shall see in later sections, the basic constituents of any machine control are a source of timing pulses, a generalized counter or sequential circuit, and some start-stop circuitry which may, if desired, be considered part of the generalized counter. The more complex the machine, the more complex the counter, but conceptually the operation is the same as in the elementary examples we have presented.

8-4. Simple Arithmetic Operations. An important function of most digital machines is arithmetic calculation. In fact, most of the so-called general-purpose computers are basically arithmetic machines. This section describes some properties of arithmetic operations and the control necessary to implement these operations. This will then lead us to the discussion of arithmetic computers in Chap. 9.

Define an *arithmetic unit* to be a set of independent registers together with sufficient dependent registers to perform one or more arithmetic operations. An arithmetic unit plus its control may be an entire system in its own right or a part of a larger system. We consider first the simple example of a serial sum mod 2 adder, similar in many respects to the example of Sec. 8-3. This is shown in Fig. 8-5. The unit consists of the six-cell registers A and B together with the necessary dependent registers to describe the shift operation, i.e., the functions first introduced in Sec. 6-3, the cyclic shifted register

$$\rho(A) = (A_6, A_1, \ldots, A_5)$$

and the subregisters A_1, $R(A)$, and $L(A)$ defined by

$$R(A) = (A_2, A_3, \ldots, A_6)$$
$$L(A) = (A_1, A_2, \ldots, A_5)$$

Further, as in Sec. 6-4 define

$$S_6{}^H(A,B) = A_6 \oplus B_6$$

The control for this computer is identical to that of Fig. 8-4, and consists of the clock pulse generator, the counter P, and start-stop flip-flop Q. The operation of the computer is then defined by the sequence

$$Q'p_0| \quad \rho(B) \to B \qquad L(A) \to R(A) \qquad S_6{}^H(A,B) \to A_1$$
$$Q'p_1| \quad \rho(B) \to B \qquad L(A) \to R(A) \qquad S_6{}^H(A,B) \to A_1$$
$$\cdots \cdots \cdots \cdots \cdots \cdots \cdots \cdots \cdots \cdots \cdots \cdots$$
$$Q'p_5| \quad \rho(B) \to B \qquad L(A) \to R(A) \qquad S_6{}^H(A,B) \to A_1$$

The control is identical to that of Sec. 8-3. This operation consists in the successive shifting of each of the registers one place to the right with

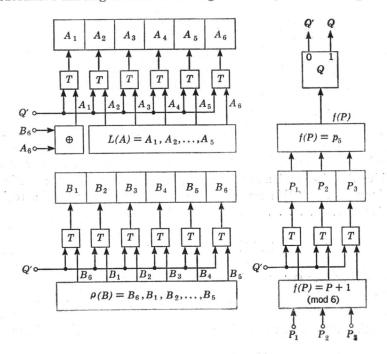

FIG. 8-5. Serial sum mod 2 adder.

the insertion into A_1 of the sum mod 2 of A_6 and B_6. After six such operations, the original value of B has been restored and the value of A has been replaced by the function $S^H(A,B) = A \oplus B$.

Now suppose that the computer of Fig. 8-5 performs componentwise multiplication instead of addition modulo 2. The operation of the computer is identical to that of the above with the function $A_6 \oplus B_6$ replaced by $A_6 \cdot B_6$. Similarly, if the operation between registers A and B were the logical sum, the operation would again be identical, with $A_6 \oplus B_6$ replaced by $A_6 + B_6$. For a little more complexity let us assume that the computer can perform all three operations, $A \oplus B$, $A \cdot B$, and $A + B$. To specify which of the three operations is to be performed, an additional register F is needed in control. Let F be a two-cell register with characteristic functions $f_0, \ldots, f_3$. Let us designate the operation to be performed by the values of F. In particular, let the functions f_1, f_2, and f_3 signify, respectively, the operations $\oplus$, $\cdot$, and $+$. Thus, the three different sets of transfers representing the three different operations may be designated by $Q' \cdot f_1 p_j$, $Q' \cdot f_2 p_j$, and $Q' \cdot f_3 p_j$, respectively. In order to specify that the computer perform one of these operations, say $A \cdot B$, it is necessary first to set the initial control conditions so that register P has the value 0, F has the value 2, and finally that Q has the value 0. At this point, the desired sequence of transfers occurs and the computer halts its operation.

The preceding operations are all arithmetic operations of a special kind (i.e., operations in which each component is independent of all other components), often called *logical* register operations. A discussion of more conventional arithmetic operations performed by digital computers is in the same spirit as for the logical operations, but with the added complexity inherent in such operations. Before one can speak of these operations is it necessary, as in previous chapters, to define a numerical representation for the register. For our first examples we consider the 2's-complement binary fraction representation described in the Appendix. In this representation, we label the cells of a register, say A, as $A_0, A_1, \ldots, A_n$ and define the representation by

$$\delta_{2F}(A) = -A_0 + \sum_{j=1}^{n} A_j 2^{-j}$$

The range of $\delta_{2F}(A)$ is from -1 to $1 - 2^{-n}$, inclusive. A_0 is referred to as the sign digit; if $A_0 = 0$, $\delta_{2F} \geq 0$, while if $A_0 = 1$, $\delta_{2F} < 0$. As is shown in the Appendix, the negative of $\delta_{2F}(A)$ is obtained by complementing each digit and adding 1 in the least significant digit, or

$$-\delta_{2F}(A) = -A_0' + \sum_{j=1}^{n} A_j' 2^{-j} + 2^{-n}$$

We shall first consider an arithmetic unit which computes the arithmetic sum of two registers A and B. If, for the moment, we disregard questions of overflow, the operation computes the fractional part of the sum (the sole exception being the number -1). The sum of registers A and B in the above representation is a register $S^2(A,B)$ with components $S_0{}^2, S_1{}^2, \ldots, S_n{}^2$, where

$$\delta_{2F}(S^2) = \delta_{2F}(A) + \delta_{2F}(B)$$

the sum designating addition in the sense of this paragraph. It is shown in the Appendix that the ith digit of S^2 is given by

$$S_i{}^2 = A_i \oplus B_i \oplus K_i{}^2(A,B)$$

where $K_i{}^2$, the carry into the ith digit in the 2's-complement system, is given by

$$K_i{}^2 = A_{i+1}B_{i+1} + K_{i+1}^2(A_{i+1} + B_{i+1}) \qquad i = 0, 1, \ldots, n-1$$
$$K_n{}^2 = 0$$

Or equivalently we may write

$$S^2(A,B) = A \oplus B \oplus K^2(A,B) = S^H(A,B) \oplus K^2(A,B)$$

The latter expression indicates immediately how the serial half adder of Fig. 8-5 may be modified to become a serial adder. We show this in Fig. 8-6, which differs from Fig. 8-5 in that an additional memory cell (carry flip-flop) C has been added. (Again the control is identical.) If we assume as an additional initial condition that C has value 0, the sequence of transfers for the addition becomes

$$\begin{array}{lllll} Q'p_0| & \rho(B) \to B & L(A) \to R(A) & S_6{}^H(A,B) \oplus C \to A_1 & K_5{}^2(A,B,C) \to C \\ Q'p_1| & & & S_6{}^H(A,B) \oplus C \to A_1 & \\ \cdots & \cdots & \cdots & \cdots & \cdots \\ Q'p_5| & & & S_6{}^H(A,B) \oplus C \to A_1 & \end{array}$$

where $K_5{}^2(A,B,C) = A_6B_6 + C(A_6 + B_6)$. At each step of the operation, C stores the carry digit (compare with Sec. 7-10).

With a slight modification, the serial adder may become a serial subtractor. As in the Appendix, let the difference function $D(A,B)$ be a register $D_0, D_1, \ldots, D_n$, where

$$\delta_{2F}(D) = \delta_{2F}(A) - \delta_{2F}(B)$$

If we again neglect questions of overflow, the digits D_i are given by

$$D_i = A_i \oplus B'_i \oplus \tilde{K}_i(A,B')$$

where $\quad \tilde{K}_i(A,B) = A_{i+1}B'_{i+1} + \tilde{K}_{i+1}(A_{i+1} + B'_{i+1})$
$$i = 0, 1, \ldots, n-1 \qquad \tilde{K}_n = 1$$

Thus, subtraction in this representation is similar to addition with the digits of B replaced by their complements and the carry into the least significant digit always a 1. This is, of course, equivalent to first computing the negative of the representation of B and then adding the negative of B to A.

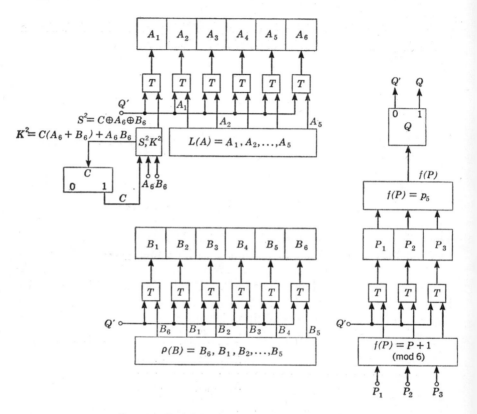

FIG. 8-6. Serial 2's-complement binary adder.

Let us next consider the operations of addition and subtraction in the 1's-complement fractional representation which is also described in the Appendix. This representation δ_{1F} is defined by

$$\delta_{1F}(A) = -A_0(1 - 2^{-n}) + \sum_{i=1}^{n} A_i 2^{-i}$$

or in terms of the 2's-complement representation,

$$\delta_{1F}(A) = \delta_{2F}(A) + A_0 \cdot 2^{-n}$$

In this representation, the range is $-(1 - 2^{-n})$ to $(1 - 2^{-n})$ inclusive, and the negative of a number is obtained by simply complementing the number. The number 0 has two representations, where $A_i = 0$ for all i and $A_i = 1$ for all i.

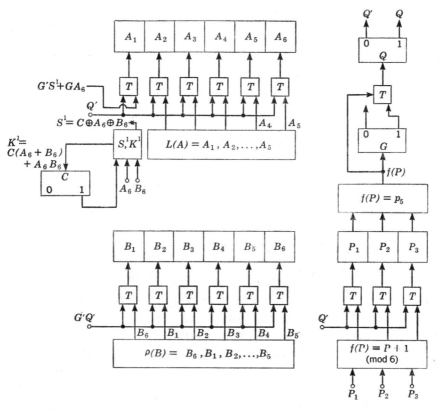

FIG. 8-7. Serial 1's-complement binary adder.

A 1's-complement adder differs from a 2's-complement adder only in the method of handling the carry into the least significant digit. The sum function $S^1(A,B)$ has components

$$S_j^1 = A_j \oplus B_j \oplus K_j^1(A,B)$$

with
$$K_j^1(A,B) = A_{j+1}B_{j+1}$$
$$+ K_{j+1}^1(A_{j+1} + B_{j+1}) \qquad j = 0, 1, \ldots, n-1$$
$$K_n^1(A,B) = A_0B_0 + K_0^1(A_0 + B_0)$$

The function $K_n^1(A,B)$ is the so-called *end-around carry*. A serial 1's-complement adder requires, therefore, an additional cycle of the register A to add in any possible carries of this type. We show a serial 1's-complement adder in Fig. 8-7. Because of the extra cycle of operation an additional control flip-flop G is required. Let us assume the initial conditions $P = 0$, $Q = 0$, $G = 0$, and $C = 0$. Then the computer proper operates according to the sequence

$$
\begin{array}{lllll}
Q'G'p_0| & \rho(B) \to B & L(A) \to R(A) & S_6{}^H(A,B) \oplus C \to A_1 & K_5{}^1(A,B,C) \to C \\
Q'G'p_1| & \rho(B) \to B & L(A) \to R(A) & S_6{}^H(A,B) \oplus C \to A_1 & K_5{}^1(A,B,C) \to C \\
\cdots & & & & \\
Q'G'p_5| & \rho(B) \to B & L(A) \to R(A) & S_6{}^H(A,B) \oplus C \to A_1 & K_5{}^1(A,B,C) \to C \\
Q'Gp_0| & & L(A) \to R(A) & A_6 \oplus C \to A_1 & A_6C \to C \\
Q'Gp_1| & & L(A) \to R(A) & A_6 \oplus C \to A_1 & A_6C \to C \\
\cdots & & & & \\
Q'Gp_5| & & L(A) \to R(A) & A_6 \oplus C \to A_1 & A_6C \to C
\end{array}
$$

while simultaneously the control is following the sequence

$$
\begin{array}{lll}
G'p_0| & P + 1 \to P & \\
\cdots & & \\
G'p_4| & P + 1 \to P & \\
G'p_5| & 0 \to P & 1 \to G \\
Gp_0| & P + 1 \to P & \\
\cdots & & \\
Gp_4| & P + 1 \to P & \\
Gp_5| & 0 \to P & 0 \to G \quad 1 \to Q
\end{array}
$$

The first cycle of operation labeled G is identical with that for 2's-complement addition. During the second cycle any end-around carry terms are added into A. It follows immediately from the definition of the 1's-complement representation that a subtractor in this system is identical to an adder, with B_j replaced by B_j'. Thus a subtraction can be performed by first complementing register B and then performing an addition as described above. Notice that the use of the 1's-complement representation for negative numbers required an extra cycle through the adder for each addition. If the 2's-complement system is used, this extra cycle is not required. When the 2's-complement system is used, however, the 2's complement of a number cannot be formed by simply complementing each binary value representing the number (see the Appendix), but a 2's complement can be formed serially by a rather simple sequential circuit (see Ref. 3). For this reason, the 2's-complement representation is often used in serial systems. This illustrates the effect of the choice of the number representation on the over-all structure of the computer.

We have shown in this section a few examples of simple arithmetic operations typical of those performed by arithmetic computers, together with the control necessary to implement these operations. These examples, together with others to follow in this chapter, are clearly not meant to be exhaustive. They do illustrate, however, the techniques by which a computer may be designed to implement one or more arithmetic algorithms in a relatively systematic way.

8-5. Arithmetic Operations: Parallel Transfers. In the previous section we presented some simple arithmetic operations and showed how one might implement these with serial arithmetic units and associated control. Each example demonstrated a sequence of transfers providing as the ultimate result the transfer

$$f(A,B) \rightarrow A$$

where the function $f(A,B)$ was in the class of arithmetic or logical functions of registers A and B such as $A \oplus B$, $A \cdot B$, and $S^2(A,B)$. As was pointed out in Chap. 6, transfers may be implemented in series, in parallel, or in some combination of the two. Consider an arithmetic unit performing the operation $A \oplus B \rightarrow A$ in parallel (refer to Fig. 6-3). If A and B are n-cell registers, there are n circuits computing the functions $A_i \oplus B_i$, $i = 1, 2, \ldots, n$; i.e., the n components of the function $A \oplus B$ are generated simultaneously. Furthermore, these n components are simultaneously transferred into register A.

We next consider an adder performing the transfer

$$S^2(A,B) \rightarrow A$$

in parallel by generating the n components $S_j^2(A,B) = A_j \oplus B_j \oplus K_j^2$ in parallel. It is necessary here to be careful in using the word parallel, since the function $S_j(A,B)$ depends in general on all the A_i and B_i, $i \geq j$. The carry function $K^2(A,B)$ may itself be generated in several ways within the parallel adder. To be more precise, let A and B be four-cell registers labeled by the subscripts 0 to 3. The least significant digit of the sum is given by

$$S_3^2 = A_3 \oplus B_3$$

and the carry from this stage is given by

$$K_2^2 = A_3 B_3$$

Similarly, the next digit is the function

$$S_2^2 = A_2 \oplus B_2 \oplus K_2^2$$

which in terms of the operations $+$ and $\cdot$ is given by

$$S_2^2 = K_2^{2\prime}(A_2 B_2' + A_2' B_2) + K_2(A_2 B_2 + A_2' B_2')$$

If both A_2 and B_2 are conventional flip-flops, A_2' and B_2' are both available as flip-flop outputs. Since $K_2^{2\prime}$ is dependent, it does not exist directly.   It may be generated as an explicit function of A_3 and B_3, i.e., by generating

$$K_2{}^{2\prime} = A_3' + B_3'$$

or it may be obtained as a function of $K_2{}^2$, that is, as the output of an inverter with $K_2{}^2$ as input.   When the first method is used, both $K_2{}^2$ and $K_2{}^{2\prime}$ are simultaneously available.   In the second method, however, $K_2{}^{2\prime}$ is available slightly later than K_2 because of the circuit delay in the inverter. The effect of these delays in operation is shown more clearly when we consider the more significant digits of this adder.   The carry $K_1{}^2$ is given by

$$K_1{}^2 = A_2B_2 + K_2{}^2(A_2 + B_2)$$

and the sum digit $S_1{}^2$ is the function

$$S_1{}^2 = A_1 \oplus B_1 \oplus K_1{}^2$$

Finally, the carry into the sign digit is

$$K_0{}^2 = A_1B_1 + K_1{}^2(A_1 + B_1)$$

and the sign digit of the result is

$$S_0{}^2 = A_0 \oplus B_0 \oplus K_0{}^2$$

To generate each component $K_i{}^2$ and its complement $K_i{}^{2\prime}$, one has the choice of using a set of identical adders with inputs A_i, B_i, and $K_i{}^2$ which realize

$$K_{i-1}^2 = A_iB_i + K_i{}^2(A_i + B_i)$$

or of generating explicit functions of A_j and B_j for $j > i$, for instance,

$$K_0{}^2 = A_1B_1 + A_1A_2B_2 + A_2B_1B_2 + A_1A_2A_3B_3 + A_1A_3B_2B_3 \\ + A_2A_3B_1B_3 + A_3B_1B_2B_3$$

obtained by combining the above equations.   It is also quite evident that combinations of the two methods may be employed.   In the first method each carry is formed from the previous one; hence the signals representing the sum $S_j{}^2$ are not available simultaneously.   In the second method all the digits $S_j{}^2$ are available essentially at the same time.   If in the first method the various components of the transfer $S_j{}^2 \rightarrow A_j$ are caused to take place as soon as the signals for $S_j{}^2$ have been generated, we have a form of asynchronous operation mentioned in Chap. 6.   If, on the other hand, the time interval allowed for the transfer is long enough to allow time for generation of the most significant digit $S_0{}^2$ in the worst case, then all the components may be clocked in simultaneously.   We shall refer to all the above types of transfers as parallel, even when some of the dependent registers are generated using an iterative circuit structure. This designation, of course, applies to functions other than arithmetic functions. The specification of the transfer

$$p_i | S^2(A,B) \rightarrow A$$

during some interval of time for which p_i is the characteristic function is, therefore, a concise notation for a basically complex operation involving the generation of the complex function $S^2(A,B)$ in one of a large variety of methods.

The list of references at the end of the chapter includes several detailed examinations of both serial and parallel adders, the effects of the choice of number representation on system complexity and speed, and other considerations. Broad treatments of some of these matters may be found in Refs. 3, 6, 19, and 23, while detailed description in specific areas may be found in Refs. 7, 8, 12, 14, 15, 18, and 20.

8-6. Arithmetic Operations: Complex Operations. As we have seen in the previous sections, the addition operation may be performed in a variety of ways. The method chosen for a given system depends upon such considerations as the speed with which the addition must be performed and the kind of electronic equipment available for the instrumentation. For a given class of circuitry, the fastest method is the parallel adder, where all the carry functions are generated independently, and, of those considered, the slowest was the serial adder. There are, of course, no hard and fast rules for designing an adder or any other piece of data-processing equipment. In general, the more parallel the system, the faster the execution time for an operation, but the more equipment is required.

Usually, an arithmetic unit consists of an adder of the kind we have been discussing plus a certain amount of additional equipment to implement one or more additional operations. From the discussion in Sec. 8-5, it is clear that subtraction differs from addition only trivially. Many computers, in fact, simply take the negative of the subtrahend and perform an addition. Multiplication, division, and still more complex arithmetic operations are most commonly implemented by appropriate sequences of additions or subtractions. It is not our purpose to describe the various algorithms which have been used in computers to implement arithmetic operations. (The interested reader is referred to the references.) Our purpose here is to show how an arithmetic unit can implement an algorithm by means of a set of transfers. As an example of a complex operation, let us consider multiplication, using what is often called the *add-shift* algorithm. Let us assume the 1's-complement binary fraction representation is used. If either multiplier or multiplicand is negative, we shall compute the negative of that number and adjust the sign of the product accordingly. To perform multiplication, we shall use an arithmetic unit consisting of three n-cell independent registers A, B, and R and a flip-flop C. Registers A and B are connected in tandem to form a $2n$-cell shift register. This double-length register is required to store a complete product with no loss in precision, since multiplication of two n-bit numbers may yield a $2n$-bit result. The addition function $S^1(A,R)$ is provided with an associated transfer of this function into A. Let us assume that the control contains a clock pulse generator, a four-cell counter P, and the start-stop cell Q. We assume further that all transfers can be performed in a single pulse interval. Suppose that initially the multiplicand is in R, and the multiplier in A. For simplicity let $n = 4$. The sequence of transfers is then as follows:

$$
\begin{array}{l|lll}
p_0 & A_0 \oplus R_0 \to C & A_0' \cdot A + A_0 \cdot A' \to A & R_0' \cdot R + R_0 \cdot R' \to R \\
p_1 & A \to B & 0 \to A \\
p_2 & B_3' \cdot A + B_3 \cdot S^1(A,R) \to A \\
p_3 & L(A,B) \to R(A,B) & A_0 \to A_0 \\
p_4 & B_3' \cdot A + B_3 \cdot S^1(A,R) \to A \\
p_5 & L(A,B) \to R(A,B) & A_0 \to A_0 \\
p_6 & B_3' \cdot A + B_3 \cdot S^1(A,R) \to A \\
p_7 & L(A,B) \to R(A,B) & A_0 \to A_0 \\
p_8 & C' \cdot A + C \cdot A' \to A & C'B + C \cdot B' \to B
\end{array}
$$

As in the previous examples Q and P initially have the values 0; P cycles through the count of 8 and then is reset to 0 as Q is set, stopping the operation. During the interval p_0 the sign digit of the product is stored in C and the multiplier and multiplicand are complemented (negated) if they are negative. During the interval p_1 the multiplier is transferred to B and A is cleared. At this point, the initial bookkeeping has been completed and the algorithm may begin. During p_2, cell B_3 contains the least significant digit of the multiplier, and, based upon that digit, register A is either left undisturbed or is added to the multiplicand, with the result stored in A (as before, we neglect questions of overflow). During p_3 the composite register A,B is shifted right by one digit. In this process the previous digit in B_3 is lost, and the second least significant digit of the multiplier is transferred to B_3. This process is repeated twice, and by the end of p_7, the magnitude of the result is stored in registers A and B. During the last interval, this result is either negated or not, according to the sign digit of the product stored in C.

For a specific example, let the multiplicand be 1.001 $(-\frac{3}{4})$ and the multiplier be 0.101 $(+\frac{5}{8})$. By the end of p_1 the registers have the values

$$R\ \boxed{0\ \ 1\ \ 1\ \ 0}\qquad\qquad\boxed{1}\ C$$

$$A\ \boxed{0\ \ 0\ \ 0\ \ 0}\qquad\boxed{0\ \ 1\ \ 0\ \ 1}\ B$$

Registers A and B assume the following values at the end of the specified intervals:

Time Interval	A	B
$p_2\vert$	0110	0101
$p_3\vert$	0011	0010
$p_4\vert$	0011	0010
$p_5\vert$	0001	1001
$p_6\vert$	0111	1001
$p_7\vert$	0011	1100
$p_8\vert$	1100	0011

At the end of p_7 register A,B contains the eight-digit number 0.0111100, which in this representation is $+15\!\!/_{32}$, the magnitude of the product. By the end of p_8 this number is negated.

There are a number of algorithms for multiplication and division, and the choice of algorithm obviously has a profound effect on system complexity and speed. A great many factors must be considered in this choice. References 12, 17, 21, and 22 give detailed accounts of some specific techniques and their advantages.

Thus far, all the arithmetic operations that we have described have employed fixed-point representations. In a more sophisticated arithmetic unit, the arithmetic operations often allow floating-point representations. To implement such operations, the pattern of transfers is much the same as for fixed-point operations, only somewhat more complex. We illustrate with the following example. Let A and $\hat{A}$ be registers of n and m cells, respectively. Let A have a binary fraction representation and $\hat{A}$ a binary integer representation. In order to be specific, let numbers be stored in 2's-complement form, and let δ_F and δ_I designate these representations respectively. Their ranges are evidently given by

$$-1 \leq \delta_F \leq 1 - 2^{-(n-1)}$$
$$-2^{(m-1)} \leq \delta_I \leq 2^{m-1} - 1$$

We define the representation $\beta(A,\hat{A})$ by

$$\beta(A,\hat{A}) = \delta_F(A) \times 2^{\delta_I(\hat{A})}$$

Let the floating-point adder consist of two register pairs $A,\hat{A}$ and $B,\hat{B}$, where A and B are 15-cell registers and $\hat{A}$ and $\hat{B}$ are four-cell registers. Also, a four-cell register C and a one-cell register D are provided for auxiliary storage. The first step that this adder must perform is to compare the two exponents stored in $\hat{A}$ and $\hat{B}$ and then shift the binary point of the number with smaller exponent to the left a sufficient number of times to make the exponents equal. For example, if $\beta(A,\hat{A}) = 0.5 \times 2^2$ and $\beta(B,\hat{B}) = 0.75 \times 2^1$, then $\beta(B,\hat{B})$ must be changed to 0.375×2^2 before the addition is performed. One method of implementing this operation is first to perform the transfers

$$\hat{A} - \hat{B} \to \hat{A} \qquad \hat{A} \to C$$

If, at the conclusion of these transfers, $\hat{A}$ is negative, then A is to be shifted; if $\hat{A}$ is positive, then B is to be shifted. Thereupon we may perform the transfers

$$\hat{A}_0' \cdot \hat{A} + A_0 \cdot (\hat{A}' + 1) \to \hat{A} \qquad \hat{A}_0 \to D$$

which replace A by its absolute value, preserving the sign of $\hat{A}_0$ in D. Finally, if at this point $\delta_I(\hat{A}) = n$, the transfers

$$D \cdot [A_0, L(A)] + D' \cdot A \to A \qquad D' \cdot [B_0, L(B)] + D \cdot B \to B$$

must be performed n times, shifting either A or B n times to the right, according to the value of D. Now the addition

$$A + B \to A$$

may be performed together with the transfer

$$D \cdot \hat{B} + D' \cdot C \to \hat{A}$$

which inserts the correct exponent into $\hat{A}$.

The control necessary to implement this sequence of transfers illustrates a situation that we have not met previously. Let $\hat{a}_0$ be the 0'th characteristic function of $\hat{A}$. For simplicity, let $\mu(A)$ designate the 15-cell register $A_0, L(A)$. Then the transfer

$$D\hat{a}_0' \cdot \mu(A) + (D\hat{a}_0')'A \to A$$

specifies that A is to be shifted right if and only if function D has the value 1 and register $\hat{A}$ does not have the value 0. This provides a mechanism for shifting register A or B the required number of times. If the shift is made conditional upon $\hat{A}$ not containing 0, then accompanying each shift by the transfer

$$\hat{A} - 1 \to \hat{A}$$

effectively counts the number of shifts.†

† If the magnitude of the difference in values in $\hat{A}$ and $\hat{B}$ exceeds 15, an error will occur when $\hat{B}$ is subtracted from $\hat{A}$. This error is due to an overflow of $\hat{A}$, and in this case, the number with the positive exponent should be simply stored in A and no addition performed. Techniques for dealing with an adder overflow will be treated in Chap. 9.

As in the previous examples, let p_j designate the characteristic functions of the control register P. The floating-point addition operation may be specified by the sequence

$$
\begin{array}{ll}
p_0| & \hat{A} - \hat{B} \to \hat{A} \qquad \hat{A} \to C \\
p_1| & \hat{A}_0' \cdot \hat{A} + A_0 \cdot (\hat{A}' + 1) \to \hat{A} \qquad \hat{A}_0 \to D \\
p_2| & \hat{a}_0'(\hat{A} - 1) + \hat{a}_0\hat{A} \to \hat{A} \qquad D\hat{a}_0' \cdot \mu(A) + (D\hat{a}_0')'A \to A \\
& \qquad\qquad\qquad\qquad\qquad\qquad D'\hat{a}_0' \cdot \mu(B) + (D'\hat{a}_0')' \cdot B \to B \\
p_3| & A + B \to A \\
p_4| & D \cdot \hat{B} + D' \cdot C \to \hat{A}
\end{array}
$$

Evidently the register P must remain with $p_2 = 1$ long enough to complete the necessary number of shifts of A or B. Hence, the sequence describing control is

$$
\begin{array}{ll}
p_0| & P + 1 \to P \\
p_1| & P + 1 \to P \\
p_2| & \hat{a}_0 \cdot (P + 1) + \hat{a}_0' \cdot P \to P \\
p_3| & P + 1 \to P \\
p_4| & 0 \to P
\end{array}
$$

where the advancing of P from p_2 to p_3 is made conditional upon the function $\hat{a}_0$.

8-7. Arithmetic Operation by Table Look-up. In the previous sections we have illustrated arithmetic operations with increasing degrees of complexity. All involved, however, the computation of some arithmetic function or functions. More specifically, in each case the operands were stored in independent registers, and the arithmetic functions were obtained as dependent registers before being stored in independent registers. An alternative method of obtaining functions, particularly of a single variable, is to store all the possible values of the function in memory cells and then to select the appropriate value of the function.

For example, consider the generation of the function A^2 of the three-cell register A, where A has the binary integer representation $\delta_I(A)$. Let M designate the set of 2^3 six-cell registers $M^0, M^1, \ldots, M^7$ with binary integer representations

$$\delta_I(M^i) = i^2$$

If a_i designates the ith characteristic function of register A, then the transfer

$$a_0 \cdot M^0 + a_1 \cdot M^1 + \cdots + a_7 M^7 = M\langle A \rangle \to B$$

describes the insertion into B of the square of the number stored in A. An alternate way of obtaining this function is to multiply the number by itself using an algorithm similar to that of Sec. 8-6. The computational method uses fewer independent registers but may be considerably slower. On the other hand the devices used to implement the registers M^i can be quite simple, since they are storing constants.   The number of registers in the set M evidently is 2^k, where k is the number of cells in register A. This number can be prohibitively large in most practical situations.   As a result, a combination of the two methods is often employed in which some fraction of the 2^k possible values of the function is stored in the table with intermediate values obtained by interpolation.   We illustrate this with the problem of obtaining the square of a six-bit number with a thirty-two-register table instead of a sixty-four-register table.   The interpolation method is based upon the relationship

$$x^2 = \tfrac{1}{2}[(x + 1)^2 + (x - 1)^2] - 1$$

As above, let A be a six-cell register with binary integer representation $\delta_I(A) = j, j = 0, 1, \ldots, 63$, and let the table M contain the registers M^i, $i = 0, 1, \ldots, 31$, having binary integer representations $\delta_I(M^i) = (2i)^2$. That is, the table stores the squares of the even integers between 0 and 63. The arithmetic unit then performs the transfers

$$L(A) \rightarrow R(A) \qquad 0 \rightarrow A_1 \qquad A_6 \rightarrow S$$

having the effect of shifting A to the right by one digit and storing the least significant bit in an auxiliary cell S.   At this point the transfer

$$a_0 \cdot M^0 + a_1 \cdot M^1 + \cdots + a_{31} \cdot M^{31} = M\langle A \rangle \rightarrow B$$

inserts the desired result into B if the original number in A was even, and inserts the square of the original number minus one if this number was odd.   The transfers

$$S' \cdot A + S \cdot (A + 1) \rightarrow A$$
$$S' \cdot B + S \cdot (M\langle A \rangle + B) \rightarrow B$$

replace B with the sum of the squares of the previous and following integers if the original integer in S was odd.   The final result is obtained by the transfers

$$S' \cdot B + S \cdot [B_1, L(B)] \rightarrow B$$
$$S' \cdot B + S \cdot (B - 1) \rightarrow B$$

In this particular example, the interpolation algorithm seems more complex than the addition of another thirty-two registers in the table M. In practical cases, the designer must decide between the size of the table and the complexity of the interpolation algorithm.   Such decisions must always be based upon the available equipment and the speed and accuracy requirements of the machine.

8-8. Computing Systems. In the preceding sections we described a number of simple idealized computing machines. These machines serve to introduce methods of machine design which are relatively straight-forward and systematic. They also serve as examples of the viewpoint stated in Sec. 8-1 that one may consider a digital system to be composed of a set of independent and dependent registers interacting through transfers specified by a control unit.

The machines that we described are idealized from two points of view. First, almost all the independent registers referred to explicitly or otherwise are flip-flops, whereas in a real system most of the independent registers are realized by cells other than flip-flops. Also, no mention has been made of how information is introduced into a system and how results are obtained from the system. Considerations of these additional complexities lead to the viewpoint that a realistic computing system is best regarded as a set of interacting computing units. These units are composed of independent and dependent registers, and perhaps terminal devices. Some of the units may contain relatively few independent registers but a large number of dependent registers with a complex set of interactions. Some units may contain large numbers of independent registers with very few dependent registers and may have basically simple interactions. A terminal unit will in general contain both registers and the terminal equipment itself. For the remainder of this chapter, we shall expand the techniques used up to this point in describing relatively simple machines to introduce the design of realistic systems, which are nothing more than sets of these simple machines.

8-9. Registers and Memories. When the cell and the register were defined in Chap. 2, several examples were described. These different realizations of the cell have widely varying properties. Indeed the only property common to all is the basic ability to store a binary value. A single computing system may very likely contain several different kinds of cells, each type being used for a different purpose. It is convenient to divide the different types of registers into two general classes, which we may call *computational* or *operation* registers and *storage* registers. Since any register, regardless of its construction, performs a storage function by its very definition, the designation computational implies, as an additional property, the ability to drive combinational circuits or, equivalently, to generate dependent registers. In most of the preceding examples in this chapter, the independent registers were computational.

There is no functional reason why all registers in computing systems cannot be constructed of high-speed flip-flops, except that in most machines thousands of registers of storage are required, and use of the flip-flop would make the machine large and prohibitively expensive. Consequently, for large-scale storage units, registers are used which can do little else but store, but which are physically small and inexpensive. Familiar examples are magnetic cores and magnetic drums and tapes. Hence, in practice we find that the bulk of the independent registers in computing machines are of the storage variety, with only a relatively small number of computational registers. Then, when computations are to be performed, the values in the selected registers of the large-scale memory must first be transferred into the computational registers. For example, let M^1, M^2, and M^3 be storage registers, and let A and B be flip-flop registers with the dependent register $S(A,B) = A + B$. The transfer

$$M^1 + M^2 \rightarrow M^3$$

cannot be performed directly, but rather through the use of intermediate transfers such as

$$M^1 \rightarrow A \qquad M^2 \rightarrow B$$
$$S(A,B) \rightarrow A$$
$$A \rightarrow M^3$$

Thus, the small number of computational registers are time shared by the large number of storage registers for the computation of functions.

Storage registers are most often grouped together in so-called *storage units* or *memories*. We shall designate such units by a Roman letter, say M, and the registers in the set M by M^1, M^2, ... Memory registers, like all other registers, may have either serial or parallel interactions with one another and with other registers. Very often a memory is constructed in such a way that only a small number of its registers may communicate with registers external to the memory at any one time, and quite often information can be transferred from only a single memory register at a given time. In such a memory, an *address register* C may be used to select the memory register from which information is to be read by a transfer, such as

$$c_0M^0 + c_1M^1 + \cdots + c_nM^n = M\langle C \rangle \rightarrow N$$
or
$$N \rightarrow M\langle C \rangle$$

where N is an external register. This symbology describes a memory in which the selection process is electronic. In memories constructed of magnetic drums or magnetic tapes in which the selection depends upon the position of the storage medium with respect to a driving or sensing device (write and read heads), there may be no need for the existence of an address register. A memory is said to be *sequential-access* if the memory registers interact with external registers in some fixed order. Drums and tapes are by their very nature sequential-access. A memory is said to be *random-access* if the order in which its registers communicate is arbitrary. An electronically addressed memory constructed of mag-

netic cores, magnetic films, storage tubes, etc., may be either sequential-
or random-access.

It is worthwhile remarking at this point that flip-flop registers are
often used simply for their storage property, although usually not in the
large numbers in which other storage registers are employed. Such
flip-flop storage registers may also be arranged in memory units in
which only one or two registers communicate outside the memory
simultaneously, as determined by one or more address registers. Similarly, fixed registers may be used as computational or storage registers.
As demonstrated in Chap. 2, a fixed register behaves like an arbitrary
register constrained to store a constant Boolean function. In particular,
a set of fixed registers may be arranged in the form of a memory, with
a particular register selected by an address register.

Another property which, *on occasion*, distinguishes computational
registers from storage registers is speed. In a given machine, the computational registers are *usually* faster than the storage registers. But
this property too is generally governed by economic considerations, since
speed and cost go hand in hand. One is usually willing to use fast, but
costly, flip-flops provided the number required is not excessive. On the
other hand, where the amount of storage is of prime importance, speed
requirements are often relaxed in the interest of economy.

8-10. Terminal Equipment. By terminal devices we mean equipment
for communicating between a machine and its environment. Such
devices are in reality divided into two classes: devices for communication between the machine and other machines or sensing devices, and
devices for communication between the computer and human operators.
Devices of the first class are often registers of the kinds we have considered previously, together with communications equipment if the
computers under consideration are at remote locations. Devices of
the second class include card punches and readers, paper tape punches
and readers, printers, cathode-ray tube displays, electric typewriters,
etc. These all consist of registers plus appropriate transducers. Some
details of the many types of input-output equipments may be found in
Refs. 3, 6, 13, 19, and 24. For the purposes of this chapter we assume
that information can be stored in suitable media, such as paper tape or
cards, and later read by some sensing device. An example of the design
for a paper tape reader and punch may be found in Chap. 9.

8-11. System Characteristics: Instructions and Programs. In the
first sections of this chapter we described the design of computing units,
giving independent and dependent registers as the elementary components. We now turn our attention to the next stage of design, i.e., the
design of computing systems which are themselves made up of the
arithmetic units, storage units, and terminal units previously described.

In the description of the operation of arithmetic units it was not necessary to discuss the detailed logical statements applicable to each flip-flop. By the same token, in a discussion of system characteristics it is neither necessary nor desirable to use a description which is so detailed as to specify each register and transfer.

Let us define a machine *instruction* as one of the elementary operations that can be specified to a machine from *outside* the machine. An instruction must therefore consist of some set of transfers specified as a whole external to the machine to designate some over-all operation. According to this definition, an instruction may consist of a single transfer or a large number of transfers; it may involve more than one unit. Suppose for example that the machine consists of two registers A and B, and the instruction specifies that the sum of the two registers with respect to some representation is to be inserted in A. As we have shown, this instruction may be implemented by a single parallel transfer or by a sequence of serial transfers. External to the computer it is irrelevant how the transfers are implemented, provided that the initial and final values of pertinent registers are specified. A *program* is a set of instructions for the performance of an over-all data-processing problem. It is evident that a program is related to the instructions contained within it as an instruction is related to the transfers comprising it. Just as a given instruction may be implemented by different sets of transfers, so a program may be implemented by different sets of instructions.

Computers are categorized in several ways. One particular category is that of special-purpose computers. Roughly speaking, a special-purpose computer is a data-processing system designed to solve a single problem or a small number of related problems. Often a machine of this type contains a single instruction and executes a program by iterating that single instruction. Another designation for such a machine is *fixed-program*, since only a single basic mode of operation exists. Another category is that of general-purpose computers. A general-purpose computer, as the name implies, can do a large variety of things. It usually has a number of instructions of various kinds which can be associated together into an enormous variety of programs. The most common method of specifying a program in such a machine is to store symbols designating the set of instructions comprising the program in the machine. Thus, such machines are often called *stored-program* computers.

It is evident that the most efficient machine for solving a particular problem is a machine designed specifically for the problem; such a machine can then solve only problems similar to the original problem. A general-purpose computer may be programmed to solve any problem. A general-purpose machine is less efficient, given a problem, than a special-purpose machine, but in a sense more versatile. In order to change the function of a special-purpose machine, the internal electronics must be altered. Obviously, no such electronic modifications are needed to modify the program of a general-purpose machine.   It turns out, however, that the problem of writing and modifying programs for general-purpose machines is by no means trivial.   So-called *automatic* programming techniques have been devised, enabling the programmer to construct his program using *pseudo-instructions* which then refer to auxiliary programs or subroutines made up of sets of machine instructions.   Thus, another level is introduced to simplify the task of programming, and we arrive at the following hierarchy:

$$\text{Program} \rightarrow \text{pseudo-instruction} \rightarrow \text{instruction} \rightarrow \text{transfer}$$

Still another method of adapting general-purpose machines to programs is to design the machine to be internally programmable, that is, to construct the machine in such a way that the sequence of transfers performed can be changed, thus making the internal structure of the machine adaptable to differing problems.   Some consideration will be given to such machines in Chap. 9.   In the following two chapters, we shall illustrate the methods of this chapter in the design of real data-processing systems.

From the above discussion, it is clear that the designation of a given set of transfers as an instruction says nothing about the complexity of that set of transfers.   The instructions simply define the elementary computer operations from the point of view of someone external to the computer.   In a special-purpose computer where the entire operation may consist of a single instruction, that instruction may be very complex.   It may include transfers involving several arithmetic units, memories, and terminal devices.   Since a general-purpose computer is designed for versatility, its instructions are usually quite simple.   Thus, a single complex instruction in a special-purpose computer may be represented by a program consisting of a large number of simple instructions in a general-purpose machine.   Although the instructions of a general-purpose machine differ widely in detail from one computer to another, they may be categorized broadly as *arithmetic* instructions, *program* instructions, and *in-out* or *terminal* instructions.   Loosely speaking, arithmetic instructions involve the computation of functions of variables and the general manipulation of data.   Program instructions are concerned with the sequencing of the instructions which comprise the program.   In-out instructions implement the methods by which data are transferred between the computer proper and terminal equipment.   From what we have said previously, a single instruction of a special-purpose machine may include all three of these functions.

It is now clear that to "design" a digital system it is necessary to define by sets of registers and by transfers the details of each instruction included in the system, the method for sequencing instructions, and the control necessary to initiate all transfers.   In a special-purpose system the first two are combined in a single complex algorithm.   The design process itself consists of three steps: (1) a statement of the desired sequence of operation of the system, (2) a translation of 1 into sets of registers and transfers together with the necessary control, and (3) the final detailed design which translates 2 into the Boolean equations governing each logical element.   This particular division of the design process is, of course, somewhat arbitrary, but extremely useful.   A more usual division is to group 1 and part of 2 together as "system design" and the rest of 2 together with 3 as "logical design."   This latter division can become unwieldy for very complicated systems where the over-all structure can be masked by consideration of logical details simultaneously with system problems.

PROBLEMS

8-1. Design a machine control which cycles a six-bit shift register three bits to the left in time T, halts for time T, and then cycles the shift register three bits to the right in time $T/2$.

8-2. A sequence of pulses is defined as follows: At time $t_j = j\tau$, $j = 1, 2, \ldots, 64$, a pulse is either present or not. Design a machine which behaves as follows: At time $t = 0$ a counter is set to 0. Thereupon the counter is incremented by 1 at $t = t_j$ if a pulse is present, or decremented by 1 if no pulse is present. If the count reaches $+16$ by t_{64}, a flip-flop F_1 is set; if the count reaches -16 by t_{64}, a flip-flop F_2 is set; if neither bound is reached by t_{64}, a flip-flop F_3 is set. F_1, F_2, and F_3 are at 0 initially.

8-3. A 30-cell register A_0, A_1, $\ldots$, A_{29} has an *arbitrary* value.   Design a machine which shifts the register to the left preserving the original value of A_0 until (1) the leftmost 1 is shifted into A_1 when $A_0 = 0$, or (2) the leftmost 0 is shifted into A_1 when $A_0 = 1$.   The machine tabulates the necessary number of shifts.

8-4. Given two numbers x and y, stored in registers A and B respectively.   Design an arithmetic unit which computes $|x| - |y|$.

8-5. Given an arithmetic unit with 1's-complement fractional representation. Before two numbers may be divided, a test must be performed to determine whether the quotient $(Q = N/D)$ is a representable number.

(*a*) Show that the overflow condition is given by

$$\lambda_{OF} = N_0'Z_0'z_0' + N_0Z_0 + Z_0z_0$$

where N_0 = sign bit of the dividend

Z_0 = sign bit of $Z = \begin{cases} |N| - |D| & N_0 = 0 \\ |D| - |N| & N_0 = 1 \end{cases}$

$z_0 = \begin{cases} 1 & Z = 0 \\ 0 & Z \neq 0 \end{cases}$

Note: Assume that whenever an addition or subtraction gives a result equal to 0, the negative-zero representation (1111) results.

(b) If N and D are stored in registers, design a transfer sequence to implement the test for division overflow which leaves N and D stored as originally and sets an alarm flip-flop F_D if the alarm condition exists.

8-6. Repeat Prob. 8-5 with 2's-complement fractional arithmetic unit. Here the alarm condition is

$$\lambda_{OF} = N_0' Z_0' (D_0' + z_0') + N_0 Z_0 z_0' + N_0 D_0 Z_0' z_0$$

where D_0 is the sign bit of the divisor.

8-7. A multiplication algorithm for numbers in 2's-complement fractional form is defined as follows: Suppose the multiplier contains $n + 1$ bits $b_0, b_1, \ldots, b_n$. Designate the multiplicand by a (see, for example, Ref. 47).

1. If $b_n = 1$, form the number $-a$; if $b_n = 0$, form the number 0.
2. Divide the result by 2.
3. If $b_{n-1} = 1$ and $b_n = 0$, subtract a from the previous result. If $b_{n-1} = 0$ and $b_n = 1$, add a to the previous result. If b_{n-1} and b_n are alike, do nothing.
4. Divide the result by 2.
5. Repeat step 3, substituting b_{n-2} and b_{n-1} for b_{n-1} and b_n, respectively.
6. Repeat step 4.

Continue repeating steps 3 and 4. The final operation uses b_0 and b_n as the test bits, and the final division by 2 is omitted.

(a) Prove the algorithm.

(b) Design an arithmetic unit containing an adder-subtracter to implement the algorithm.

(c) What is the machine result when both multiplicand and multiplier have values of -1?

REFERENCES

1. Alt, F. L.: "Electronic Digital Computers," Academic Press, Inc., New York, 1958.
2. Alt, F. L.: "Advance in Computers," Academic Press, Inc., New York, 1960.
3. Bartee, T. C., "Digital Computer Fundamentals," McGraw-Hill Book Company, Inc., New York, 1960.
4. Berkeley, E. C., and L. Wainwright: "Computers, Their Operation and Applications, "Reinhold Publishing Corporation, New York, 1956.
5. Blaauw, G. A.: Indexing and Control-word Techniques, *IBM J. Research and Develop.*, vol. 3, pp. 288–301, July, 1959.
6. Bloch, E.: The Engineering Design of the Stretch Computer, *Proc. Eastern Joint Computer Conf.*, Boston, Mass., pp. 48–58, Dec. 1–3, 1959.
7. Brooks, E. P.: The Execute Operations—A Fourth Mode of Instruction Sequencing, *Communs. Assoc. for Computing Machinery*, vol. 3, pp. 168–170, March, 1960.
8. Brooks, E. P., G. A. Blaauw, and W. Buchholz: Processing Data in Bits and Pieces, *IRE Trans. on Electronic Computers*, vol. EC-8, pp. 118–124, June, 1959.
9. Burks, A. W., H. Goldstine, and J. von Neumann: "Preliminary Discussion of the Logical Design of an Electronic Computing Instrument," The Institute for Advanced Study, Princeton, N.J., 1947.
10. Burtsev, V. S.: Accelerating Multiplication and Division Operations in High-speed Digital Computers, "Exact Mechanics and Computing Technique," Acad. Sci. USSR, Moscow, 1958.
11. Campbell, S. J., and G. H. Rosser, Jr.: An Analysis of Carry Transmission in Computer Addition, presented at the 13th National Meeting of the Association for Computing Machinery, University of Illinois, Urbana, June 11–13, 1958.

12. Dreyfus, P.: System Design of the Gamma-60, *Proc. Western Joint Computer Conf.*, pp. 130–132, May 6–8, 1958.
13. Eckert, J. P., J. C. Chu, A. B. Tonil, and W. F. Schmitt: Design of Univac-Larc System I, *Proc. Eastern Joint Computer Conf.*, pp. 59–65, Dec. 1–3, 1959.
14. Engineering Research Associates, Inc.: "High-speed Computing Devices," McGraw-Hill Book Company, Inc., New York, 1950.
15. Estrin, B., B. Gilchrist, and J. H. Pomerene: A Note on High-speed Digital Multiplication, *IRE Trans. on Electronic Computers*, vol. EC-5, p. 140, September, 1956.
16. Flores, I.: "Computer Logic," Prentice-Hall, Inc., Englewood Cliffs, N.J., 1960.
17. Foss, F. A.: Use of a Reflected Code in Digital Control Systems, *IRE Trans. on Electronic Computers*, vol. EC-3, no. 4, pp. 1–6, December, 1954.
18. Frankel, S. P.: Logical Design of a Simple General-purpose Computer, *IRE Trans. on Electronic Computers*, vol. EC-6, pp. 5–14, March, 1957.
19. Frankovitch, J. M., and H. P. Peterson: A Functional Description of the Lincoln TX-2 Computer, *Proc. Western Joint Computer Conf.*, pp. 146–155, Feb. 26–28, 1957.
20. Frieman, C. V.: A Note on Statistical Analysis of Arithmetic Operations in Digital Computer, *Proc. IRE*, vol. 49, no. 1, pp. 91–103, January, 1961.
21. Garner, H. L.: Generalized Parity Checking, *IRE Trans. on Electronic Computers*, vol. EC-7, no. 3, pp. 207–213, September, 1958.
22. Garner, H. L.: The Residue Number Systems, *IRE Trans. on Electronic Computers*, vol. EC-8, no. 2, pp. 140–147, June, 1959.
23. Gilchrist, B., J. H. Pomerene, and S. Y. Wong: Fast Carry Logic for Digital Computers, *IRE Trans. on Electronic Computers*, vol. EC-4, pp. 133–136, 1955.
24. Hendrickson, H. C.: Fast High-accuracy Binary Parallel Addition, *IRE Trans. on Electronic Computers*, vol. EC-9, pp. 465–468, December, 1960.
25. Holland, J.: A Universal Computer Capable of Executing an Arbitrary Number of Subprograms Simultaneously, *Proc. Eastern Joint Computer Conf.*, pp. 108–113, Dec. 1–3, 1959.
26. Hollingdale, S. H.: "High Speed Computing," The Macmillan Company, New York, 1959.
27. Ivall, T. E.: "Electronic Computers," Philosophical Library, Inc., New York, 1959.
28. Jeenel, J.: "Programming for Digital Computers," McGraw-Hill Book Company, Inc., New York, 1959.
29. Kampe, T. W.: The Design of a General-purpose Microprogram-controlled Computer with Elementary Structure, *IRE Trans. on Electronic Computers*, vol. EC-9, pp. 208–212, June, 1960.
30. Kautz, W. H.: Optimized Data Encoding for Digital Computers, *IRE National Convention Record*, part 4, pp. 47–57, 1954.
31. Lawless, W. J., Jr.: Developments in Computer Logical Organization, in "Advances in Electronics and Electron Physics," vol. 10, pp. 153–183, Academic Press, Inc., New York, 1959.
32. Lehman, M.: High Speed Multiplication, *IRE Trans. on Electronic Computers*, vol. EC-6, pp. 204–205, 1957.
33. Lehman, M.: Shortcut Multiplication and Division in Automatic Binary Digital Computers, *Proc. IEE*, vol. 105B, pp. 496–504, September, 1958.
34. Leiner, A. L., W. A. Notz, J. L. Smith, and A. Weinberger: Pilot—A New Multiple Computer System, *J. Assoc. for Computing Machinery*, vol. 6, pp. 313–335, July, 1959.
35. Leiner, A. L., J. L. Smith, and A. Weinberger: "System Design of Digital Com-

puter at the National Bureau of Standards," National Bureau of Standards Circular 591, Feb., 1958.

36. Lenaerts, E. H.: Automatic Square Rooting, *Elec. Eng.*, vol. 27, pp. 287–289, July, 1955.

37. Lourie, N., H. Schrimpf, R. Reach, and W. Kahn: Arithmetic and Control Techniques in a Multiprogram Computer, *Proc. Eastern Joint Computer Conf.*, pp. 75–81, Dec. 1–3, 1959.

38. Maclean, M. A., and D. Aspinall: A Decimal Adder Using a Stored Addition Table, *Proc. IEE*, vol. 105B, pp. 129–135, 144–146, March, 1958.

39. MacSorley, O. L.: High Speed Arithmetic in Binary Computers, *Proc. IRE*, vol. 49, no. 1, pp. 67–91, January, 1961.

40. McCracken, D. D.: "Digital Computer Programming," John Wiley & Sons, Inc., New York, 1959.

41. Metropolis, N., and R. L. Ashenhurst: Significant Digit Computer Arithmetic, *IRE Trans. on Electronic Computers*, vol. EC-7, pp. 265–267, December, 1958.

42. Montgomerie, G. A.: "Digital Calculating Machines," D. Van Nostrand Company, Inc., Princeton, N.J., 1956.

43. Phister, Montgomery: "Logical Design of Digital Computers," John Wiley & Sons, Inc., New York, 1958.

44. Reed, I. S.: Symbolic Synthesis of Digital Computers, *Proc. of the 1952 Meeting, Assoc. for Computing Machinery*, pp. 90–94, Richard Rimbach Associates, Pittsburgh, Pa., 1952.

45. Reed, I. S.: "The Symbolic Design of Digital Computers," MIT Lincoln Laboratory, Lexington, Mass., Tech. Memorandum no. 23, 1953.

46. Reitwiesner, G. W.: The Determination of Carry Propagation Length for Binary Arithmetic, *IRE Trans. on Electronic Computers*, vol. EC-9, pp. 35–38, March, 1960.

47. Richards, R. K.: "Arithmetic Operations in Digital Computers," D. Van Nostrand Company, Inc., Princeton, N.J., 1955.

48. Richards, R. K., "Digital Computer Components and Circuits," D. Van Nostrand Company, Inc., Princeton, N.J., 1957.

49. Robertson, J. E.: A New Class of Digital Division Methods, *IRE Trans. on Electronic Computers*, vol. EC-7, pp. 218–222; September, 1958.

50. Scott, N. R.: "Analog and Digital Computer Technology," McGraw-Hill Book Company, Inc., New York, 1960.

51. Shimshoni, M.: An Improved Technique for Fast Multiplication on Serial Digital Computers, *Elec. Eng.*, vol. 30, pp. 504–505, August, 1958.

52. Smith, J. L., and A. Weinberger: "Shortcut Multiplication for Binary Digital Computers," p. 22, National Bureau of Standards Circular 591, Feb. 14, 1958.

53. Stevens, W. Y.: "A Study of Decision Operations in Digital Computers," Ph.D dissertation, Cornell University, Ithaca, N.Y., September, 1958.

54. Tocher, T. D.: Techniques of Multiplication and Division for Automatic Binary Computers, *Quart. J. Mech. Appl. Math.*, vol. XI, part 3, pp. 364–384, 1958.

55. Weinberger, A., and J. L. Smith: A One-microsecond Adder Using One-megacycle Circuitry, *IRE Trans. on Electronic Computers*, vol. EC-5, pp. 65–73, June, 1956.

56. Weinberger, A., and J. L. Smith: "Logic for High-speed Addition," National Bureau of Standards Circular 591, Feb. 14, 1958.

57. Wilkes, M. V.: "Micro-programming," *Proc. Eastern Joint Computer Conf.*, pp. 18–20, Dec. 3–5, 1958.

58. "On the Design of a Very High-speed Computer," pp. 180–187, Digital Computer Laboratory Rept. no. 80, University of Illinois, Urbana, October, 1957.

9

General-purpose Computers

9-1. Introduction. This chapter begins the discussion of realistic computing machines, with Chap. 8 serving as the foundation. Chapter 9 describes general-purpose computers and Chap. 10 is devoted to special-purpose machines. As we have already pointed out, the only essential difference between the two classes of machines is one of versatility: as the names imply, a general-purpose machine can solve a large variety of problems and a special-purpose machine a restricted class of problems. Other than that, the two classes are, in principle, the same, being composed of sets of independent and dependent registers. The general-purpose computer, however, has come to have a characteristic structure. Although individual computers vary widely in size, speed, and over-all capabilities, they all retain certain elements of this structure. Most of this chapter is devoted to a design of a simple computer which might be called typical of the class of general-purpose computers despite its simplicity. That is to say, the computer designed here was made just complex enough to illustrate many of the design problems associated with modern computers. (By definition, no special-purpose computer can be typical of anything.) The machines described in Chap. 10 have been chosen to illustrate the same design techniques in their application to machines which are not of the general-purpose class.

Common to the design of all digital machines, elementary as in Chap. 8, general-purpose as in this chapter, and special-purpose as in Chap. 10, are the problems of system design, structural design, and logic design. By way of repetition, system design involves the specification of the general properties of the system; structural design defines the system on the transfer level; finally, logic design translates the structural design into Boolean equations. In this chapter we assume a system design and proceed from there to develop a complete machine design for a general-purpose computer. We then translate portions of this into a logic design to demonstrate the general techniques involved. No circuit minimization, other than the most obvious, is attempted.

It should be emphasized that no attempt is made to present anything approaching an exhaustive description of general-purpose machines. The bibliography at the end of the chapter contains references to many varieties of computers. Our emphasis throughout is on machine-design techniques, with specific structures serving as examples. These techniques have been described in Refs. 4, 14, 20, and 21.

9-2. The Basic Subdivisions of a Computer. The general-purpose computer was defined in Chap. 8 as a machine with a number of different instructions which could be arranged in arbitrary sequences to perform a large variety of computations. It will be recalled that an instruction was defined as an elementary machine operation, as specified external to the machine, and that a set of instructions realizing some computation is called a program. It was also noted in Chaps. 1 and 8 that when the instruction words comprising a program can be stored in a given computer's memory, the machine is generally referred to as a stored-program computer.

Let us now consider the constituent elements of a general-purpose stored-program computer. The machine will contain one or more sets of storage registers, one or more arithmetic units, a program unit, terminal equipment, and one or more control units. The sets of storage registers, or memories, are required to store both programs and data. It is not uncommon for a machine to have tens of thousands of registers in such memories. The arithmetic units are similar to those described in Chap. 8. They contain relatively small numbers of independent and dependent registers and are used for the computation of functions of the variables stored in the memories. The program unit is a specialized arithmetic unit which has as its function the initiation of each instruction at the proper time. The terminal devices, are, of course, the means by which information is directed to the machine by the operator and vice-versa. The control unit specifies the timing of the individual transfers in the machine, as described in Chap. 8.

9-3. Instruction and Data Storage. Most of the memory capacity of a general-purpose computer is for the storage of the program and the data to be processed. Occasionally, separate memory units are provided for program and data storage. Most often, however, both instructions and data may be stored in the same set of registers.

Whenever information is stored in a register, a representation must be defined by which a correspondence is established between the values to be stored and the values of the register. To represent an instruction, a register must be able to specify, first, which of the possible machine operations is required, and second, which register(s) contains the data to be operated upon. To clarify this: suppose that the computer contains a 128-register memory and two computational registers A and B in an arithmetic unit. A typical instruction might require the transfer of the value of register X to register A, where X is one of the 128 memory registers. The register storing this instruction must specify the operation $X \rightarrow A$, and it must also define which of the memory registers is represented by X. Another instruction might specify the addition of the values of registers X and Y and the storage of the sum in register Z, where X, Y, and Z are all memory registers. To specify any one of the above instructions, a register is divided into several subregisters, one subregister specifying the nature of the operation and the other subregisters designating the memory registers containing the operands. If the machine has the ability to perform k distinct operations, then at least $\log_2 k$ bits of information are necessary to specify each operation uniquely. Thus, the *operation subregister* must contain at least $\log_2 k$ cells. If the operand storage capacity is m registers, then $\log_2 m$ bits are needed to specify uniquely each operand, and at least $\log_2 m$ cells are needed to specify each operand in memory. Correspondingly, the *instruction word* itself contains an *operation part* and one or more *address parts*. Each of the subregisters specifying operand locations in memory is called an *address subregister* of the instruction.

An instruction specifying n registers in memory is called an *n-address* instruction. A computer in which the *maximum* number of addresses specified by any instruction is m is called an *m-address* computer. In an m-address computer all of the instructions need not specify m addresses. Thus, in the computer of the previous paragraph an instruction may require the adding of the values of arithmetic registers A and B and storing of the result in A. No memory registers are specified and the address parts of the instruction are ignored. Another instruction may specify that register A be shifted j bits to the left. In this case, one of the address parts of the instruction stores the number j in some representation; the operation specification tells the computer to regard the value of the address part of the instruction as a number rather than as an address.

The performance of an instruction by a computer involves two distinct phases: (1) reading the instruction from memory and interpreting it, and (2) the execution of the instruction. If the instruction is an n-address instruction, then $n + 1$ registers of memory must be referenced. In most computers, only a single memory register may be read from or written into at a given time; hence, $n + 1$ accesses to memory in sequence are required in the performance of an n-address instruction.

9-4. The Program Unit. A program is a set of instructions to the machine specifying the computation to be performed. The program unit of the computer has as its function the selection from the memory of the correct set of instructions. In this sense, the program unit as it relates to the execution of instructions is analogous to the control unit directing the execution of transfers in the proper sequence. From the examples of simple computing systems presented in Chap. 8 it becomes quite clear that there is often a degree of arbitrariness in the arrangement of the transfers. In some cases, transfers which could have been specified simultaneously are indicated as occurring sequentially, with a corresponding increase in the execution time of the operation. Conversely, in some of the examples, the execution time can be shortened by permitting simultaneous transfers wherever possible. In general, the same kind of "serial versus parallel" arguments that were presented in discussing simple transfers apply to transfers within an instruction, and to instructions within a program. It follows that the more parallelism at any level of the computing hierarchy, the more equipment is involved, but the faster the operation.

In most present-day computers, the execution of the program is sequential; i.e., the program consists of a sequence of instructions which the computer executes one at a time. Here the function of the program unit is to initiate the correct sequence of instructions from memory. Usually, the instructions are selected for execution according to some built-in sequence, with deviations from the sequence specified by program-type instructions, as mentioned in Sec. 8-11. The most common method of accomplishing this is to store the instructions of a program in sequentially numbered registers in the computer memory. Here the program unit sees to it that the instructions are taken from memory in numerical order, unless a program instruction specifies an alternative.

We saw in Chap. 8 that in describing transfers there were many possible gradations between completely serial and completely parallel operation. Similarly, in describing programs there are many degrees of departure from purely serial operation. In a strictly sequential computer no instruction is begun until the preceding instruction has been completed. There are many drawbacks to such strictly serial operation. Chief among these is the fact that the entire machine is held back by the speed of its slowest instruction, which usually involves some terminal device. In particular, a very fast arithmetic unit may be kept idle because of a slow card punch. Thus, many computers allow some degree of overlap among the instructions of a program. Such machines are still sequential in the sense that the instructions are initiated serially, but some overlap is permitted in the execution of the instructions. The overlap feature is not necessarily restricted to terminal instructions, but it is here that the value of overlapping instructions becomes most apparent. For a familiar example, an instruction directing the printing of a single character on an electric typewriter may require a major fraction of a second. During this time, the computer may be performing thousands of arithmetic or program-type instructions, or indeed any instructions in its repertoire, provided that the typewriter itself is not demanded before its previous operation has been completed.

This brings us to the essential point in discussing any kind of instruction overlap that a computer may have: the concurrent instructions cannot have any conflicting transfers. In the previous example, it is quite clear that an electric typewriter cannot print two characters concurrently. By the same token, any other part of the computer involved in the printing operation cannot be used simultaneously for another instruction.

Another common kind of overlap allows an instruction to be interpreted while the previous instruction is still being executed. The interpretation of an instruction obviously requires as a first step the receipt of the instruction word from memory. If the memory is such that only a single register may be referenced at any time, then the obtaining of an instruction from memory cannot overlap that part of a preceding instruction which itself requires a reference to memory.

The examples cited above are all of basically sequential machines in the sense that a single program is executed by the performance of a sequence of instructions which are initiated one after another. The next step in program parallelism is to allow the simultaneous execution of more than one program. At this point, we are content to mention the existence of such computing structures and restrict our discussion in this chapter to sequential computers with, at best, some overlap in instruction execution.

9-5. The Control Unit. The elements of the computer control were described in Chap. 8. The control was defined to be a set of registers which assumed certain states which in turn specified the transfers in the machine. What transfers must the control of a general-purpose machine specify? First of all, it must designate the transfers needed to implement each instruction in the repertory of the machine. Second, it must specify the program transfers. Finally, it must control itself.

In the remainder of this chapter, we shall design a simple general-purpose computer. We shall proceed in a manner similar to that of the previous chapter. The sample computer will have a small number of instructions, each of which will be described by a set of transfers connecting independent and dependent registers in various parts of the computer. As in Chap. 8, a control statement will be necessary to initiate each of these transfers. The transfers defining the program operation of the computer will be stated together with the control for initiating them. The machine is complete with the specification of how the control specifies its own transfers and hence its own operation. In Sec. 9-17 some general statements about control units will be presented.

9-6. An Example: Basic Elements.   In order to illustrate the design procedure for a general-purpose machine we shall choose a rather arbitrary system design with a number of component specifications that seem realistic.   The machine will have a small list of instructions, selected, for the most part, to illustrate the operation of typical instructions.   It is clearly not our attempt to design a complete machine, but rather to demonstrate with this example some basic principles, common

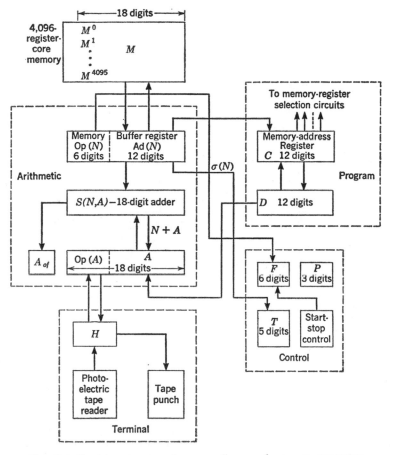

Fig. 9-1. Register structure for a small general-purpose computer.

to the design of all machines, which may be employed for any specific machine, however complex.   Figure 9-1 shows the basic register structure of the machine and will be used as a reference during the description.

First, let us assume the machine will have a memory unit containing 4,096 eighteen-cell registers.   These numbers are, of course, arbitrary, but are fairly typical for small machines. We further assume that the memory unit is constructed of ferrite cores and is random-access. This memory will be used for both program and data storage. Symbolically we shall refer to the memory as M and to its individual registers as M^j, $j = 0, 1, \ldots, 2^{12} - 1$. We shall denote the register used to select one of the registers M^j as the *memory-address* register C. Following the notation of Chap. 6, we shall denote by $M\langle C \rangle$ the memory register selected by C. That is, $M\langle C \rangle = M^j$ where j is the value of C. Finally, we let N denote the 18-cell *memory-buffer* register used to transmit information between M and the rest of the computer.

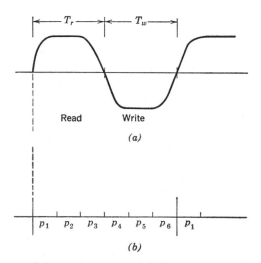

FIG. 9-2. Fundamental timing intervals. (a) Memory cycle; (b) timing intervals.

Let us specify the operation of M by the timing diagram in Fig. 9-2a. The memory cycle of operation contains two equal parts devoted, respectively, to reading out the current value of $M\langle C \rangle$ and to writing a new value into $M\langle C \rangle$. The read operation is destructive by its very nature; hence, if it is desired to preserve the original value of $M\langle C \rangle$, then this value must be rewritten during the write interval. We may summarize this operation with two pairs of transfers. Let Ω be a Boolean function which, when 1, signifies that a new value is to be inserted in $M\langle C \rangle$ during the write interval, and, when 0, signifies that the old value is to be preserved. We then have for the read interval T_r,

$$T_r|\quad \Omega'M\langle C \rangle + \Omega X \to N \qquad 0 \to M\langle C \rangle$$

where X designates some register external to M whose value is to be stored in M. During the write interval T_w we have,

$$T_w| \quad N \to M\langle C \rangle \qquad Y \to C$$

where Y designates a register external to M storing the next address in M to be referenced. The function Ω is specified by control, and its value for each cycle must be specified prior to the beginning of the cycle. Another way of expressing the transfers which will be used in the description of the computer treats Ω explicitly as a control function:

$$\Omega' T_r| \quad M\langle C \rangle \to N \qquad 0 \to M\langle C \rangle$$
$$\Omega' T_w| \quad N \to M\langle C \rangle \qquad Y \to C$$
$$\Omega T_r| \quad X \to N \qquad 0 \to M\langle C \rangle$$
$$\Omega T_w| \quad N \to M\langle C \rangle \qquad Y \to C$$

The arithmetic unit of the computer consists of two independent registers N and A, each containing 18 cells, and the one-cell register A_{of}. N is, of course, the memory-buffer register, and A is an *accumulator*. Several dependent registers, functions of N and A, will allow parallel arithmetic operations to be performed between the two registers.

The program unit contains the independent registers C and D, each 12 cells in length. C is, of course, the memory-address register, and D is the *program counter*.

The control unit contains a six-cell register F, a three-cell register P, and a five-cell register T. These registers and the remainder of the control unit will be specified later.

For terminal equipment our computer will contain a paper tape punch and a tape reader with a buffer register H.

9-7. Fundamental Timing. For simplicity, the computer (except for the terminal devices) will operate completely synchronously, with the memory cycle divided into six basic intervals as defined by Fig. 9-2b. The memory will operate continuously, alternating its read and write intervals T_r and T_w. The memory cycle is synchronized to the period of a clock pulse generator and timing counter in such a way that T_r and T_w each contain exactly three clock intervals. More precisely, a three-cell timing counter P continuously cycles through the values 1 through 6 with the characteristic functions $p_1 + p_2 + p_3$ and $p_4 + p_5 + p_6$ defining T_r and T_w, respectively. Thus, a memory cycle time and a cycle time of P are equivalent. Following the notation of Chap. 8, we assume that all transfers are gated by one of the functions p_j ($j = 1, \ldots, 6$), or the logical sum of several p_j. Typically, the duration of each p interval might be 1 μsec, with a corresponding memory cycle time of 6 μsec.

Each instruction will require an integral number of memory cycles for its execution.   During each cycle, the memory operation will be defined by the set of transfers of Sec. 9-6, which in the new timing notation are written as

$$
\begin{array}{ll}
p_1| & \Omega' M\langle C\rangle + \Omega X \to N \qquad 0 \to M\langle C\rangle \\
p_2| & \qquad\qquad\downarrow \qquad\qquad\qquad\qquad \downarrow \\
p_3| & \qquad\qquad\downarrow \\
p_4| & \qquad\quad N \to M\langle C\rangle \\
p_5| & \qquad\qquad\downarrow \\
p_6| & \qquad\qquad\downarrow \qquad\qquad\qquad\qquad Y \to C
\end{array}
$$

where the vertical arrows indicate that the transfers occur at some time during the interval.   For instance, $0 \to M\langle C\rangle$ occurs during the interval from p_1 to p_3.

9-8. Register Representations.   The 18-cell registers in the memory and arithmetic unit will store both data and instructions.   Accordingly, two representations must be defined, one for data and one for instructions. In all cases, we shall label the cells of a register with the subscripts 0 to 17 from left to right.

Numbers will be stored in the 1's-complement fractional representation defined in the Appendix by

$$
\delta_f(X) = -X_0(1 - 2^{-17}) + \sum_{i=1}^{17} X_i 2^{-i}
$$

The range of $\delta_f(X)$ is from $-(1 - 2^{-17})$ to $1 - 2^{-17}$, and $\delta_f(X) = 0$ for two values of X: $X_i = 0$ and $X_i = 1$, $i = 0, 1, \ldots, 17$.

To represent instructions, we divide the register X into two subregisters: Op $(X) = (X_0, X_1, \ldots, X_5)$ and Ad $(X) = (X_6, X_7, \ldots, X_{17})$. The six-cell subregister Op (X) will be referred to as the *operation subregister* of X and will define one of up to 64 instructions by establishing a correspondence between each instruction and a unique value of Op (X). The 12-cell register Ad (X) is called the *address subregister* of X.   For single-address instructions [i.e., when Op (X) specifies a single-address instruction] the values of Ad (X) are in one-to-one correspondence with the 2^{12} registers in M.   In particular, when Ad (X) has the integer representation j it will refer to register M^j.   In *zero-address instructions*, the value of Ad (X) has a special meaning defined by each such instruction.[†]   As indicated in the previous section, the 12-cell address register C and program counter D refer to memory locations and hence use the same representation as Ad (X), where X is any 18-cell register.

† A zero-address instruction is one in which the address part does not refer to an address in memory.

9-9. Arithmetic Instructions.   The arithmetic unit of the computer was defined in Sec. 9-6 as consisting of two 18-cell independent registers A and N.   To perform addition we assume the dependent register†

$$S(A,N) = A + N$$

which computes the sum of the numbers (in 1's-complement fractional representation) stored in registers A and N.   In order to store the result of the addition, we assume the transfer $S(A,N) \rightarrow A$.   The fact that a clock pulse generator and an interval counter P have been defined to operate at a fixed rate (say 1 M pulse/sec) does not in any sense imply that all transfers must be accomplished between successive clock pulses. In fact, in Sec. 9-7, the transfers defining the read and write operations of the core memory required three intervals for execution.   Clearly the transfer of a complex function such as $S(A,N)$, which may involve many cascaded logic levels in its carry network, will result in more delay than the transfer of a simple function.   In this machine, we shall assume that simple transfers (e.g., $N \rightarrow A$) may be accomplished in a single pulse interval, but the transfer $S(A,N) \rightarrow A$ requires two pulse intervals. Thus, we shall always demand that A and N receive their values at least two intervals prior to the transfer (that is, by the end of p_4).

A typical addition instruction reads as follows: add the number in memory register X to that in register A, storing the sum in A.   If an overflow occurs as the result of the addition (i.e., the sum of the two numbers has magnitude 1 or greater), store a 1 in flip-flop A_{of}.   Otherwise store a 0 in A_{of}.   We shall abbreviate this instruction by ad X and assign to it the integer 1 as its representation in the operation subregister.   Hence, when a memory register stores the instruction ad X, its operation subregister has the value 1 and its address subregister the value X ($0 \leq X \leq 4{,}095$).

The execution of the instruction ad X will take two memory cycle intervals, the first to obtain and interpret the instruction itself and the second to perform the actual addition.   If we assume that the instruction is stored in register Y, then clearly the address register C must have the value Y during the cycle in which the instruction is obtained from the memory.   The method by which this is accomplished will be deferred to Sec. 9-10, where program transfers are discussed.   Furthermore, this instruction interpretation cycle is one in which information is obtained from memory and not transferred to memory; hence, the function Ω has the value 0 during this cycle.   A discussion of how this is accomplished will be deferred to Sec. 9-11, where the control is discussed.

Some of the transfers necessary to the instruction interpretation cycle

† Since only 1's complements will be used in this chapter, we shorten S^1 to S.

are:

$$\alpha p_1| \quad M\langle C\rangle \to N \qquad 0 \to M\langle C\rangle$$
$$\alpha p_2|$$
$$\alpha p_3| \qquad \downarrow \qquad\qquad \downarrow$$
$$\alpha p_4| \quad N \to M\langle C\rangle$$
$$\alpha p_5|$$
$$\alpha p_6| \qquad \downarrow \qquad \text{Ad } (N) \to C \qquad \text{Op } (N) \to F$$

These transfers define a memory access in which the instruction is transferred to N by p_3 and is restored in memory by p_6. During p_6, the value designating the particular operation (in the case of ad X, the value 1) is transferred to control register F; also during p_6 the address X is transferred from Ad (N) to C. Hence, during the next cycle of the machine, the control "knows" that an addition is to be performed next, and the memory is accessed at address X.

The above list of transfers is incomplete in that the control is only partially defined. The control function α has the value 1, identifying the fact that this is an instruction interpretation cycle; among other things it must guarantee that $\Omega = 0$. We shall assume for now that the presence of the value 1 in control register F is sufficient to cause the control function f_1 to have the value 1 during the following cycle of operation, specifying uniquely that an addition is to be performed. This next cycle then contains the set of transfers:

$$f_1 p_1| \quad M\langle C\rangle \to N \qquad 0 \to M\langle C\rangle$$
$$f_1 p_2|$$
$$f_1 p_3| \qquad \downarrow \qquad\qquad \downarrow$$
$$f_1 p_4| \quad N \to M\langle C\rangle$$
$$f_1 p_5|$$
$$f_1 p_6| \qquad \downarrow \qquad S(N,A) \to A \quad \lambda(N,A) \to A_{of}$$

Since C contains the address X, the value of register X is transferred to N by the end of p_3, and the addition is complete by the end of the p_6 interval. The function $\lambda(N,A)$ is the overflow function defined for the 1's-complement fractional representation by

$$\lambda(N,A) = N_0 A_0 K_0' + N_0' A_0' K_0$$

which states that an overflow exists if initially N and A have negative values and there is no carry into the sign digit, or if N and A are positive and there is a carry into the sign digit.

A subtraction instruction is almost exactly the same. We define it precisely by: subtract the number in memory register X from the number in A, storing the difference in A. If an overflow occurs, store a 1 in A_{of}.

We abbreviate the instruction by su X and designate it by the number 2 as its operation representation. Its instruction interpretation cycle is identical to that described for ad X. The transfer Op $(N) \rightarrow F$ inserts the value 2 in F, which in turn assures that the control function f_2 is 1 during the execution cycle. The latter is defined by

$$
\begin{array}{lll}
f_2 p_1| & M\langle C\rangle \rightarrow N & 0 \rightarrow M\langle C\rangle \\
f_2 p_2| & \quad\bigg\downarrow & \\
f_2 p_3| & & \\
f_2 p_4| & N \rightarrow M\langle C\rangle & N' \rightarrow N \\
f_2 p_5| & \quad\bigg\downarrow & \\
f_2 p_6| & & S(N,A) \rightarrow A \qquad \lambda(N,A) \rightarrow A_{\text{of}}
\end{array}
$$

This cycle is identical with that for addition except for the additional transfer $N' \rightarrow N$ during p_4, which negates the number in N and which, when combined with the addition, produces the desired subtraction.

For a final example of an arithmetic-type instruction, we describe the instruction "store in X," defined by: transfer the value of A to memory register X, preserving the value of A. We abbreviate this instruction by st X and assign to it the representation 3. The instruction acquisition cycle of st X is again identical to that previously described, and the execution cycle is defined by

$$
\begin{array}{lll}
f_3 p_1| & & 0 \rightarrow M\langle C\rangle \\
f_3 p_2| & & \\
f_3 p_3| & A \rightarrow N & \quad\bigg\downarrow \\
f_3 p_4| & N \rightarrow M\langle C\rangle & \\
f_3 p_5| & \quad\bigg\downarrow & \\
f_3 p_6| & &
\end{array}
$$

Since, during this latter cycle a new value is being transferred to $M\langle C\rangle$, the control function f_3 must imply the function Ω.

9-10. Program Operation. The computer will execute a program by taking consecutive instructions from consecutive memory registers unless told otherwise by a program-type instruction. The transfers needed to realize this mode of operation are rather simple. For example, during the execution of an instruction interpretation cycle, the address register C contains the location of the current instruction. This address is advanced by 1 and stored in program register D. Thus, by the end of this cycle, the D register contains the address of the next instruction in the absence of a program instruction which breaks the sequence. With these modifications, the instruction interpretation cycle now becomes:

$$
\begin{array}{llll}
\alpha p_1| & M\langle C\rangle \to N & 0 \to M\langle C\rangle & C \to D \\
\alpha p_2| & \downarrow & \downarrow & D + 1 \to D \\
\alpha p_3| & \downarrow & \downarrow & \\
\alpha p_4| & N \to M\langle C\rangle & & \\
\alpha p_5| & \downarrow & & \\
\alpha p_6| & \downarrow & \mathrm{Ad}\,(N) \to C & \mathrm{Op}\,(N) \to F
\end{array}
$$

The two new transfers were located arbitrarily at p_1 and p_2. They could have been located during any pair of time intervals which would have provided the desired result by the end of the cycle.

During the execution cycle of any instruction in which the normal program sequence is not broken, the transfer $D \to C$ is performed, thus readying the memory for its next instruction interpretation cycle. Thus, the ad X instruction is modified to provide:

$$
\begin{array}{llll}
f_1 p_1| & M\langle C\rangle \to N & 0 \to M\langle C\rangle & \\
f_1 p_2| & \downarrow & \downarrow & \\
f_1 p_3| & \downarrow & \downarrow & \\
f_1 p_4| & \mathrm{V} \to M\langle C\rangle & & \\
f_1 p_5| & \downarrow & & \\
f_1 p_6| & \downarrow & S(N,A) \to A & \lambda(N,A) \to A_{\mathrm{of}} \quad D \to C
\end{array}
$$

The transfer $D \to C$ is, of course, constrained by the basic memory operation to occur during p_6.

A rather common class of program instructions is that of the *conditional jump*. In an instruction of this kind, a binary decision is made which is based upon a specified Boolean function. This decision specifies that the program is to continue in its normal consecutive sequence if the function has one value, or it is to jump elsewhere in memory for the next instruction if the function has the other value. A typical example is the instruction "jump, if negative, to X," abbreviated jn X and defined by the statement: if the sign digit of the A register is 1, take the next instruction from register X; if it is 0, follow the normal sequence. We designate jn X by the representation 4. The execution cycle is made up of the transfer sequence:

$$
\begin{array}{lll}
f_4 p_1| & M\langle C\rangle \to N & 0 \to M\langle C\rangle \\
f_4 p_2| & \downarrow & \downarrow \\
f_4 p_3| & \downarrow & \downarrow \\
f_4 p_4| & N \to M\langle C\rangle & \\
f_4 p_5| & \downarrow & \\
f_4 p_6| & \downarrow & A_0 C + A_0' D \to C
\end{array}
$$

The conditional transfer at p_6 leaves C unchanged (with value X) if

$A_0 = 1$ or inserts the value of D (the address of the next instruction in sequence) if $A_0 = 0$. Again, this transfer is constrained to occur during p_6 by the basic memory operation. The normal memory read-write operation is indicated even though no use is made at this time of the output of the memory. It is interesting to observe that the value of N during the last half of the cycle is that of the next instruction to be executed if $A_0 = 1$. We could have taken advantage of this fact to shorten the execution time of the instruction when $A_0 = 1$. This, of course, would only have been at the cost of more equipment and complexity.

A jump instruction, conditional or otherwise, is a perfectly general way of causing a program to depart from a sequence of consecutive registers. It is often desirable for a program to jump out of a sequence to some sub-program located elsewhere in memory and then to return to the original sequence at the point of departure. Moreover, it is often desirable to jump to the subprogram from an arbitrary point in the main sequence and always return to the sequence from the point of departure. Sub-programs having this property are often used. To obtain the desired operation, an additional transfer may be included in the jump instructions. This transfer, $D \rightarrow \text{Ad}\ (A)$, preserves in the address subregister of A the address of the next instruction in the main program. The subprogram itself will begin with a special program-type instruction called "return to the main program from X" (abbreviation rf X) where X is the address of the jump instruction concluding the subroutine. The instruction rf X transfers the value of Ad (A) to Ad (X). Hence, the final instruction of the subprogram is a jump instruction referring back to the proper point of the main program.

The instruction rf X requires the modification of the address sub-register X while leaving the remainder of the register intact. In order to accomplish this, the function Ω is replaced by two functions Ω_0 and Ω_{Ad} referring respectively to cells 0 to 6 and 7 to 18. In all the instructions considered thus far, $\Omega = 1$ implies $\Omega_0 \Omega_{\text{Ad}} = 1$ and $\Omega = 0$ implies $\Omega_0' \Omega_{\text{Ad}}' = 1$. In the instruction rf X, the control function f_5 must imply $\Omega_0' = \Omega_{\text{Ad}} = 1$. The execution cycle of this instruction then becomes:

$$
\begin{array}{l}
f_5 p_1| \\
f_5 p_2| \\
f_5 p_3| \\
f_5 p_4| \\
f_5 p_5| \\
f_5 p_6|
\end{array}
\quad
\begin{array}{c}
\text{Op }(M\langle C\rangle) \rightarrow \text{Op }(N) \\
\downarrow \\
\\
N \rightarrow M\langle C\rangle \\
\downarrow
\end{array}
\quad
\begin{array}{c}
0 \rightarrow M\langle C\rangle \\
\downarrow \\
\\
\\
D \rightarrow C
\end{array}
\quad
\begin{array}{c}
\text{Ad }(A) \rightarrow \text{Ad }(N)
\end{array}
$$

Included in the list of instructions of this computer are two additional jump instructions. The first of these, "jump on overflow to X," is designed to follow an addition or subtraction instruction to provide an alternative computation if an overflow exists. It is identical to the jn X instruction with the one obvious exception that the cell A_{of} rather than A_0 determines the decision of the instruction. The transfer sequence is as follows:

$$
\begin{array}{llll}
f_6 p_1| & M\langle C\rangle \to N & 0 \to M\langle C\rangle & D \to \text{Ad } (A) \\
f_6 p_2| & \downarrow & \downarrow & \\
f_6 p_3| & & & \\
f_6 p_4| & N \to M\langle C\rangle & & \\
f_6 p_5| & \downarrow & & \\
f_6 p_6| & & & A_{of} \cdot C + A'_{of} \cdot D \to C
\end{array}
$$

The final jump instruction will be to illustrate one of the types of instruction overlap discussed in Sec. 9-4. This instruction, "jump unconditionally to X" (abbreviated ju X), requires the following transfer sequence for its execution:

$$
\begin{array}{llll}
f_7 p_1| & M\langle C\rangle \to N & 0 \to M\langle C\rangle & D \to \text{Ad } (A) \\
f_7 p_2| & \downarrow & \downarrow & \\
f_7 p_3| & & & \\
f_7 p_4| & N \to M\langle C\rangle & & \\
f_7 p_5| & \downarrow & & \\
f_7 p_6| & & &
\end{array}
$$

It is readily observed that the only transfer required specifically for this instruction is $D \to \text{Ad } (A)$, which prepares for a possible subroutine to follow. All the other transfers pertain to the memory read and write operations which continue automatically. In this case, however, the value of C during the cycle is the address X, which is the address of the next instruction to be executed. We take advantage of this fact and interpret the next instruction during this cycle of operation. More precisely, by demanding that $\alpha = 1$ during the cycle, all the transfers occurring during an instruction interpretation cycle will occur; the condition $f_7 = 1$ causes the additional transfer $D \to \text{Ad } (A)$.

9-11. Control Operation. We have reached a point in the discussion of the computer where the general structure of the control unit can be described. It will be recalled that the control unit must establish states or characteristic functions in one-to-one correspondence with the transfers to be performed in the machine, and that it is these functions which initiate the transfers.

Several of the general features of the control have already been established: each instruction takes an integral number of cycles for its execution, where the cycle is defined to be the memory read-write time. The cycle is subdivided into six intervals which define the clock periods at which all transfers occur.

We define an *instruction interval* to begin just when the instruction is about to be implemented (i.e., just after the interpretation cycle has been completed) and to conclude at the end of the *next* interpretation cycle. Clearly, the final cycle of each instruction interval is the interpretation cycle for the next instruction of the program. Consider, for example, the program consisting of the sequence

$$\begin{aligned}
&\text{ad } X \\
&\text{su } Y \rightarrow \text{ad } A \\
&\qquad\qquad \text{ha} \\
&\text{ju } Z
\end{aligned}$$

The final instruction (halt, abbreviated ha) will be described later. It will be seen to require two cycles for its operation. This program is performed in time according to Fig. 9-3. The lower line divides the time

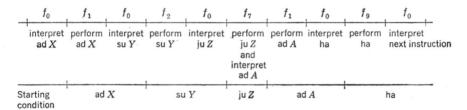

FIG. 9-3. Timing intervals for a short program.

into instruction intervals. Since the first instruction of a program is defined to begin with the execution of that instruction, the interpretation of the instruction must be performed during the starting period as an initial condition.

The functions which initiate each transfer are characteristic functions of the form $f_i p_j$, where $f_i = 1$ implies that the control is in its ith state, and $p_j = 1$ corresponds to the pulse counter P having value j. The control function f_i will never change more rapidly than once per cycle. On the upper line of Fig. 9-3 the cycles are labeled by the control functions with which they are in correspondence. We have designated by f_0 those cycles during which an instruction is interpreted and nothing else happens. It then follows that the function α, defined in Sec. 9-9 to designate the instruction interpretation cycle, is given by

$$\alpha = f_0 + f_7$$

The functions f_i are, of course, the characteristic functions of the control register F. Since these functions correspond in time to cycles, any transfers to F must be performed during the p_6 intervals. Hence, to describe completely the operation of the register F, we need only specify appropriate inputs during each p_6 interval. One such input has already been stated: during each instruction interpretation cycle ($\alpha = 1$), the value k of the operation itself is transferred to F, thus giving F the value k, or, equivalently, requiring the function f_k to have the value 1 during the next cycle. All that remains to be done for all the instructions defined thus far is to perform the transfer $0 \rightarrow F$ during the execution cycles.

There will be some departures from this simple mode of operation of control necessitated by more complex instructions to be described subsequently. These departures will be, however, only by way of introducing new inputs to F. The basic structure will not be changed.

9-12. A Shifting Instruction. We consider now an example of an instruction which demands a slightly more complex control. This instruction, "cycle the A register to the right n times," is abbreviated cr n and is assigned the value 8. Unlike the previous instructions, cr n is a *zero-address instruction;* the address subregister represents the integer n instead of an address. Since n is a variable, the number of shifts required by the instruction, and hence the execution time of the instruction, will be variable. An additional control register T will be used to keep track of the number of shifts performed. Since A contains only 18 cells, no shift need be larger than that number. Accordingly, T is a five-cell register into which n is transferred. Upon execution of each shift, T is decremented by 1; when T reaches 0, the shifting is terminated.

For convenience, we introduce the transfer of the number n to T during the instruction interpretation cycle. To do this, let $\sigma(X)$ be the rightmost five cells of register X [or of Ad (X)]; we shall assume the transfer $\sigma(N) \rightarrow T$ to occur always during p_6 of the interpretation cycle (whenever $\alpha = 1$), i.e., at the same time as the transfers Op $(N) \rightarrow F$ and Ad $(N) \rightarrow C$. The instruction is then executed according to the sequence:

$$
\begin{array}{lll}
f_8 p_1| & t_0 \cdot A + t_0' \cdot \rho(A) \rightarrow A & t_0 \cdot T + t_0' \cdot (T - 1) \rightarrow T \\
f_8 p_2| & t_0 \cdot A + t_0' \cdot \rho(A) \rightarrow A & t_0 \cdot T + t_0' \cdot (T - 1) \rightarrow T \\
f_8 p_3| & t_0 \cdot A + t_0' \cdot \rho(A) \rightarrow A & t_0 \cdot T + t_0' \cdot (T - 1) \rightarrow T \\
f_8 p_4| & t_0 \cdot A + t_0' \cdot \rho(A) \rightarrow A & t_0 \cdot T + t_0' \cdot (T - 1) \rightarrow T \\
f_8 p_5| & t_0 \cdot A + t_0' \cdot \rho(A) \rightarrow A & t_0 \cdot T + t_0' \cdot (T - 1) \rightarrow T \\
f_8 p_6| & t_0 \cdot A + t_0' \cdot \rho(A) \rightarrow A & t_0 \cdot T + t_0' \cdot (T - 1) \rightarrow T \qquad D \rightarrow C \\
& & (t_0 + t_1)' \cdot F + (t_0 + t_1) \cdot 0 \rightarrow F
\end{array}
$$

The function t_0 is, of course, the characteristic function of T, which has the value 1 if $T = 0$. A 1 is subtracted from the T register each time A

is cycled right one place. The function $\rho(A)$ was defined in Chap. 6 to be the dependent register with cells $(A_{17}, A_0, \ldots , A_{16})$, and hence $\rho(A) \to A$ represents the cycle-right operation. During p_6 the value of F is preserved at 8 if more cycling is required (i.e., if T has value 2 or greater at the *beginning* of p_6) and is cleared if the cycling is complete, thus placing the control in state f_0. Note that the second term of the input to F, $(t_0 + t_1) \cdot 0$, need not be included. The explicit statement is included for clarity.

9-13. Start-Stop Control: Initial Conditions. In Sec. 9-11 the control of the operating computer was described. In particular, the various ways in which the control register F could change its state were defined. The remaining gap in the general picture of control operation is the implementation of the start-stop control, i.e., the transition between operating and nonoperating conditions.

We use a flip-flop Q to distinguish between the two conditions: $Q = 0$ implies operation, $Q = 1$ implies inactivity. Thus, many of the transfers already described must be conditioned by the function Q'. Some transfers need not be conditioned in this way. The P register, which establishes the timing within each cycle, is most conveniently allowed to operate regardless of the value of Q. This assures a synchronous time base regardless of the mode of computer operation. It may also be convenient, from an equipment point of view, to allow the memory to operate continuously. If this is done, it is mandatory that $\Omega = 0$ whenever $Q = 1$, so that the memory keeps reading and writing the same value. We shall assume that P and M operate continuously, but with the restrictions noted. All other transfers will be considered as conditioned by the function Q'.

The instruction interval was defined to begin with the execution cycle of that instruction. Thus, the first instruction of a program must itself be interpreted during a cycle initiated by start-stop control. We shall make the restriction that the first instruction of every program will be located in register M^0. This restriction is not serious, however, since if a program begins at register X the instruction ju X inserted in M^0 will meet the requirements of the computer. Let W_p be a function which has value 1 when a "start program" button is depressed. The start-stop control operation then has the form:

$$Qp_1|$$
$$Qp_2|$$
$$Qp_3|$$
$$Qp_4|$$
$$Qp_5|$$
$$Qp_6| \quad W_p' \to Q \qquad W_p' \cdot C + W_p \cdot 0 \to C \qquad W_p' \cdot F + W_p \cdot 0 \to F$$

The three transfers occurring during p_6 are sufficient to establish the initial conditions. When the button is depressed, F is set to value 0, ensuring that the following cycle will be an instruction interpretation cycle, C is set to value 0, requiring that this instruction be taken from register M^0, and Q is set to 0, thus starting the computer.

The halt instruction itself, referred to in Sec. 9-11, is now easily defined by:

$$
\begin{array}{l}
f_9 p_1| \\
f_9 p_2| \\
f_9 p_3| \\
f_9 p_4| \\
f_9 p_5| \\
f_9 p_6| \quad 1 \to Q \qquad D \to C \qquad 0 \to F
\end{array}
$$

The transfers $D \to C$ and $0 \to F$ prepare for a following instruction interpretation cycle as in many other instructions (for example, ad X). The transfer $1 \to Q$ places the computer in its inactivity state.

The halt instruction is another example of a zero-address instruction. Here the address subregister is not used. If a number, say Z, is stored in the address subregister, the memory reads and rewrites the value of register M^Z during the execution cycle.

It is often desirable to start the computer from a stopping point, i.e., to start at the address following that of a halt instruction. To do this, another start button, labeled "restart," is provided. This initiates the following sequence:

$$
\begin{array}{l}
Q p_1| \\
Q p_2| \\
Q p_3| \\
Q p_4| \\
Q p_5| \\
Q p_6| \quad W_R' \to Q
\end{array}
$$

W_R is a function having value 1 when the "restart" button is depressed. Unlike the "start program" condition, the "restart" condition needs only to set $Q = 0$; all other initial conditions are correct.

9-14. Terminal Instructions. The final two instructions are terminal instructions which employ the tape reader and the tape punch. We shall not describe these instructions in complete detail, since to do so would involve a discussion of the detailed properties of two complex electromechanical devices. Both the instructions illustrate instruction overlap.

Information is placed on tape in rows of six digits each. A 1 is represented by a hole and a 0 by the absence of a hole. The value of a computer register, for example, could be stored in three rows of tape. A row of tape is, of course, a register in the general sense.

The photoelectric tape reader contains six photosensitive devices, each of which emits a pulse when it "sees" a hole (1) and no pulse when it sees no hole (0). These pulses are applied to a six-cell register H. If the row of tape under the photosensitive detectors is designated by Y, then the reading operation may be represented by $Y \rightarrow H$.

The rt instruction starts the tape unit by clearing the tape-reader start-stop flip-flop J_R. Thereupon it performs the transfer $H \rightarrow$ Op (A) which represents the value of the row of tape read previously. The computer continues with the rest of its program, and some time later another rt instruction is given. While the program is being executed, the tape is advanced until the next row of information reaches the sensors. At this time, the transfer $Y \rightarrow H$ occurs (under the direction of the reader), and a 1 is inserted in J_R, stopping the motion. If the second rt instruction occurs before this operation is complete, the computer is forced to halt and wait for its completion. The transfers are as follows:

$$f_{10}p_1| \quad J'_R \rightarrow Q \qquad 0 \rightarrow J_R$$
$$f_{10}p_2|$$
$$f_{10}p_3|$$
$$f_{10}p_4|$$
$$f_{10}p_5|$$
$$f_{10}p_6| \quad Q \cdot F + Q' \cdot 0 \rightarrow F \qquad Q \cdot \text{Op } (A) + Q' \cdot H \rightarrow \text{Op } (A)$$
$$Q \cdot C + Q' \cdot D \rightarrow C$$

As remarked in the previous section, all control functions $f_i p_k$ up until now have implied $Q = 0$. The functions $f_{10}p_i$ are independent of Q, since the dependence on Q is given explicitly in the transfers themselves.

When the instruction is executed, J_R is first tested: if $J_R = 0$, the tape reader is busy; if $J_R = 1$, it has completed its previous instruction. If $J_R = 0$, the computer is halted, thus inhibiting the transfers at p_6. Once the reader is idle, the computer restarts and the tape reader is restarted. During p_6, the information transfer is performed and control is advanced to f_0, thus continuing with the program.

The punch tape instruction, abbreviated pt, is treated in very much the same manner as the rt instruction. The punch has its own start-stop flip-flop J_P, which is set to 0 by the computer when a punch operation is initiated and set to 1 by the punch when the operation is complete. The data are transferred to the punch by the transfer Op $(A) \rightarrow H$. The transfer sequence is as follows:

$$f_{11}p_1| \quad 0 \rightarrow J_P \qquad J'_P \rightarrow Q$$
$$f_{11}p_2| \quad Q' \cdot \mathrm{Op}\ (A) + Q \cdot H \rightarrow H$$
$$f_{11}p_3|$$
$$f_{11}p_4|$$
$$f_{11}p_5|$$
$$f_{11}p_6| \quad Q' \cdot 0 + Q \cdot F \rightarrow F \qquad Q' \cdot D + Q \cdot C \rightarrow C$$

As in the rt instruction, the control functions $f_{11}p_i$ do not imply $Q = 0$.

9-15. Implementation of the Computer: Logic Design. The computer is now completely defined on the transfer level. A summary of the transfers is presented in Table 9-1. The value of this method of description is shown by the fact that the entire machine is described with a relatively small amount of symbology. To complete the design, it is necessary only to collect the input transfer relations for each register and translate these into flip-flop input equations. This translation is, of course, dependent upon the nature of the flip-flop and gating circuitry to be used for the machine. Up to now, the assumptions that have been made about the machine have been rather general. We have assumed, for example, that the machine is synchronous with six clock pulses per memory cycle.

We shall assume further that the flip-flop is a set-reset (RS) flip-flop, as defined in Sec. 7-5, which operates according to the relation

$$S + XR' \rightarrow X$$

where S and R represent pulses at the set and reset inputs respectively. Let us assume that each flip-flop in the computer will be preceded by a so-called *steering circuit*, consisting of an inverter and two AND gates. The inputs to the particular steering circuit we shall use are the clock pulse train E and the functions G and K. The outputs are the functions

$$S = KGE \qquad \text{and} \qquad R = KG'E$$

That is to say, a clock pulse appears at S whenever both K and G are 1, at R when $K = 1$ and $G = 0$, and nowhere when $K = 0$. The transfer relation satisfied by the flip-flop X then becomes

$$KG + X(KG')' \rightarrow X$$

which reduces to

$$KG + K'X \rightarrow X$$

The steering circuit, therefore, causes the transfer of the function G into X whenever $K = 1$ and leaves X undisturbed whenever $K = 0$. For simplicity, we shall refer to the RS flip-flop combined with the above steering circuit as a GK flip-flop, and shall use the notation $G(X)$ and $K(X)$ to refer to the G and K inputs of flip-flop X.

Let us consider first the implementation of the control register F with GK flip-flops. In Table 9-1, two functions are indicated as being transferred into F under various conditions. These functions are Op (N) and 0. Thus, the general transfer equation defining the inputs to F may be written in the form

$$\lambda \cdot \text{Op } (N) + \mu \cdot 0 + (\lambda + \mu)' \cdot F \rightarrow F$$

where λ is a scalar function including all control functions which specify Op $(N) \rightarrow F$, and μ is a similar function specifying the transfer $0 \rightarrow F$. The final term states explicitly that whenever a transfer is unspecified F is to remain undisturbed. This transfer may be realized by setting

$$\begin{aligned} K(F_j) &= \lambda + \mu \\ G(F_j) &= \lambda N_j \end{aligned} \qquad j = 0, 1, \ldots, 5$$

The functions λ and μ are obtained immediately from Table 9-1:

$$\lambda = Q'\alpha p_6 = Q'p_6(f_0 + f_7)$$
$$\mu = Q'p_6[f_1 + f_2 + f_3 + f_4 + f_5 + f_6 + f_9 + f_8(t_0 + t_1) + f_{10} + f_{11}]$$
$$+ Qp_6(W_p + W_r)$$

It follows that

$$\lambda + \mu = p_6\{Q(W_p + W_r) + Q'[f_8' + (t_0 + t_1)']\}$$

The above translation of transfer equations into flip-flop equations may be generalized quite simply. Let B^i, $i = 1, 2, \ldots, n$, be a set of functions to be transferred into a register A. Further, let λ_i be the control function selecting B^i. It then follows that the general input transfer equation for register A is

$$\lambda_1 B^1 + \lambda_2 B^2 + \cdots + \lambda_n B^n + (\lambda_1 + \lambda_2 + \cdots + \lambda_n)'A \rightarrow A$$

Then if A is realized with GK flip-flops, we have

$$\begin{aligned} K(A_j) &= \lambda_1 + \lambda_2 + \cdots + \lambda_n \\ G(A_j) &= \lambda_1 B_j{}^1 + \lambda_2 B_j{}^2 + \cdots + \lambda_n B_j{}^n \end{aligned}$$

The functions λ_i $(i = 1, 2, \ldots, n)$ need be generated only once for an entire register and then used as the inputs to the combinational networks realizing $G(A_j)$ for each j.

The realization of the F register with GK flip-flops is, of course, just one of the many possible alternatives. We present it to illustrate the general method of translating transfer relations into flip-flop equations. The procedures using other kinds of flip-flops are quite similar.

Table 9-1. Summary of Computer Transfer Relations

Memory Operation Common to All Instructions

$\Omega'_O p_1$	Op$(M\langle C\rangle) \to$ Op(N)	$\Omega'_{Ad} p_1$	Ad$(M\langle C\rangle) \to$ Ad(N)	p_1	$0 \to M\langle C\rangle$
$\Omega'_O p_2$		$\Omega'_{Ad} p_2$		p_2	
$\Omega'_O p_3$		$\Omega'_{Ad} p_3$		p_3	
$\Omega'_O p_4$		$\Omega'_{Ad} p_4$		p_4	$N \to M\langle C\rangle$
$\Omega'_O p_5$		$\Omega'_{Ad} p_5$		p_5	
$\Omega'_O p_6$		$\Omega'_{Ad} p_6$		p_6	

Control Operation Common to All Instructions

Subinterval Counter:

	Start Control:	
p_1	$P+1 \to P$	Qp_1
p_2	$P+1 \to P$	Qp_2
p_3	$P+1 \to P$	Qp_3
p_4	$P+1 \to P$	Qp_4
p_5	$P+1 \to P$	Qp_5
p_6	$1 \to P$	Qp_6

$$(W_P + W_R)' \cdot Q \to Q$$
$$W_P \cdot 0 + W'_P \cdot C \to C \qquad W_P \cdot 0 + W'_P \cdot F \to F$$

$$\Omega_O\Omega_{Ad} = f_3 \qquad \Omega'_O\Omega_{Ad} = f_5 \qquad \alpha = f_0 + f_7$$
$$\Omega' = f_0 + f_1 + f_2 + f_4 + f_6 + f_7 + f_8 + f_9 + f_{10} + f_{11} = (f_3 + f_5)' = f'_3 f'_5$$

Instruction Interpretation

$Q'\alpha\Omega' p_1$	$C \to D$	$Q'f_1\Omega' p_1$
$Q'\alpha\Omega' p_2$	$D+1 \to D$	$Q'f_1\Omega' p_2$
$Q'\alpha\Omega' p_3$		$Q'f_1\Omega' p_3$
$Q'\alpha\Omega' p_4$		$Q'f_1\Omega' p_4$
$Q'\alpha\Omega' p_5$		$Q'f_1\Omega' p_5$
$Q'\alpha\Omega' p_6$	Ad$(N) \to C \qquad$ Op$(N) \to F \qquad \sigma(N) \to T$	$Q'f_1\Omega' p_6$

ad X

$$S(N,A) \to A \qquad \lambda(N,A) \to A_{of} \qquad 0 \to F \qquad D \to C$$

su X

$$Q'f_2\Omega'p_1|$$
$$Q'f_2\Omega'p_2|$$
$$Q'f_2\Omega'p_3| \qquad N' \to N$$
$$Q'f_2\Omega'p_4|$$
$$Q'f_2\Omega'p_5|$$
$$Q'f_2\Omega'p_6| \qquad S(N,A) \to A \qquad \lambda(N,A) \to A_{of} \qquad 0 \to F \qquad D \to C$$

st X

$$Q'f_3\Omega p_1|$$
$$Q'f_3\Omega p_2|$$
$$Q'f_3\Omega p_3| \qquad A \to N$$
$$Q'f_3\Omega p_4|$$
$$Q'f_3\Omega p_5|$$
$$Q'f_3\Omega p_6| \qquad 0 \to F \qquad D \to C$$

jn X

$$Q'f_4\Omega'p_1| \qquad D \to \mathrm{Ad}(A)$$
$$Q'f_4\Omega'p_2|$$
$$Q'f_4\Omega'p_3|$$
$$Q'f_4\Omega'p_4|$$
$$Q'f_4\Omega'p_5|$$
$$Q'f_4\Omega'p_6| \qquad 0 \to F \qquad A_0 \cdot C + A'_0 \cdot D \to C$$

rf X

$$Q'f_5\Omega'_0\Omega_{\mathrm{Ad}}p_1|$$
$$Q'f_5\Omega'_0\Omega_{\mathrm{Ad}}p_2|$$
$$Q'f_5\Omega'_0\Omega_{\mathrm{Ad}}p_3| \qquad \mathrm{Ad}(A) \to \mathrm{Ad}(N)$$
$$Q'f_5\Omega'_0\Omega_{\mathrm{Ad}}p_4|$$
$$Q'f_5\Omega'_0\Omega_{\mathrm{Ad}}p_5|$$
$$Q'f_5\Omega'_0\Omega_{\mathrm{Ad}}p_6| \qquad 0 \to F \qquad D \to C$$

jo X

$$Q'f_6\Omega'p_1| \qquad D \to \mathrm{Ad}(A)$$
$$Q'f_6\Omega'p_2|$$
$$Q'f_6\Omega'p_3|$$
$$Q'f_6\Omega'p_4|$$
$$Q'f_6\Omega'p_5|$$
$$Q'f_6\Omega'p_6| \qquad 0 \to F \qquad A_{of} \cdot C + A'_{of} \cdot D \to C$$

ju X

$$Q'f_7\Omega'p_1| \qquad D \to \mathrm{Ad}(A)$$
$$Q'f_7\Omega'p_2|$$
$$Q'f_7\Omega'p_3|$$
$$Q'f_7\Omega'p_4|$$
$$Q'f_7\Omega'p_5|$$
$$Q'f_7\Omega'p_6|$$

Table 9-1. Summary of Computer Transfer Relations (Continued)

cr n

$Q'f_8\Omega'p_1\|$	$t_0 \cdot A + t'_{0P}(A) \to A$	$t_0 \cdot T + t'_0(T-1) \to T$
$Q'f_8\Omega'p_2\|$	$t_0 \cdot A + t'_{0P}(A) \to A$	$t_0 \cdot T + t'_0(T-1) \to T$
$Q'f_8\Omega'p_3\|$	$t_0 \cdot A + t'_{0P}(A) \to A$	$t_0 \cdot T + t'_0(T-1) \to T$
$Q'f_8\Omega'p_4\|$	$t_0 \cdot A + t'_{0P}(A) \to A$	$t_0 \cdot T + t'_0(T-1) \to T$
$Q'f_8\Omega'p_5\|$	$t_0 \cdot A + t'_{0P}(A) \to A$	$t_0 \cdot T + t'_0(T-1) \to T$
$Q'f_8\Omega'p_6\|$	$t_0 \cdot A + t'_{0P}(A) \to A$	$t_0 \cdot T + t'_0(T-1) \to T$

$$D \to C \qquad (t_0 + t_1)' \cdot F + (t_0 + t_1) \cdot 0 \to \boldsymbol{F}$$

ha

$Q'f_9\Omega'p_1\|$	
$Q'f_9\Omega'p_2\|$	
$Q'f_9\Omega'p_3\|$	
$Q'f_9\Omega'p_4\|$	
$Q'f_9\Omega'p_5\|$	
$Q'f_9\Omega'p_6\|$	$1 \to Q \qquad D \to C \qquad 0 \to F$

rt

$f_{10}\Omega'p_1\|$	$0 \to J_R$	$J_R \to Q$	
$f_{10}\Omega'p_2\|$			
$f_{10}\Omega'p_3\|$			
$f_{10}\Omega'p_4\|$			
$f_{10}\Omega'p_5\|$			
$f_{10}\Omega'p_6\|$	$Q \cdot F + Q' \cdot 0 \to F$	$Q' \cdot H + Q \cdot \text{Op}(A) \to \text{Op}(A)$	$Q' \cdot D + Q \cdot C \to C$

pt

$f_{11}\Omega'p_1\|$	$J'_P \to Q$	$0 \to J_P$
$f_{11}\Omega'p_2\|$	$Q' \cdot \text{Op}(A) + Q \cdot H \to H$	
$f_{11}\Omega'p_3\|$		
$f_{11}\Omega'p_4\|$		
$f_{11}\Omega'p_5\|$		
$f_{11}\Omega'p_6\|$	$Q' \cdot 0 + Q \cdot F \to F$	$Q' \cdot D + Q \cdot C \to C$

Another example, exhibiting somewhat more complexity, occurs in the realization of the A register. Collecting together all of the transfers into A, we obtain the sequence:

$$Q'p_1| \quad (f_4 + f_6 + f_7) \cdot D \to \text{Ad} \ (A) \qquad f_8 t'_0 \rho(A) \to A$$
$$Q'p_2| \qquad\qquad\qquad\qquad\qquad\qquad f_8 t'_0 \rho(A) \to A$$
$$Q'p_3| \qquad\qquad\qquad\qquad\qquad\qquad f_8 t'_0 \rho(A) \to A$$
$$Q'p_4| \qquad\qquad\qquad\qquad\qquad\qquad f_8 t'_0 \rho(A) \to A$$
$$Q'p_5| \qquad\qquad\qquad\qquad\qquad\qquad f_8 t'_0 \rho(A) \to A$$
$$Q'p_6| \quad (f_1 + f_2) \cdot S(N,A) \to A \qquad f_8 t'_0 \rho(A) \to A \qquad f_{10} H \to \text{Op} \ (A)$$

Since the transfers into A are partially dependent upon breaking A into subregisters, it is most convenient in deriving the flip-flop equations to treat A as the subregisters Op (A) and Ad (A). Considering Op (A), we obtain

$$\lambda_1 \cdot \text{Op} \ [\rho(A)] + \lambda_2 \cdot \text{Op} \ [S(N,A)] + \lambda_3 \cdot H + (\lambda_1 + \lambda_2 + \lambda_3)' \cdot \text{Op} \ (A)$$
$$\to \text{Op} \ (A)$$

Similarly, for Ad (A) we obtain

$$\lambda_1 \cdot \text{Ad} \ [\rho(A)] + \lambda_2 \cdot \text{Ad} \ [S(N,A)] + \lambda_4 \cdot D + (\lambda_1 + \lambda_2 + \lambda_4)' \cdot \text{Ad} \ (A)$$
$$\to \text{Ad} \ (A)$$

where
$$\lambda_1 = f_8 t'_0 Q'$$
$$\lambda_2 = (f_1 + f_2) p_6 Q'$$
$$\lambda_3 = f_{10} p_6 Q'$$
$$\lambda_4 = (f_4 + f_6 + f_7) p_1 Q'$$

It therefore follows that if A is realized in GK flip-flops, the input equations are:

$$K(A_j) = \lambda_1 + \lambda_2 + \lambda_3 \qquad\qquad j = 0, 1, \ldots, 5$$
$$G(A_j) = \lambda_1 \cdot A_{j-1} + \lambda_2 S_j(N,A) + \lambda_3 \cdot H_j \qquad j = 0, 1, \ldots, 5$$
$$A_{-1} \equiv A_{17}$$

and
$$K(A_j) = \lambda_1 + \lambda_2 + \lambda_4 \qquad\qquad j = 6, 7, \ldots, 17$$
$$G(A_j) = \lambda_1 A_{j-1} + \lambda_2 S_j(N,A) + \lambda_4 D_j \qquad j = 6, 7, \ldots, 17$$

To complete the logic design of the computer, it is necessary only to combine the input relations for each register in a similar fashion. Once this step has been completed, the equations may be translated into hardware. If a particular logical expression yields electronic circuitry which is awkward or excessively complex, an alternate method of arranging the transfers may be attempted, in order to find a simpler circuit configuration. In this way, hardware considerations usually strongly influence the logic and machine design. In fact a system design parameter which results in an excessively complex implementation is often relaxed or eliminated to yield a more reasonable circuit realization.

9-16. Program Units. The discussion of the preceding sections has been concerned with the design of a rather simple machine. The machine chosen as the example was kept simple to illustrate design techniques. There are, however, certain features of general-purpose computers, alluded to at the beginning of the chapter but omitted from the example, which are important enough to warrant further discussion. These relate to the program and control units.

First, let us introduce the notion of indexing and index registers. Almost all modern computers have index registers because their use greatly facilitates the writing of iterative programs. Suppose that the computer in the preceding sections of this chapter were to be used to add 100 numbers stored in registers 1001 through 1100. The simplest program consists of the sequence:

$$
\begin{array}{ll}
\text{st} & 1000 \\
\text{su} & 1000 \\
\text{ad} & 1001 \\
\text{ad} & 1002 \\
\multicolumn{2}{c}{\cdots\cdots} \\
\text{ad} & 1099 \\
\text{ad} & 1100 \\
\text{ha} &
\end{array}
$$

While this program is straightforward, it is lengthy.† Another sequence performing the same computation is as follows:

500	st	1000			
501	su	1000	600	temporary storage (initial value 0.000–0)
502	ad	600	601	0.00 · · · 1
503	ad	1001	602	ad	1100
504	st	600			
505	su	600			
506	ad	503			
507	ad	601			
508	st	503			
509	su	602			
510	jn	500			
511	ha				

This program requires 12 instructions stored in registers 500 to 511 and three constants as opposed to the preceding program requiring 103 instructions and no constants. The program, while shorter, is considerably more complex. Register 600 is used to store the sum. Register 503 is modified by the program to contain successively ad 1001, ad 1002, . . . , ad 1100, thus performing the iterated program.

---

† Note the awkward method of clearing the A register. See Prob. 9-1 for a more convenient method.

The use of an index register allows the iteration of a program without modifying an instruction in memory as was done in the above program. Suppose the computer contains an additional 12-cell register S, which we refer to as an *index* register. We define two new instructions which make use of S: "set index to n" (in n), and "jump on index to X" (ji X). The first instruction simply inserts the integer n into S. The second instruction is a conditional jump; S is first counted down by 1, and then the jump occurs if the value of S after the subtraction is not 0.

Any machine instruction which defines an operation involving register X is *indexed* by referring the operation, not to X, but rather to X minus the value m of register S. Thus "indexed ad X" means add the value of register $X - m$ to A and store the result in A. With the index register, the sample program simplifies to seven instructions.

500	in	100
501	st	1000
502	su	1000
503	ad(index)	1101
504	st	600
505	ji	503
506	ha	

The first time the program is performed, the indexed instruction is interpreted as ad $(1101 - 100) = $ ad 1001; the 100th time, S has value 1, the indexed instruction is interpreted as ad 1100, the jump on index instruction counts S from 1 to 0, and the program halts.

In order to implement the indexing of an instruction, an additional bit is necessary in the instruction format. Let $B(X)$ designate the fact that the instruction stored in register X is to be indexed. Then during p_6 of the instruction interpretation cycle, the transfer Ad $(N) \to C$ is replaced by

$$B'(N) \text{ Ad } (N) + B(N)\{\text{Ad } (N) - S\} \to C$$

The original transfer remains if $B(N)$ has the value 0; if $B(N) = 1$ the indicated subtraction is performed.

The simplification of the sample program, made possible by the incorporation in the machine of an index register, was quite dramatic. Index registers have, in general, been found to be so useful that several are provided in most computers. Here the index subregister of an instruction register must, of course, be large enough to address the set of index registers. Further, it is often convenient to change the value of an index register by integers other than 1, the value used above, and some machines provide such variability.

9-17. Generalized Control Centers. In Secs. 8-1 and 9-5 the control unit was defined in a general way. We have seen examples of control units in the elementary machines of Chap. 8 and in the general-purpose computer of this chapter. Further examples will be shown in the special-purpose computers described in Chap. 10. In this section, we shall expand the discussion of control units to include more general structures. The starting point of this discussion will be the computer of this chapter. Many of the remarks, however, are applicable to special-purpose as well as to general-purpose computers.

As we have pointed out previously, the control unit assumes a sequence of states which are in correspondence with transfers or sets of transfers in the computer. One class of transfers determines the operation of the machine proper, and another class defines the operation of the control itself. Thus in the computer of this chapter, the F and P registers form the control unit, and all the remaining registers the machine proper. The states of F and P define the transfers for the rest of the machine and for themselves. The machine, however, is quite simple, and the control unit is correspondingly simple. In more complex machines, it may be desirable to build more flexible control units, units which permit additions and changes to be incorporated more readily. We shall examine methods of achieving this by expanding our discussion of the control of the computer of this chapter.

The implementation chosen for the control unit of the computer was simple and straightforward. Each instruction had the duration of an integral number of memory cycle intervals, and each of the latter was further subdivided into six subintervals. The memory cycle intervals were in correspondence with control states, and the subintervals with substates. In fact, all instructions except the unconditional jump (ju) made use of two control states, one of which was distinct for each instruction and the other of which was common to all. The ju instruction used just one control state.

In Fig. 9-4 we show the control state diagram, omitting for simplicity the start-stop control.

When in state 0 (characteristic function f_0), a transition is possible to all other states. This must be true since state 0 corresponds to the instruction interpretation cycle, after which any instruction may be initiated. State 7 corresponds to the unconditional jump instruction, during the performance of which the next instruction in sequence is interpreted, and hence any of the states 1 to 11 inclusive may follow. States 1 to 6 and 9 are always followed unconditionally by state 0. States 8, 10, and 11 are either repeated or followed by state 0. When in state 0 or 7, the next state is either state 0 or a repetition of the same state. All of this was summarized in Sec. 9-15 in the general transfer equation

$$\lambda \cdot \mathrm{Op}\ (N) + \mu(F,X) \cdot 0 + (\lambda + \mu)' \cdot F \rightarrow F$$

where the next state is shown explicitly to be Op (N), 0, or F, a repetition of the current state. The functions λ and μ are scalar functions of F and certain other cells collectively designated by X.

Thus, to construct the control, the F register required $1 + k$ states, where k is the number of instructions, or correspondingly $\log_2 (1 + k)$ bits of storage. The substate register P required γ states or $\log_2 (\gamma)$ cells. To add another instruction of the same complexity as the existing instructions, it is necessary to add a single control state to F and use the

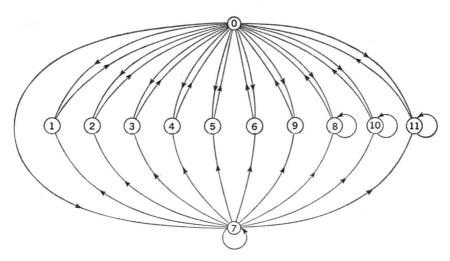

FIG. 9-4. Control state diagram.

characteristic function of that state together with those of P to initiate the transfers for the new instructions. Suppose, however, that a new instruction is to be added which requires the use of three distinct states, say 12, 13, and 0 in that order, following the state sequence

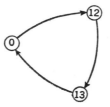

This requires a modification of the F register input equation to include the constant 13 as a possible input. In the general case where each instruction requires an arbitrary number of states and any given state may, in general, be used in any instruction, the state diagram becomes complex, with possibly all connections allowable, and F becomes a general class B sequential circuit with n storage cells and 2^n states. The transfer relation is, then [temporarily letting Op (N) be of dimension n],

$$\lambda \cdot \text{Op } (N) + \sum_{j=0}^{2^n-1} \mu_j \cdot K_j + \Big(\sum_{j=0}^{2^n-1} \mu_j + \lambda \Big)' \cdot F \to F$$

where the possible inputs to F are seen to be Op (N), F, and the states K_j, $j = 0, 1, \ldots, 2^n - 1$. The most straightforward method of implementing this circuit is to store the 2^n n-bit constants K_j in a combinational network together with the control functions μ_j. With the above relation each instruction is now a "program" consisting of a sequence of states K_j (the "instructions" of the program), with λ and μ_j specifying the program sequence. Storage of the K_j and μ_j in combinational circuits results in a fixed control program. If we now imagine the state n-tuples K_j to be stored in a memory which we designate Z, then the above relation becomes

$$\lambda \cdot \text{Op } (N) + \mu \cdot Z(F,X) + (\lambda + \mu)'F \to F$$

where
$$\mu = \sum_{j=0}^{2^n-1} \mu_j$$

The relation, when written in this way, shows three possible sources of the next state of F: Op (N), the set Z addressed by F and X, and F itself.

The set $Z(F,X)$ contains up to 2^n dependent registers, all of which are the distinct values which F can assume. The selection process represented by $Z(F,X)$ is potentially a very complex many-to-one configuration. A simpler selection process results if the number of registers in Z is allowed to increase. Suppose that the machine has k instructions $I_1, I_2, \ldots, I_k$, with instruction I_j having r_j states, resulting in a total of $r = \sum_{j=1}^{k} r_j$ states. Let Z be a set of at least r registers storing the r states and let G be the address register of the memory Z. The r states are not in general distinct. Thus, Z may contain many more than 2^n registers and G more than n cells. Part of G, which we call Op (G), contains $\log_2 k$ cells and receives operation codes from Op (N). Let $Y(G)$ be the remainder of G. Then the transfers governing control are

$$\lambda\{\text{Op } (N), 0\} + \mu Z\langle G\rangle + (\lambda + \mu)'\{\text{Op } (G), Y(G)\} \to [\text{Op } (G), Y(G)]$$

Thus G is either changed by the occurrence of a new instruction or receives information from the memory $Z\langle G\rangle$. Hence each instruction is executed by the performance of the sequence of control states stored in the control memory Z. The transfers in the remainder of the machine are governed by the set of states stored in Z. Thus, the internal structure of the machine is stored in the control memory Z. This structure may be modified by modifying the program of microinstructions stored in Z, that is, by constructing Z in such a way that its contents may be changed.

To take best advantage of such a programmable control, the structure of the words in Z can be made to correspond to the transfers in the machine. Thus, one subregister of each register Z^i in Z, called $G(Z^i)$, determines control operation. Other cells in Z^i either singly or in groups are in one-to-one correspondence with transfers. Thus, for example, one cell, say Z_j^i, may be placed in correspondence with a particular transfer, say $A \rightarrow N$. With this association, to select the transfer $A \rightarrow N$ in any instruction, the cell Z_j^k of register Z^k for that instruction takes on the value 1. Hence, to create a new instruction which makes use of existing transfers, one need only set up the appropriate registers in Z in the configuration to initiate the transfers for the instruction.

All the generalizations of control described here have the effect of making the machine more easily modified, and hence more flexible. This is achieved at the cost of more equipment. Thus, for example, a control memory configuration which is programmable is effective only if the transfers to be used in a modified instruction are already built in.

These remarks on flexible control units and microprogramming have been made within the context of a general-purpose computer. The notions, however, are quite general, and are applicable to any digital machine. More details and examples may be found in Refs. 14 and 21 to 23.

PROBLEMS

9-1. Design a "clear and add" instruction for the computer which transfers the value of a memory register to the accumulator.

9-2. The transfer $D \rightarrow \text{Ad}\,(A)$ occurs during the execution of each jump instruction of the computer designed in Chap. 9. This has the effect of destroying the previous contents of A. Indicate another method of preparing for subroutines which leaves A undisturbed.

9-3. Write a set of transfers defining the instruction "shift right n," having the effect of shifting A to the right n times, preserving the sign of A.

9-4. Indicate how the punch tape instruction might be simplified to eliminate the overlap feature.

9-5. Describe the modifications necessary to incorporate a single index register composed of flip-flops into the computer.

9-6. Suppose that it is desired to incorporate an index register into the computer without adding an additional flip-flop register. Describe a method for doing this which uses a fixed location in the main memory as an index register.

9-7. Consider a "subtract magnitudes X" instruction which computes $|a| - |x|$, where a and x are the numbers stored in A and X, respectively. This instruction requires an initial comparison of A and X, followed in some cases by a change of sign of the result (see Prob. 8-4).

(a) Design a "subtract magnitudes" instruction which requires two cycles. The first cycle is used for the initial comparison, the second cycle for the possible change of sign as well as for interpretation of the next instruction.

(b) Design a "subtract magnitudes" instruction which requires three cycles. This differs from part a in that the sign change and interpretation of the next instruction occur in consecutive cycles.

9-8. Describe the input equations to the D register under the assumption that RS flip-flops are used.

9-9. Consider a somewhat more flexible control for the computer of this chapter. Let $Z(F)$ be a function of control register F and let the characteristic function $z_i p_j$ be used rather than the $f_i p_j$ as the control functions for the machine. Assume that each transfer is initiated by a distinct bit of Z.

(a) Select a representation of Z to initiate the transfers necessary for the ad X instruction.

(b) Design a combinational network to realize the representation of (a).

(c) Extend (a) and (b) to include the su X instruction.

(d) Discuss how $Z(F)$ may be transformed into a memory W with output register Z and address register F.

REFERENCES

1. Bartee, T. C.: "Digital Computer Fundamentals," McGraw-Hill Book Company, Inc., New York, 1960.
2. Blaauw, G. A.: Indexing and Control-word Techniques, *IBM J. Research and Develop.*, vol. 3, pp. 288–301, July, 1959.
3. Brooks, E. P., G. A. Blaauw, and W. Buchholz: Processing Data in Bits and Pieces, *IRE Trans. on Electronic Computers*, vol. EC-8, pp. 118–124, June, 1959.
4. Dinneen, G. P., I. L. Lebow, and I. S. Reed: The Logical Design of CG24, *Proc. Eastern Joint Computer Conf.*, July, 1959.
5. Dreyfus, P.: System Design of the Gamma-60, *Proc. Western Joint Computer Conf.*, pp. 130–132, May 6–8, 1958.
6. Eckert, J. P., J. C. Chu, A. B. Tonil, and W. F. Schmitt: Design of Univac-Larc System I, *Proc. Eastern Joint Computer Conf.*, pp. 59–65, Dec. 1–3, 1959.
7. Flores, I.: "Computer Logic," Prentice-Hall, Inc., Englewood Cliffs, N.J., 1960.
8. Frankel, S. P.: On the Minimum Logical Complexity Required for a General Purpose Computer, *IRE Trans. on Electronic Computers*, vol. EC-7, pp. 282–285, December, 1958.
9. Frankel, S. P.: "Logical Design of a Simple General Purpose Computer, *IRE Trans. on Electronic Computers*, vol. EC-6, pp. 5–14, March, 1957.
10. Frankovich, J. M., and H. P. Peterson: A Functional Description of the Lincoln TX-2 Computer, *Proc. Western Joint Computer Conf.*, pp. 146–155, Feb. 26–28, 1957.
11. Holland, J.: A Universal Computer Capable of Executing an Arbitrary Number of Subprograms Simultaneously, *Proc. Eastern Joint Computer Conf.*, pp. 108–113, Dec. 1–3, 1959.
12. Jeenel, J.: "Programming for Digital Computers," McGraw-Hill Book Company, Inc., New York, 1959.
13. Lawless, W. J., Jr.: Developments in Computer Logical Organization, in "Advances in Electronics and Electron Physics," vol. 10, pp. 153–183, Academic Press, Inc., New York, 1959.

14. Lebow, I. L.: Communications in Digital Systems, *Proc. Fourth Intern. Symposium on Inform. Theory*, Butterworth & Co. (Publishers), Ltd., London, 1961.

15. Lourie, N., H. Schrimpf, R. Reach, and W. Kahn: Arithmetic and Control Techniques in a Multiprogram Computer, *Proc. Eastern Joint Computer Conf.*, pp. 75–81, Dec. 1–3, 1959.

16. McCracken, D. D.: "Digital Computer Programming," John Wiley & Sons, Inc., New York, 1959.

17. Mercer, R. J.: Micro-programming, *J. Assoc. for Computing Machinery*, vol. 4, pp. 151–157, April, 1957.

18. Phister, Montgomery: "Logical Design of Digital Computers," John Wiley & Sons, Inc., New York, 1958.

19. Reed, I. S.: "Some Mathematical Remarks on the Boolean Machine," MIT Lincoln Laboratory, Lexington, Mass., Rept. TR-2, Dec. 19, 1951.

20. Reed, I. S.: "Symbolic Design of Digital Computers," MIT Lincoln Laboratory, Lexington, Mass., Rept. TM-23, Jan. 19, 1953.

21. Reed, I. S.: "Symbolic Design Techniques Applied to a Generalized Computer," MIT Lincoln Laboratory, Lexington, Mass., Tech. Rept. 141, Jan. 3, 1957.

22. Wilkes, M. V., and J. B. Stringer: Micro-programming and the Design of Control Circuits in an Electronic Digital Computer, *Proc. Cambridge Phil. Soc.*, April, 1953.

23. Wilkes, M. V.: Micro-programming, *Proc. Eastern Joint Computer Conf.*, pp. 18–20, Dec. 3–5, 1958.

24. "On the Design of a Very High-speed Computer," pp. 180–187, Digital Computer Laboratory Rept. 80, University of Illinois, Urbana, October, 1957.

10

Special-purpose Computers

10-1. Introduction. In previous chapters, especially Chaps. 7 and 8, a number of special-purpose computing devices were discussed. However, these machines were generally elements of some larger computing machine. For example, the parallel adder, as described in Sec. 8-4, is a special-purpose computer which can be considered to be a component of a large class of potential computers.

In this chapter the concept of a special-purpose computer is extended to include an entire machine which is capable of computing some special classes of mathematical problems or of handling the data from some larger system in some specialized manner. Usually, it is a component of some control or data-processing system; it performs a routine class of tasks as an integral member of the system.

For the sake of economy, light weight, small size, or simplicity in design, the special-purpose machine is often fixed-program. If the machine is changeable, it is general-purpose only in the sense that it could be programmed to accomplish a particular task within some specialized class of functions. In many cases a change of purpose can be achieved only by a physical change of the connections within the machine, e.g., by plug boards or by an actual rewiring of connections.

It would be difficult to list or even classify the variety of possible special-purpose computers. If one includes electromechanical or digital-analog hybrid machines, the class becomes very large. For then the class includes such diverse machines as automatic or semiautomatic telephone exchanges, missile guidance computers, automatic milling machines, computers, oil refinery control computers, accounting and banking computers, automatic radar detection and processing machines, and so on.

Because of the size of this class of possible specialized computers, we limit our attention in this chapter to two examples. The first is an automatic radar detection and processing machine. This machine was designed and constructed by members of the MIT Lincoln Laboratory in 1954 and early 1955. Its purpose was to detect automatically aircraft returns seen by an air search radar, to estimate the position of a return, and finally to prepare the information concerning aircraft radar returns for transmittal over a digital data link. This machine was called the Fine Grain Data (FGD) computer; it was an operational prototype of part of the FST-2 system, the radar-site computer of the SAGE (Semiautomatic Ground Environment) system.

The second example of a special-purpose computer is the Digital Differential Analyzer (DDA). This class of machines was first developed at Northrop Aircraft, Inc., during the years 1947 through 1950.[7,13,21] As a computing machine to analyze ordinary nonlinear differential equations, this machine has shown little advantage over the more versatile high-speed general-purpose computer, such as discussed in the last chapter. However, as an accurate control element for the automatic guidance of aircraft or missiles, the DDA is often superior to the more general machine. For the same number of components, the DDA, as a control element, can be designed usually to have a shorter response time and a greater capacity than the general-purpose computer would have for a similar application. Consequently, for an application which involves only the solution of ordinary differential equations (requiring an accuracy greater than can be achieved with an analog differential analyzer), the DDA has possibly the least weight and complexity of any machine which can be developed for the purpose.

The first example of a special-purpose computer will be discussed in more detail than the second. In the first example, we develop the design details of a particular machine that was actually constructed and operated. For the second example we design only a hypothetical example from the class of possible DDA's and demonstrate the method of encoding such a machine to analyze ordinary differential equations.

10-2. Automatic Radar Target Detection. The FGD is a system which automatically detects radar targets and transmits their polar coordinates to a remote location over a digital data link. The somewhat peculiar name of the system stems from the fact that the azimuth information obtained is considerably more accurate than that achieved in an earlier automatic radar detection system.†

The radar is assumed to be an air search type having an antenna beam with narrow azimuth cross section and broad elevation cross section, thus producing the fanlike pattern shown in Fig. 10-1. The antenna is assumed to rotate continuously at a fixed rate, f_a rps. Accordingly, the azimuth angle θ varies with time as in Fig. 10-2a, where θ is defined to have the value 0° when the center of the beam points to the north.

† The principals of this development were E. W. Bivans, J. V. Harrington, and I. S. Reed.

The radar transmits narrow pulses ($\sim$1 μsec in duration) every T seconds. It therefore transmits $1/(360f_aT)$ pulses per degree of antenna rotation. If the width of the beam in azimuth is $B°$, then $B°/(360f_aT)$ pulses are transmitted in the time required for the antenna to rotate

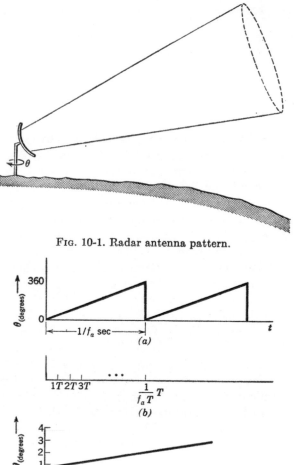

Fig. 10-1. Radar antenna pattern.

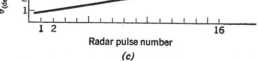

Fig. 10-2. Radar azimuth angle and pulse patterns.

one beam width. In a typical case, $B = 2.3°$, $f_a = \frac{1}{6}$ rps, and $T = 2.4$ msec, resulting in 16 radar pulses per beamwidth of rotation. The radar trigger pulses are plotted in Fig. 10-2b. Because of the correspondence in time between radar trigger pulses and antenna azimuth, one can measure azimuth by pulse number. An expanded section of Fig. 10-2*a* and *b* is shown in Fig. 10-2*c*. Here we see the angle θ plotted against pulse number, with about seven pulses per degree of rotation.

Suppose now that a target is strong enough to produce a detectable echo each time it is hit by a radar pulse. If it produces such echoes on 16 successive pulses as the antenna scans past the target, then the azimuth angle θ corresponding to the eighth such pulse is a good estimate of the true azimuth angle of the target. This technique of obtaining azimuth accuracy greater than that inherent in the antenna beamwidth by observing echoes from the several pulses occurring during a beamwidth has been called "beam splitting."

The other target coordinate measured by the radar is range. This is, of course, given by the time t at which the echo is received. A target at a range r will produce an echo at time $t_p + 2r/c$, where t_p is the time at which the pulse is transmitted and the time interval $2r/c$ is approximately $12r$ μsec if r is measured in nautical miles.

Only a very strongly reflecting target will result in an echo detectable by the radar receiver on every impinging pulse. This is due to variations in the effective reflectivity of targets and noise fluctuations in the receiver. In general, the output of the radar receiver is a function V, the value of which at time t is a function of the sum of the received signal from a possible target occurring at the range corresponding to that time and the receiver noise occurring at that time. The decision as to whether a target is present at time t (or range r) is based upon whether the value of $V(t)$ is sufficiently greater than its expected value in the presence of noise alone. The function $V(t)$ is often called the radar *video* output.

Thus the detection procedure in the radar receiver is basically that of discrimination between signal plus noise and noise alone. If a target is hit by n successive radar pulses and consequently produces echoes on these pulses, the receiver may base its detection procedure on the returns from the n pulses, with a corresponding improvement in signal detectability. Thus the fact that several returns are received from a single target is used both as an aid in target detection and as a means of improving the azimuth-angle accuracy of the observation.†

Given this brief background we are now in a position to describe the FGD detection and beam-splitting process. The signal $V(t)$ is "quantized" to obtain the Boolean function $V_j(r_i,\lambda)$ shown in Fig. 10-3 and defined as follows: the time base is divided into 6-μsec ($\frac{1}{2}$ nautical mile)

---

† We sketch here only enough of the qualitative notions of radar detection to motivate the description of the FGD computer. The classical paper in radar detection theory is that of Marcum (Ref. 11). Some insight into digital techniques is presented in Refs. 6 and 9. A basic paper describing the theory of beam-spliting is that of Swerling (Ref. 23). For the modern theory of radar detection see Ref. 10.

intervals, each of which is labeled by the range corresponding to the beginning of the interval. Then the function $V_j(r_i,\lambda)$ in the ith interval r_i has value 1 if V_j exceeds a threshold λ during that interval and has value zero otherwise. The index j designates the radar pulse for which V_j is the output signal. This *binary quantization* process, therefore, replaces the function $V(t)$ by a sequence of 0's and 1's which represent *tentative* decisions as to the absence or presence respectively of targets in $\frac{1}{2}$-mile *range boxes*. Such a sequence is obtained for each radar trigger pulse. The determination of the threshold λ will be discussed later.

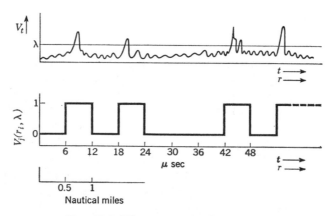

FIG. 10-3. Binary quantization process.

The final detection decision is based upon n successive sequences, where n is the number of pulses per beamwidth. Consider now a fixed range box r_i in which a very strongly reflecting target is located, so that $V_j(r_i,\lambda)$ has the value 1 for 16 successive values of j or on 16 successive pulses. This is plotted in Fig. 10-4a.

We also assume no target or noise returns on either side of the target in question. In Fig. 10-4b is shown the function $Y_j(r_i,\lambda)$, defined by

$$Y_j(r_i,\lambda) = \sum_{k=j-15}^{j} V_k(r_i,\lambda)$$

Y_j is simply the sum of the 16 successive values of V_k, ending with V_j. Also shown in the figure is a second threshold μ used to define another Boolean function $S_j(r_i,\lambda,\mu)$, which has the value 1 whenever $Y_j(r_i,\lambda) \geq \mu$ and value 0 if $Y_j(r_i,\lambda) < \mu$. In Fig. 10-4c μ was chosen to be 8, and S_j is 1 whenever Y_j is 8 or greater, and 0 otherwise. The function $S_j(r_i,\lambda,\mu)$ represents a final decision based upon 16 samples of the function V_j. Let S_j have the value 1 between radar pulses j_b and j_e respectively. Then the pulse defining the center of this interval is given by $\frac{1}{2}(j_b + j_e)$, and the azimuth angle θ of the target center is $\frac{1}{2}(\theta_b + \theta_e) - 8\Delta\theta$, where θ_b and θ_e are the antenna angles corresponding to j_b and j_e and $\Delta\theta$ is the increment of antenna rotation between radar pulses. $8\Delta\theta$ represents the fixed bias introduced by the detection process.

The above description is, of course, highly idealized. In a more realistic case V_j contains 1's due to noise alone and 0's in the target region. The thresholds λ and μ are chosen to maximize the probability of making a final detection on a true target, while keeping the probability of a final

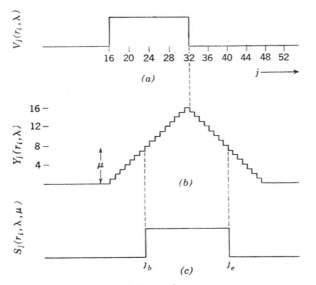

FIG. 10-4. Detection process.

detection of "noise" targets acceptably low. We shall assume, from now on, that λ and μ have been chosen and shall suppress explicit reference to them.

The operation of the FGD may now be summarized as follows: the returned signal $V_j(t)$ for each radar pulse is quantized in $\frac{1}{2}$-mile range boxes. In each range box r_i, the quantized signals are examined by a "moving-window" detector which measures target beginnings and endings as described above. From such measurements the azimuths of the target centers are computed. The target coordinates (range and azimuth) are then stored temporarily and are finally transmitted over a digital data link.

10-3. The FGD Detector. The first step in the design of any system is the specification of the basic timing of the major components of the system. This problem is somewhat more complex in a real time system such as the FGD, where the input and output data rates are determined by the environment of the computing system, and consequently the computing rate is subservient to these externally controlled factors.

The fastest clock necessary to the system is a "range mark" clock which produces pulses at 6 μsec intervals corresponding to $\frac{1}{2}$–nautical mile range boxes. To keep track of the range marks we utilize a register $R = (R_1, \ldots, R_9)$ which is counted according to the sequence

$$
\begin{aligned}
r_0| \quad & R + 1 \to R \\
r_1| \quad & R + 1 \to R \\
\cdots & \cdots \cdots \cdots \\
r_{398}| \quad & R + 1 \to R \\
r_{399}| \quad & 0 \to R
\end{aligned}
$$

The functions r_j are, of course, the characteristic functions of register R. The radar pulses will be assumed to occur at the end of the interval r_{399}. Thus the interval during which r_j has value 1 represents the range box between $j/2$ and $(j + 1)/2$ nautical miles. The range-mark counter R represents, therefore, a short-term control register for the FGD computer.

The detection process requires the comparison of the returns from the current radar pulse with those from the preceding 15 pulses. This comparison is performed in a register $W = (W_0, W_1, \ldots, W_{15})$. Suppose that the computer is now examining range box r_i on radar pulse j. Then, using the notation of the previous section, $V_j(r_i)$ has value 1 if the quantizer judges that the receiver output was sufficiently large during the range box r_i and 0 otherwise. Letting the quantizer be designated by V, the transfer

$$
r_i| \quad V \to W_0
$$

has the effect of inserting in W_0 the quantizer decision for box r_i on the current (jth) radar pulse. In order to perform the comparison with returns from the preceding 15 pulses, the above transfer must be accompanied by the transfers

$$
r_i| \quad V_{j-1}(r_i) \to W_1 \qquad V_{j-2}(r_i) \to W_2 \quad \cdots \quad V_{j-15}(r_i) \to W_{15} \qquad (1)
$$

The storage required to preserve the past values of quantized video necessary for these transfers is provided on the FGD by a magnetic drum which behaves like a set of shift registers and is synchronized to the radar pulses.[2] The drum, sketched schematically in Fig. 10-5, contains a number of channels, some of which are designated "short" channels and the remainder of which are referred to as "long" channels. Each channel contains a "read" and a "write" head positioned such that the drum must rotate about 90° to bring a spot from the write head to the read head on a short channel, and 270° on a long channel.

Fifteen of the short channels designated C^1, C^2, . . . , C^{15} are used to provide the necessary video storage. Each channel C^k, $k = 1, 2, . . . ,$ 15, is provided with an input flip-flop $\hat{W}_k$ and an output flip-flop W_k (the 15 rightmost cells of the decision register W). The channels are connected effectively in series by the transfers

$$r_i| \quad W_0 \rightarrow \hat{W}_1 \qquad W_1 \rightarrow \hat{W}_2 \qquad \cdots \qquad W_{14} \rightarrow \hat{W}_{15} \qquad (2)$$

or

$$r_i| \quad L(W) \rightarrow \hat{W}$$

where $L(W)$ is the 15-cell register W_0, W_1, . . . , W_{14}. The timing requirement of Eq. (1) is then satisfied if each video channel of the drum

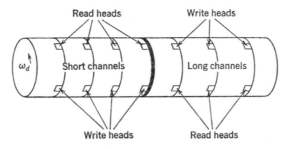

Fig. 10-5. Drum layout for FGD.

(including the input and output flip-flops) behaves like a delay line of length 400 range boxes (2.4 msec). This condition is fulfilled if the drum speed is synchronized to the radar pulse frequency in such a manner that $\frac{1}{4}$ revolution requires 2.4 msec, corresponding to a speed of approximately 6,250 rpm.

The drum operation may, therefore, be summarized by

$$r_i| \quad V \rightarrow W_0 \qquad C \rightarrow R(W) \qquad L(W) \rightarrow \hat{W} \qquad \hat{W} \rightarrow C$$

where the transfers $C \rightarrow R(W)$ and $\hat{W} \rightarrow C$ symbolize the read and write operations of the drum video channels respectively [$R(W) = (W_1, . . . , W_{15})$].

The threshold μ defined in the previous section is obtained as a function of the W register. At any given time register W contains the 16 latest samples of quantized video from the range box corresponding to that time. Let $\sigma(W)$ define a Boolean function which has value 1 if the integer value of W is μ or greater and has the value 0 at other times. Then at any time t the function $\sigma(W)$ indicates whether or not the range box in question contains a target. In particular a transition of $\sigma(W)$ from 0 to 1 indicates a target beginning and the transition from 1 to 0 specifies a target ending.

10-4. Azimuth Computation and Storage. Azimuth information is obtained directly from the antenna pedestal in the form of a sequence of *azimuth change pulses.* A total of 4,096 such pulses are generated for each revolution of the antenna, resulting in a conversion precision of 12 bits or an angular precision of 0.088° in transforming from antenna shaft position to a quantized value. The angular precision of the beam-splitting process is at best 0.16°, with a 2.3° beam and 16 radar pulses per beamwidth. Thus the 12-bit conversion represents sufficient precision for the measurement of angle. The azimuth change pulses are accumulated in a 12-cell register A which is cleared once per revolution as the antenna passes north. Thus the integer values of A are in correspondence with antenna bearing θ.

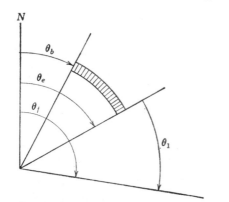

Fig. 10-6. Azimuth location of target.

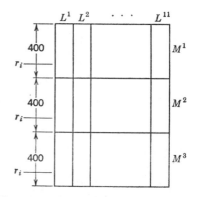

Fig. 10-7. Drum buffer-channel format.

The method of storing azimuth information and performing the beam-splitting computation takes advantage of the particular characteristics of the drum storage system. In Fig. 10-6 we show a target, detected between the angles θ_b and θ_e, having a measured bearing $\frac{1}{2}(\theta_e + \theta_b)$. When the target is first detected at θ_b, a register is set to value 0. Thereupon the register accumulates a value proportional to the angle of rotation until the target ending θ_e is detected. The register value then continues to increase but now proportionally to twice the angle of rotation. This continues until the target coordinates are ready for transmission. If the bearing is θ_f at this time, then the register has a value proportional to $2\theta_1 + \theta_e - \theta_b$. Finally this value is divided by 2 and subtracted from that of a register storing the azimuth $\theta_f - 8\Delta\theta$. The result of the subtraction is the true bearing of the target center defined in Sec. 10-2.

The accumulation and storage of the angle $\theta = 2\theta_1 + \theta_e - \theta_b$ is performed using 11 of the long channels of the drum, each of which is equivalent to a shift-register delay line of length $3T = 7.2$ msec, or $3 \times 400 = 1,200$ cells. This storage will be treated as three blocks

M^1, M^2, and M^3, each containing 400 eleven-cell registers, one cell from each of the 11 channels L^1, L^2, . . . , L^{11} (Fig. 10-7). Moreover, the ith register in each block will correspond to the range box r_i. Thus storage is provided for three targets for each range box.

When a target beginning is detected in range box r_i, one of the three azimuth registers at range r_i is cleared and assigned to that target. Thereupon each time that register is read out of the drum (every $3T$ seconds) the target to which it is assigned is examined. The register value is incremented by the number of azimuth change pulses occurring in the last $3T$ seconds until a target ending is detected. After that the value is incremented by twice the number of azimuth pulses occurring until the target is finally removed from memory to be transmitted over the data link.

10-5. Target Processing. We have, up until now, specified the use of 15 short channels for video storage and 11 long channels for target azimuth storage. In addition to these, 3 short channels and 2 long channels are used as status channels for the bookkeeping purposes of recording target beginnings and endings, assigning azimuth storage to targets, and controlling the readout operation. Let $\hat{S}_j$ and S_j ($j = 1, 2, 3$) denote the input and output buffer flip-flops, respectively, of the short status channels. We further designate the register $(\hat{S}_1, \hat{S}_2, \hat{S}_3)$ by $\hat{S}$ and (S_1, S_2, S_3) by S. The timing is such that status information regarding range box r_i appears in S at the same time as the video information for r_i appears in W. We further let $\hat{B} = (\hat{B}_1, \hat{B}_2)$ and $B = (B_1, B_2)$ be the input and output registers, respectively, for the long status channels and let $\hat{L} = (\hat{L}_1, \hat{L}_2, . . . , \hat{L}_{11})$ and $L = (L_1, L_2, . . . , L_{11})$ be similar buffers for the azimuth storage channels. The characteristic functions of S and B serve as control functions for the operation of the machine. The states of S are indicative of the status of incomplete targets, and those of B represent the status of the azimuth buffer storage.

Suppose, first, that the current range box does not contain a detected target, and let s_0 specify that condition. If no target beginning is detected, we then have the transfers

$$s_0 r_i \sigma'(W)| \quad L(W) \to \hat{W} \qquad S \to \hat{S}$$

which have the effect of advancing the video and maintaining the status information in S. If, on the other hand, a target beginning is detected, an azimuth storage channel must be assigned to the target. Let b_0 designate the fact that the current azimuth register is available for a new target. We then have

$$s_0 r_i \sigma(W) b_0| \quad 0 \to \hat{L} \qquad 4 \to S$$

where, clearly, the state s_4 will designate that a target was detected and an azimuth register assigned. If the current azimuth register is not available, we have

$$s_0 r_i \sigma(W) b_0' | \quad 5 \to \hat{S}$$

where the value 5 in S records this fact. In summary, the S register is governed by

$$s_0 r_i | \quad \sigma'(W) \cdot S + \sigma(W) b_0 \cdot 4 + \sigma(W) b_0' \cdot 5 \to \hat{S}$$

during state s_0.

Now consider the state s_5, designating the fact that a target has been detected but no azimuth register assigned. T seconds later this range box is again examined, and simultaneously the second azimuth register for the range box is examined. The transfer governing operation is now

$$s_5 r_i | \quad \sigma'(W) \cdot 0 + \sigma(W) b_0 \cdot 4 + \sigma(W) \cdot b_0' \cdot 6 \to \hat{S}$$

The condition $\sigma'(W)$ denotes that the target has ended before its azimuth could be recorded; it is discarded and S is cleared. If $\sigma(W) = 1$ and a register is available, $\hat{S}$ assumes stage 4, while if no register is available, $\hat{S}$ assumes state 6. Similarly, in state s_6, if an azimuth register is found, the machine goes to s_4, otherwise it goes back to s_0.

Thus the machine is in state s_4 while waiting for a target ending. Once an ending is found, the machine returns to s_0, directly if it finds the correct azimuth register immediately, or via states s_1 or s_1 and s_2 if it takes T or $2T$ seconds respectively to locate the appropriate azimuth register.

The detection and azimuth assignment are summarized by the flow chart of Fig. 10-8. The transfers defining this flow chart are

$$s_0 r_i | \quad \sigma b_0 \cdot 4 + \sigma b_0' \cdot 5 + \sigma' S \to \hat{S}$$
$$s_1 r_i | \quad b_1 \cdot 0 + b_1' \cdot 2 \to \hat{S}$$
$$s_2 r_i | \quad 0 \to \hat{S}$$
$$s_4 r_i | \quad \sigma \cdot S + \sigma' b_1 \cdot 0 + \sigma' b_1' \cdot 1 \to \hat{S}$$
$$s_5 r_i | \quad \sigma b_0 \cdot 4 + \sigma b_0' \cdot 6 + \sigma' \cdot 0 \to \hat{S}$$
$$s_6 r_i | \quad \sigma b_0 \cdot 4 + (\sigma' + b_0') \cdot 0 \to \hat{S}$$

Next we consider the azimuth storage registers themselves. Their operation is simply defined. Let D be a three-cell register which accumulates azimuth change pulses over three radar pulse intervals. When an azimuth register is assigned to a target, it is first cleared. This must occur on one of the three possible transitions into state 4, and is given by

$$b_0 \sigma (s_0 + s_5 + s_6) r_i | \quad 0 \to \hat{L}$$

Under the same control condition the transfer $1 \rightarrow B$ specifies that the register in azimuth storage has been assigned. Each time thereafter L is incremented by the value of D until the target end is reached; i.e.,

$$b_1 r_i| \quad L + D \rightarrow \hat{L}$$

Register B itself is placed in state 3, when a target ending is detected, by the transfer

$$b_1 \sigma'(s_4 + s_1 + s_2) r_i| \quad 3 \rightarrow B$$

As indicated above, once the target end has been reached, L is counted up at twice the azimuth pulse rate by the transfer

$$b_3 r_i| \quad L + 2D \rightarrow \hat{L}$$

Finally, when a target is removed from memory, the transfer $0 \rightarrow B$ indicates that the register is available for a new target.

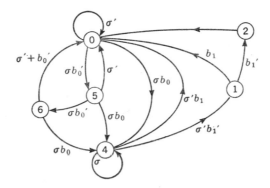

Fig. 10-8. Flow diagram of detection and azimuth assignment.

10-6. Azimuth Synchronization. To complete the description of the target detection and azimuth storage, we must indicate the operation of the D and A registers which count azimuth change pulses over different periods of time. All the transfers specified up to this point have been synchronous with the range mark clock. The antenna rotation rate, and correspondingly the azimuth change pulse rate, is not synchronous with the range mark clock. It turns out to be most convenient to synchronize the azimuth change pulses with the range mark clock before making use of these pulses. The result of the synchronization process is to delay each azimuth pulse until the occurrence of the first (or perhaps second) range mark following the azimuth pulse. To do this reliably and ensure that one and only one synchronized pulse is generated, two flip-flops Z_0 and Z are used. Suppose that an azimuth change pulse derived from the antenna pedestal initiates the transfer $1 \rightarrow Z_0$. Then the transfers

$$r_k| \quad Z_0 \to Z \qquad Z_0 \cdot Z' \to Z_0$$

result in the flip-flop Z having the value 1 for exactly one range interval shortly after the occurrence of an azimuth change pulse. The azimuth counter A is incremented after each azimuth change pulse by

$$Z| \quad A + 1 \to A$$

A is reset to 0 by a "north mark" pulse generated at the antenna pedestal whenever the antenna points close to north, as indicated in Sec. 10-4.

The D register stores azimuth change pulses over a $3T$-second interval. To do this we first need a counter $J = (J_1, J_2)$ which counts up to three radar pulses. This is accomplished by the transfers

$$r_{399}| \quad J + 1 \to J$$

where the addition is to be regarded as modulo 3. This may be written explicitly as

$$r_{399}| \quad J_1' \cdot (J + 1) + J_1 \cdot 0 \to J$$

Finally a counter $\hat{D}$ is incremented by

$$\begin{aligned}
r_k j_0'| &\quad Z \cdot (\hat{D} + 1) + Z' \cdot \hat{D} \to \hat{D} &\quad \text{all } k \\
r_k j_0| &\quad Z \cdot (\hat{D} + 1) + Z' \cdot \hat{D} \to \hat{D} &\quad k \neq 399 \\
r_{399} j_0| &\quad Z \cdot 1 + Z' \cdot 0 \to \hat{D}
\end{aligned}$$

These transfers specify that $\hat{D}$ is incremented by 1 each time a quantized azimuth pulse Z occurs and is reset after the J counter is reset. If a reset pulse and azimuth pulse should coincide ($r_{399} j_0 Z = 1$), $\hat{D}$ is reset to value 1 rather than to value 0. The maximum value of $\hat{D}$ is stored in D every 3 radar pulses by

$$r_{399} j_0| \quad \hat{D} \to D$$

10-7. The FGD Output. There remains to be discussed the method of removing detected targets from storage on the drum for transmission over a telephone line. Information from each target is encoded into a 32-bit word or message which is transmitted at a rate of 1,300 bits/sec. The format of the message is shown below.

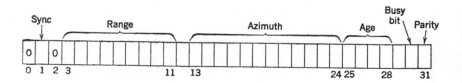

Bits 3 to 11 contain the range and 13 to 24 the azimuth of the target. In bits 25 to 28 an indication of the "age" of the target is transmitted. Bit 30 contains a 1 if the message contains target information, and 0 otherwise. Bit 1 contains a special sync symbol protected by 0's on either side in bits 0 and 2. Bit 31 is a so-called "odd-parity" bit. It has a value such that the total number of 1's in the word is odd. Bits 12 and 29 are not used. A block diagram of the output registers is shown in Fig. 10-9.

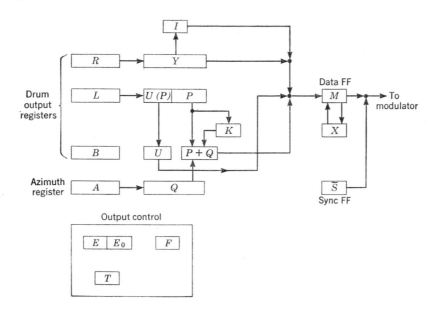

Fig. 10-9. Output registers.

The basic phone-line timing is generated by a 1,300-cps tuning-fork oscillator independent of the range mark clock. As in the case of the azimuth change pulses, the phone-line pulses are synchronized to the range marks by the use of two flip-flops E_0 and E. E_0 is set to value 1 by a phone-line pulse. Thereupon the transfers

$$r_k| \quad E_0 \to E \qquad E_0 \cdot E' \to E_0$$

ensure that E has value 1 for one and only one range interval following the occurrence of a phone-line pulse.

The synchronized phone-line pulses are counted in a five-cell register T to generate control functions for the output computations.

$$E| \quad T + 1 \to T$$

The integer values of T are associated with the digit numbers of Fig. 10-8 such that T has value j when bit j is being transmitted. Let $\alpha(T)$ be a scalar function of T having value 1 during the intervals t_{24} through t_{31} of a message and t_0 and t_1 of the following message and 0 at all other times. During the ten intervals = 7.7 msec that $\alpha(T)$ has the value 1, the drum azimuth channels are sampled and the first complete target (signified by $b = 3$) is read out of memory and prepared for transmission. Since the azimuth channels may all be sampled in 7.2 msec, the entire storage may be searched for a target during the interval $\alpha(T)$.

If a target is found in the drum, the range and azimuth information pertinent to that target is placed in temporary storage. More precisely, a flip-flop F directs this temporary storage by the transfer

$$E \cdot t_{23}| \quad 1 \to F$$

If a complete target is found during $\alpha(T)$, F is cleared; otherwise F is cleared at the end of $\alpha(T)$. Thus we have

$$r_j F(b_3 + t_1)| \quad 0 \to F$$
$$r_j F b_3| \quad 0 \to B$$

The first completed target ($b_3 = 1$) found in the buffer after F is set to 1 initiates a transfer of data to be described below. It further clears B, the latter signifying that the buffer register just sampled is now available for new information. The range information is held in an 11-cell register $Y = (Y_1, \ldots, Y_{11})$.

$$r_j F b_3| \quad R \to Y$$

The target azimuth information must be computed by subtracting one-half the value of L from that of A, the register containing the current azimuth (see Secs. 10-4 and 10-5). This is performed in two steps: first, the current azimuth is stored in a register $Q = (Q_1, \ldots, Q_{12})$ by

$$r_j F b_3| \quad A \to Q$$

and simultaneously one-half the 1's complement of L is stored in a 10-cell register $P = (P_1, \ldots, P_{10})$ by the transfer

$$r_j F b_3| \quad L(L') \to P$$

where $L(L')$ is the subregister $L_1', \ldots, L_{10}'$.

The transmission of the coordinates of this target is accomplished by shifting the appropriate functions of Y, P, and Q serially into a flip-flop M according to the timing indicated above. The phone-line modulator samples M for indication of the appropriate digit to be transmitted.

Thus the transfer of range information is governed by

$$t_2E| \quad L(Y,M) \rightarrow R(Y,M) \qquad 0 \rightarrow Y_1$$
$$t_3E| \quad L(Y,M) \rightarrow R(Y,M)$$
$$\cdots \cdots \cdots \cdots \cdots$$
$$t_{10}E| \quad L(Y,M) \rightarrow R(Y,M)$$

The azimuth is obtained by adding the values of P and Q with an initial carry of 1.   Define a carry flip-flop K set initially to 1 by

$$t_{11}E| \quad 1 \rightarrow K$$

The least significant digit of the target azimuth is obtained by

$$t_{12}E| \quad K \oplus P_{10} \oplus Q_{12} \rightarrow M$$

accompanied by the transfers

$$t_{12}E| \quad L(P) \rightarrow R(P) \qquad 0 \rightarrow P_1$$
$$L(Q) \rightarrow R(Q) \qquad 0 \rightarrow Q_1$$
$$K(P_{10} + Q_{12}) + P_{10}Q_{12} \rightarrow K$$

which prepare for computation of the next digit.   The above two sets of transfers are repeated during t_{13} through t_{23} until the entire azimuth word is transmitted.

Bits 25 through 28 are used to transmit an indication of the length of time the target remained in storage before being transmitted.   The value of the drum azimuth register is approximately proportional to twice this time, and the complement of one-half this value was transferred into P during the interval $\alpha(T)$.   Let $U(P)$ designate the leftmost four cells of P and let $U = (U_1, \ldots, U_4)$ be a four-cell register.   The transfer

$$t_2E| \quad U(P') \rightarrow U$$

recomplements the four most significant digits of this word.   These digits are transmitted after the transfers

$$t_{24}E| \quad L(U,M) \rightarrow R(U,M) \qquad 0 \rightarrow U_1$$
$$\cdots \cdots \cdots \cdots \cdots \cdots \cdots \cdots$$
$$t_{27}E| \quad L(U,M) \rightarrow R(U,M) \qquad 0 \rightarrow U_1$$

Message bit number 31, the "busy bit," is to have value 1 if the current message contains target information.   It may be determined in a number of ways.   One convenient method takes advantage of the fact that the first range box is gated out by the radar and hence never contains a target.   Thus the binary representation of the range of any target always contains at least a single 1.   Hence the transfers

$$t_2E| \quad Y_9 + I \to I$$
$$\cdots\cdots\cdots\cdots$$
$$t_{10}E| \quad Y_9 + I \to I$$
$$t_{29}E| \quad I \to M \qquad 0 \to I$$

record in cell I the presence of a single 1 in the range value when this function is transferred to M.

Finally the parity bit must be generated.   This is accomplished by the transfers

$$t_1E| \quad 1 \to X$$
$$t_3E| \quad M \oplus X \to X$$
$$\cdots\cdots\cdots\cdots$$
$$t_{29}E| \quad M \oplus X \to X$$
$$t_{30}E| \quad M \oplus X \to M$$

The synchronization bit is generated by

$$t_1E| \quad 1 \to \bar{S}$$
$$t_2E| \quad 0 \to \bar{S}$$

where flip-flop $\bar{S}$ is used to indicate the special sync pulse generator in the modulator.

10-8. Digital Differential Analyzers: The Digital Integrator.   A large class of problems involves the solution of differential equations.   Often closed-form solutions are not known and an analytic treatment involves a prohibitive amount of calculation.   A mechanical machine called a differential analyzer, specifically designed to handle such equations, was proposed by Lord Kelvin[6] in 1876, but over 50 years passed before the first successful version of a mechanical differential analyzer was designed and constructed by Vannevar Bush.[7]   Later, electronic versions of the differential analyzer, which used electrical signals continuous within intervals to represent values of the variables, were constructed, and these machines came to be loosely grouped under the title *analog computer*. In 1950 the first digital differential analyzer (DDA), the MADDIDA, was constructed.†   The DDA offers greater accuracy than the analog version; also the independent variable in an analog machine is invariably time, whereas the digital machine does not suffer from this restriction.

Our approach to the DDA will be as follows: first, the *integrator*, which is the basic unit of both digital and analog differential analyzers, is dis-

† MADDIDA was first described by F. G. Steele and D. E. Eckdahl at the ACM Meeting, Rutgers University, August, 1950.   Mathematical considerations were presented by I. S. Reed at the IRE Convention, Long Beach, Calif., December, 1950.

cussed.   A typical *digital* integrator consisting of several dependent and independent registers is then described.

After the integrator is introduced, a short description of the programming of a digital differential analyzer is presented.   Programming consists basically of three steps: (1) mapping, which concerns the interconnection of a set of integrators so that a given equation is realized; (2) scaling, which refers to the scaling of the variables in the problem and the representation of the variable values in machine form; and (3) coding, which concerns setting up the problem on a specific machine.

A simple DDA is then described using the symbolic notation of this and previous chapters.   This machine is made especially straightforward by the fact that it is a parallel machine.

The basic unit of both the analog and digital differential analyzers is the integrator.   An integrator can be defined as a device which accepts two input rates and delivers one output rate.   For analog devices, the two input rates, say dx/dt and dy/dt, are related to the output rate dz/dt by the expression

$$\frac{dz}{dt} = Cy\frac{dx}{dt}$$

where C is some constant.

The definite integral of the above relation is

$$z(t) = z(t_0) + C \int_{t_0}^{t} y(\tau)x'(\tau)\, d\tau$$

where the prime designates the time derivative.   The basic function of the integrators used in DDA's is to compute the above type of integral. Since we are now dealing with a machine that handles numbers in discrete form, we must approximate the value of a given integral using a series consisting of a finite number of terms.   The value of the integral above can be approximated using the Riemann sum:

$$z(t) = z(t_0) + C \sum_{k=1}^{n} y(t_k)x'(t_k)\, \Delta t + \epsilon_n'$$

where $$\Delta t = \frac{t - t_0}{n} \qquad t_k = t_0 + k\,\Delta t$$

and ϵ_n' is the truncation error.   If we further approximate $x'(t_k)\,\Delta t$ by the differential

$$\Delta x\,(t_k) = x(t_{k+1}) - x(t_k) \approx x'(t_k)\,\Delta t$$

we have the following elementary approximation to the integral.

$$z(t) = z(t_0) + C \sum_{k=1}^{n} y(t_k)\,\Delta x\,(t_k) + \epsilon_n$$

The values of $y(t_k) \Delta x (t_k)$ could be formed in a digital machine in a relatively straightforward manner, as could the values of the series $\sum_{k=1}^{n} y(t_k) \Delta x (t_k)$. But we would like the machine to contain a number of integrators which can be interconnected by using successive output values from a given integrator as input values to other integrators. The DDA's constructed to date transfer only "increments" instead of "entire values" and are often referred to as *incremental* machines. The output from a given integrator therefore contains fewer digits than the basic values stored in the integrator. In order to maintain precision, a number

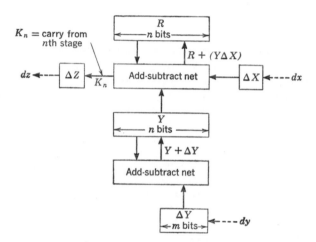

Fig. 10-10. Digital integrator.

of interesting algorithms have been invented, and we will describe a specific one, that used in MADDIDA, which permits the Δx, Δz values to be represented by single binary digits. The transferring of increments rather than entire values considerably simplifies construction of a given machine.

First, we will examine the operation of a typical integrator and then illustrate the use of a specific number system, the MADDIDA system. Figure 10-10 shows a digital integrator consisting of five independent registers Y, R, ΔY, ΔX, and ΔZ, and three dependent registers K, $R + (\Delta X \cdot Y)$, and $Y + \Delta Y$. The inputs to the ΔX register and the ΔY register are generally from other registers and are transferred into ΔX and ΔY before each iteration of the basic integration algorithm. Basically, during each iteration of the integrator algorithm the contents of ΔY are added to Y; then $\Delta X \cdot Y$ is added to R and any carry K from the addition is transferred into ΔZ. In order to facilitate description, let all the values in the Y register be nonnegative and let ΔX always have value 1. (The programmer must see that any given value does not overflow the Y register by proper scaling, which is described in the next section.) Any overflow K occurring as a result of the addition of $R + Y \cdot \Delta X$ and its insertion into R will be transferred into ΔZ. Let R and Y be n-cell registers, so that $2^n - 1$ is the maximum integer value of R and Y. Therefore, the addition into R will be modulo 2^n, and the residue after each addition will be stored in R. Let $\hat{y}_i$, $i = 0, 1, \ldots,$ $q, \ldots,$ represent the successive values stored in Y, $\Delta \hat{y}_i$ the successive values in ΔY, $\hat{r}_i$ the successive values in R, and $\Delta \hat{z}_i$ and $\Delta \hat{x}_i$ the values in ΔZ and ΔX, respectively. The values generated in the Y register are therefore determined by the relation

$$\hat{y}_i = \hat{y}_{i-1} + \Delta \hat{y}_i \tag{1}$$

Each iteration $\hat{y}_i \, \Delta \hat{x}_i$ is added to $\hat{r}_{i-1}$, forming $\hat{r}_i$, which is the *residue* of $\hat{y}_i + \hat{r}_{i-1}$, mod 2^n. Therefore, any overflow $K = \Delta \hat{z}_i$ has the value $2^n \, \Delta \hat{z}_i$, and the successive values of $\Delta \hat{z}_i$ satisfy the relation

$$2^n \, \Delta \hat{z}_i + \hat{r}_i = \hat{y}_i \, \Delta \hat{x}_i + \hat{r}_{i-1} \tag{2}$$

Summation of the successive values generated in this way yields

$$\sum_{i=1}^{k} (2^n \, \Delta \hat{z}_i + \hat{r}_i) = \sum_{i=1}^{k} (\hat{y}_i \, \Delta \hat{x}_i + \hat{r}_{i-1})$$

$$2^n \sum_{i=1}^{k} \Delta \hat{z}_i = \sum_{i=1}^{k} (\hat{y}_i \, \Delta \hat{x}_i + \hat{r}_{i-1} - \hat{r}_i) \tag{3}$$

$$2^n \sum_{i=1}^{k} \Delta \hat{z}_i = \hat{r}_0 - \hat{r}_k + \sum_{i=1}^{k} \hat{y}_i \, \Delta \hat{x}_i$$

Since $|\hat{r}_0 - \hat{r}_k|$ must be less than 2^n, the series $2^n \sum_{i=1}^{k} \Delta \hat{z}_i$ approximates the series $\sum_{i=1}^{k} \hat{y}_i \, \Delta \hat{x}_i$ with an error of magnitude less than 2^n. We now describe a number system which allows the values in R, Y, ΔX, and ΔZ to be positive or negative. This is the number system originally used in MADDIDA. Let the state of register Y with cells $Y_1, Y_2, \ldots, Y_n$ at time k be designated $Y^k = (Y_1^k, Y_2^k, \ldots, Y_n^k)$. To each value of Y^k we assign an integer value y_k as follows:

$$y_k = -2^{n-1} + \sum_{i=1}^{n} 2^{n-i} Y_i^k = \hat{y}_k - 2^{n-1}$$

The value of y_k is therefore a positive or negative integer with $2^{n-1} - 1 \geq y_k \geq -2^{n-1}$.

If $Y^k = 100 \cdots 0$, then $y_k = 0$; if $Y^k = 111 \cdots 1$, then

$$y_k = 2^{n-1} - 1$$

and if $Y^k = 000 \cdots 0$, then $y_k = -2^{n-1}$. (A 1 in the "sign position" indicates a positive number, and a 0 a negative number.) Let the same representation system be used for registers R and $Y \, \Delta X$, so that

$$r_k = \hat{r}_k - 2^{n-1} \text{ and } y_k \, \Delta x_k = \hat{y}_k \, \Delta \hat{x}_k - 2^{n-1}\dagger$$

Since ΔX is a single cell, we let ΔX^k be the state of ΔX at time k and assign a *value* Δx_k such that $\Delta x_k = 2\Delta X^k - 1 = (\pm 1)$. Similarly let Δz_k be the value of ΔZ^k at k and let $\Delta z_k = 2\Delta Z^k - 1 = 2\Delta \hat{z}_k - 1$.

The cells in the register $\Delta Y = \Delta Y_1, \Delta Y_2, \ldots, \Delta Y_q$ will, in general, contain the ΔZ outputs from q different integrators, so the value of ΔY at time k should be the sum of the individual values; therefore, let the *value* Δy_k be

$$\Delta y_k = \sum_{i=1}^{q} (2\Delta Y_i^k - 1)$$

We now describe a sequence of transfers which will cause the registers in Fig. 10-10 to realize the incremental operations

$$y_k = y_0 + \sum_{j=1}^{k} \Delta y_j$$

and

$$z_k - z_0 = \sum_{i=1}^{k} \Delta z_i = 2^{-(n-1)} \sum_{i=1}^{k} (y_i \, \Delta x_i) + \epsilon_k$$

where ϵ_k is an error of magnitude less than 2.

Let us substitute Δz_k for $\Delta \hat{z}_k$, r_k for $\hat{r}_k$, $y_k \, \Delta x_k$ for $\hat{y}_k \, \Delta \hat{x}_k$, etc., into Eqs. (1) to (3) giving

$$y_i = y_{i-1} + \Delta y_i \tag{4}$$
$$2^{n-1} \, \Delta z_i + r_i = r_{i-1} + y_i \, \Delta x_i \tag{5}$$
$$2^{n-1} \sum_{i=1}^{k} \Delta z_i = r_0 - r_k + \sum_{i=1}^{k} y_i \, \Delta x_i \tag{6}$$

Each iteration of relations (4) and (5) can be realized by the transfer

$$\Delta Y + Y \rightarrow Y$$

† This is equivalent to a representation system where a 1 in Y_1 indicates a positive number, a 0 in Y_1 indicates a negative number, and the remaining digits Y_2 through Y_n are in a normal integer representation system with negative values stored in 2's-complement form. It is also equivalent to a representation system where a given number is stored as its integer value plus 2^{n-1}.

followed by the transfers

$$\Delta X \cdot Y + R \to R \qquad K \to \Delta Z$$

where the values of ΔX determine whether y_k is to be added to r_k or subtracted from r_k, and K is the carry from the last stage of the adder for $R + \Delta X \cdot Y$. After each iteration new values will be read into ΔX and ΔY and the next iteration (set of transfers) then performed. The values of Δx_k, which are constrained to be ± 1, determine whether a given y_k is to be added to or subtracted from r_k in Eqs. (5) and (6). Also, by letting $\hat{r}_0$ be equal to 0 in Eq. (6), we can eliminate r_0 from the equation, and by summing the values of ΔZ in another integrator, we can effectively integrate. It follows from Eq. (6) that

$$p^{-1} \sum_{i=k}^{k+p} \Delta z_i \approx p^{-1} 2^{-(n-1)} \sum_{i=k}^{k+p} y_i \, \Delta x_i$$

for reasonable values of p, so that the digital integrator approximates the differential relation $dz = Cy \, dx$, where C equals $2^{-(n-1)}$. Therefore, dz can be approximated at a point indexed by k by simply summing the next $p \, \Delta x_i$'s and then dividing by p.

Figure 10-11 shows two block-diagram symbols for the integrator. It is customary to use the same symbol for integrators in DDA's as has been used in mechanical and electronic analog computers, and we have conformed to this symbology; thus, a given integrator realizes the relation $dz = Cy \, dx$. Also, the output dz may be either positive or negative; hence the relation realized is either $dz = +Cy \, dx$ or $dz = -Cy \, dx$. The signed output is useful, as will be seen in the following section, and

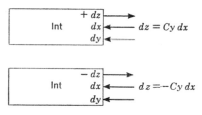

Fig. 10-11. Block-diagram symbols for integrators.

may be easily formed in the MADDIDA number system by simply complementing the output values.

10-9. Programming the Digital Differential Analyzer. This section will describe the programming of a set of integrators to yield solutions to differential equations. First we describe the mapping of a few differential equations. Mapping roughly consists of drawing a diagram showing the interconnections between a set of integrators which will realize a given equation. To facilitate description we assume a "black-box" viewpoint, with integrators which realize the relation $dz = Cy \, dx$ and interconnections indicated by lines on drawings. In actual practice these interconnections will consist of transfers, but we defer this to the next section.   The mapping procedure described here is identical with that for analog machines; any good text on analog computers will present more details on mapping differential equations.   The material on scaling which follows in this section is indigenous to the digital machine, however.

To illustrate the interconnections which must be made to solve a differential equation, consider first the simple equation

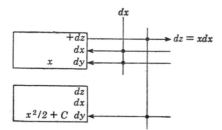

$$\frac{dy}{dx} = x$$

or its differential equivalent

$$dy = x\,dx$$

FIG. 10-12. Solution of $dy = x\,dx$.

The solution of this equation is $y = x^2/2 + C$.   Figure 10-12 shows two integrators interconnected so that the Y register of the top integrator in the figure accumulates the dx increments; therefore x is the content of the Y register.   The dz connection therefore produces the output $dz = x\,dx$.   The second integrator accumulates the dz increments of the first integrator, producing $x^2/2 + C$, where C is the content of the Y register when $x = 0$.

Now consider the equation $dy = -y\,dx$.   The solution of this equation is $y = Ae^{-x}$, where A is a constant, and a mapping of this equation can be found in Fig. 10-13.   Assume y to be stored in the Y register; the

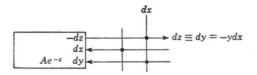

FIG. 10-13. Solution of $dy = -y\,dx$.

dx increments and the sign change then produce the output $dy = -y\,dx$. This output is fed back to the dy input and added into the Y register to produce y.   The constant A is determined by the initial contents of Y and the length of the Y register.

The solution of the differential equation

$$y'' + 2\alpha y' + \beta^2 y = 0$$

where $y' = dy/dx$ and $0 \le \alpha < \beta$ is the periodic function

$$y = e^{-\alpha x}[A \sin (\beta^2 - \alpha^2)x + B \cos (\beta^2 - \alpha^2)x]$$

where A and B are constants. With the valuations

$$dy' = y'' \, dx$$
$$dy = y' \, dx$$

and
$$dy'' = -2\alpha \, dy' - \beta^2 \, dy$$

the mapping given in Fig. 10-14 is easily obtained. The two dy connections to the first integrator are added to form the dy input to the integrator.

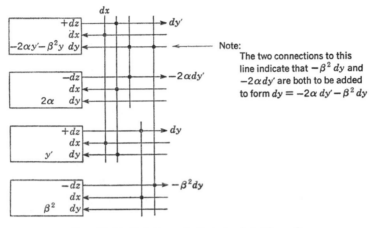

Note:
The two connections to this line indicate that $-\beta^2 \, dy$ and $-2\alpha \, dy'$ are both to be added to form $dy = -2\alpha \, dy' - \beta^2 \, dy$

Fig. 10-14. Solution of $y'' + 2\alpha y' + \beta^2 y = 0$.

As a final example of encoding the DDA, consider the Van der Pol equation

$$y'' - \mu(1 - y^2)y' + y = 0$$

where $\mu > 0$ is a constant. This equation is an example of a nonlinear differential equation. There is no known closed-form solution for this equation. However, if μ is very small, i.e., $\mu \ll 1$, the equation does have a periodic solution with range ± 2, differing very little from the solution of the previous equation if $\alpha = 0$, $\beta = 1$, and $A^2 + B^2 = 2$. In Fig. 10-15 we see the machine connections for the Van der Pol equation, resulting from the form

$$dy' = y'' \, dx$$
$$dy = y' \, dx$$
$$dy'' = \mu \, dy' - \mu y^2 \, dy' - \mu y' \, d(y^2) - dy$$

Now let us briefly consider the scaling of a problem, given a particular mapping. We adopt the convention that lower-case letters represent actual values of the variables in the problem and upper-case letters represent machine values.† Therefore, the scaling relations for a given integrator will consist in three relations:

$$dX = S_x\,dx$$
$$dY = S_y\,dy$$
$$dZ = S_z\,dz$$

where S_x, S_y, and S_z are the scale factors which convert problem values into machine values.

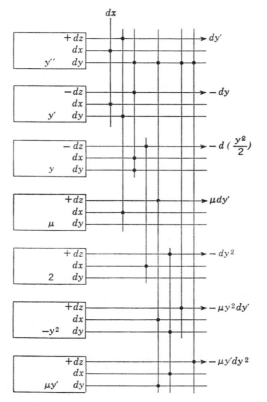

Fig. 10-15. Solution of Van der Pol's equation.

In the following discussion we shall assume the use of a binary machine; for machines using other number systems, the value 2 in the following relations need merely be changed to the value of the radix used.

Consider a problem which has been mapped and which requires the use of r integrators, which we designate I^1, I^2, . . . , I^r. Let the number of

† Please notice that this is a departure in notation from the preceding section. The use of upper-case letters for machine values and lower case for actual problem values is made to facilitate description.

cells in registers Y^i and R^i of integrator I^i be designated N_i. The basic relation yielded by integrator I^i is therefore $dZ = 2^{-(N_i-1)} Y \, dX$. The value of N_i may be chosen for each integrator; each value of N_i is set into the machine before a problem is started and remains fixed while the problem is run. However, since the maximum length of each register is fixed, and N_i must be less than or equal to this maximum, the relation $N_i \leq \max N_i$ must be satisfied, where $\max N_i$ is the total number of cells which comprise Y and R.

Since $dZ = 2^{-(N-1)} Y \, dX$ is the basic integrator relation, the following relation between scale factors for integrator I^i must also exist:

$$S_{z_i} = \frac{S_y S_{x_i}}{2^{N_i-1}}$$

This may also be written

$$\log_2 S_{z_i} = \log_2 S_{y_i} + \log_2 S_{x_i} - (N_i - 1) \tag{1}$$

Now the value of $Y^i = S_{y_i} y_i$ must not exceed the largest value which can be stored in the Y register of integrator I^i. We must therefore satisfy the relation

$$2^{N_i-1} \geq \max |Y^i| = \max S_{y_i} |y_i|$$

Or, this may be written

$$\max N_i - 1 \geq N_i - 1 \geq \log_2 \max |y_i| + \log_2 S_{y_i} \tag{2}$$

All further scale-factor relationships are obtained from the mapping for the given problem. If the dz output of integrator I^k is used as a digit of the dy input to the integrator I^i, the scale of the output increment of the integrator I^k must be the same as the scale of all other inputs which compose the dy to integrator I^i. This type of scale relationship must hold for all integrators of the problem. The interconnection scale constraints in the preceding paragraph and relations (1) and (2) constitute the entire set of scale-factor relationships which must be satisfied in order to encode an ordinary differential equation properly for solution by a DDA.

The scaling of a problem therefore consists in setting the value of the N_i and determining the values of the scale factors S_{x_i}, S_{y_i}, and S_{z_i} for all the integrators so that the scaling is consistent between integrators and so that relations (1) and (2) are satisfied. The other criteria which must be satisfied are apparent. The significance or accuracy of the solution will be determined by the scale factor chosen. Thus, the problem is to provide scale factors such that maximum accuracy of output is realized and a reasonable time for solution is obtained. To set up a problem so that a desired accuracy is realized in a reasonable operation time, the scaling between integrators is consistent, and relations (1) and (2) are satisfied also involves the particular mapping chosen to realize the equation. Generally several reasonable mappings are available, and a choice will be necessary. This choice, plus a choice of scale factors, may require trial runs of the problem, especially if the intervals in which all the variables lie are not precisely known. In this case each Y register will be checked for overflow, and if this occurs the problem must be rescaled.

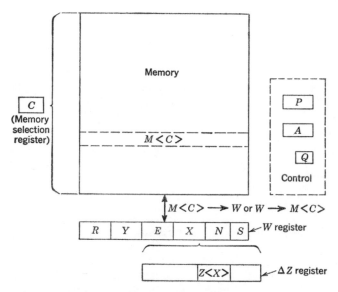

Fig. 10-16. Elementary parallel DDA.

It is possible, of course, to trade time for precision. If the scale of the independent variable is fixed, the number of iterations required is also fixed. If the independent variable is rescaled and the N_i adjusted accordingly, precision may be increased or decreased with an inverse effect on the solution time required.

The coding of the problem for a particular machine is the last step of programming a given problem. This will, of course, depend on the machine used, and is deferred to the next section, where we consider a specific machine.

10-10. A Parallel DDA. Let us now design a DDA with an arithmetic center which will be time shared by m integrators. Figure 10-16 shows a block diagram of the various registers that will be used in developing the design. M is a memory of m words. C is the address register of M, and $M\langle C\rangle$ is that register of M addressed by C. The memory-buffer register W is the register into which $M\langle C\rangle$ is transferred and is composed of six subregisters R, Y, E, X, N, and S. R corresponds to the R register of an integrator in Fig. 10-11, and similarly Y corresponds to the Y register of an integrator. E holds a *code word* for selecting the dy inputs to be added to the Y register, and X stores the code word for selecting the dx input. N stores the length of the integrator such that the scale relations derived in the last section are satisfied; hence N determines the effective length of Y and R.   Finally S, a one-bit register, stores the sign of the output dz increment, where $S = 0$ designates an integrator with a plus output, and $S = 1$ an integrator with a minus output. Register W, as well as any register $M\langle C\rangle$ in M, is capable of containing all information required for an integrator, except for input-output increments of an integrator.   All increments (the dz output increments) which are transferred between the integrators of the machine are stored in the Z register.   Since M has storage for m integrators, Z must be an m-bit register with cells Z_1, Z_2, . . . , Z_m.   Finally $Z\langle X\rangle$ is that digit of Z addressed by X, the subregister of W used to select the dx input for the integrator currently in the W register.   Therefore, the number of cells in X must be equal to or greater than $\log_2 m$, and we can let the integer value of X designate the index of the Z_i selected.   Let us assume that the code in E for dy selection from the Z register is a linear code; that is, if it is desired to select the kth bit Z_k of the Z register there is a 1 in the kth bit of the E register.   There are 0's in all bits of E where no selection is desired.   For such a code E must contain m bits, the same number of bits as the Z register.   Register E then selects the digits from register Z which are to comprise ΔY^i for a particular integrator I^i.   Let us designate the subregister E at memory address i (which corresponds to integrator I^i) as E^i, and let the cells of E^i be designated $E_1{}^i$, $E_2{}^i$, . . . , $E_m{}^i$.   Then the value of a given ΔY^i is determined by the current value of Z and E^i, and is

$$\lambda(\Delta Y^i) = \sum_{k=1}^{m} E_k{}^i(2Z_k - 1)$$

We can form a dependent register with outputs $\Delta Y_1{}^i$, $\Delta Y_2{}^i$, . . . , $\Delta Y_r{}^i$, where $r \geq \log_2 m$ and where the integer value of a given ΔY^i is $\sum_{k=1}^{m} E_k{}^i(2Z_k - 1)$ for a given E^i and Z; it is this value ΔY^i which is added to register Y each iteration.†

---

† The dependent register (combinational net) for ΔY can be formed using a symmetric network as described in S. H. Caldwell, "Switching Circuits and Logical Design," John Wiley & Sons, Inc., New York, 1958.   For most machines to date, the number of inputs comprising a given ΔY is limited to be on the order of five or ten. $\Delta Y + Y$ can be formed using the same adder as is used for $Y + R$.

The P register holds the control states of the machine. Let the Boolean functions of these states yielded by a dependent register be p_0, p_1, p_2, and p_3, so we can let the P register be a simple two-cell counter which generates p_0, p_1, p_2, p_3, p_0, p_1, p_2, ...

We now provide a simple means for controlling the number of iterations performed by the integrators, which corresponds to the number of increments which compose the range of the independent variable for the problem to be solved. Let us load the number of iterations to be performed into register A, which has p cells, hence 2^p states; we designate these states a_0, a_1, . . . , a_{2^p-1}, where the subscripts are in correspondence with the integer value of the register A. We also provide a start-stop flip-flop and a push button. Let the push button place a 0 in Q and a 0 in P, thus starting the machine. The transfer sequence for our DDA is as follows:

$$Q'p_0|\quad M\langle C\rangle \to (R,Y,E,X,N,S) = W \qquad P + 1 \to P$$
$$Q'p_1|\quad \Delta Y + Y \to Y \qquad P + 1 \to P$$
$$Q'p_2|\quad Z\langle X\rangle(R + Y) + Z'\langle X\rangle(R - Y) \to R$$
$$\qquad (K\langle N\rangle \oplus S)' \to Z\langle C\rangle \qquad P + 1 \to P \qquad C_0(A - 1) + C_0'A \to A$$
$$Q'p_3|\quad W \to M\langle C\rangle \qquad C + 1 \to C \qquad 0 \to P \qquad a_0 \to Q$$

At time p_0 the integrator at location C in M is transfered to the composite register $W = (R,Y,E,X,N,S)$. During the interval for which $p_1 = 1$, the sum of the dy increments designated by the current value of E is added to the integrand in Y; i.e., the integrand is brought up to date. At time p_2 the dx increment ΔX is selected from the Z register by X. If the increment is positive, Y is added to R; if the increment is negative, Y is subtracted from R. The result is transferred to R (refer to Fig. 10-11).

The length of the R register and of the Y register is determined by the value of register N. Let $K\langle N\rangle$ denote the binary carry at the nth stage of the above addition or subtraction, where n is the binary integer value of register N. The second transfer at time p_2 is the transfer of the output dz increment ΔZ (which has been added mod 2 to the sign value in S) into that bit of register Z which is addressed by C (that bit of Z wherein the output increment of the current integrator must be deposited). Finally, at time p_3 the updated R and Y registers of the present integrator are returned to storage, the C register is advanced by one step, and the P register is reset to state 0. As a consequence the machine will proceed again to time p_0 and then update the next integrator of M in sequence. Assume, finally, that C is reset to zero when the last integrator in the file has been updated. By this means, every integrator is updated periodically, with a period corresponding to the amount of time required to update sequentially all integrators of M once in review.

The above design of a rudimentary DDA is complete except for a means for entering the problem into the machine and reading out the final values. These operations could be performed via punched cards or punched tape as described in Chap. 9. DDA's are often used in real-time control systems where the input data are from analog-to-digital converters, and rather specialized problems are included in reading in such data.

A few remarks about incremental machines, of which the DDA represents an example, are in order. Most of the incremental machines to date have been serial machines, which require fewer components than the parallel versions, and magnetic drums are often used. The basic characteristic of the incremental machine, the fact that only changes of variables are transferred and not entire values, sometimes leads to what is called the "slewing problem." In real-time systems in which sensory or tracking devices must "hunt" a target, the response time of the incremental machines may be slower than might be desirable, since variable values can be changed at a maximum rate which is dependent upon the size of the increments which can be transferred, and hence the machine cannot slew quickly. To alleviate this, hybrid machines can be used which are primarily incremental but which have the ability to transfer entire machine words between certain registers when this becomes necessary.

Other problems arise due to the fact that errors which occur are propagated throughout the computations. Some information on this subject as well as a discussion of integrator techniques and their results on errors may be found in Refs. 1 and 18. The effects of quantization are often important, as is the basic cycle time of the machine. Despite these limitations, however, incremental machines are often quite suitable for specialized computing functions because of their inherent simplicity, light weight, size, low power consumption, etc.

PROBLEMS

10-1. (a) Write the complete transfer equations for registers $\hat{S}$, $\hat{B}$, and $\hat{L}$ of Sec. 10-5 for *any* range box r_i.

(b) Draw the state diagrams for registers $\hat{B}$ and $\hat{L}$ in accordance with part a.

(c) By use of the RS flip-flop of Sec. 7-5, write the minimal input equations to the flip-flops of registers $\hat{S}$, $\hat{B}$, $\hat{L}$.

(d) In the sequential circuits for the short status channels, when implemented with flip-flops, there are two unused states, $S = 3$ and 7. Similarly for the long status channels there is the single unused state, $B = 2$. Now if it happened that a circuit transient pulse kicked one of these sequential circuits into an unused state, there is no guarantee that the circuit would recover, that is, eventually return to one of the normally used states. Introduce transitions in the state diagrams for $\hat{S}$ and $\hat{B}$ which will guarantee recovery from all unused states in such a manner that no false target is generated at some range r_i.

(*e*) Incorporate the transitions from the unused states developed in part *d* into the flip-flop equations of part *c* where necessary.

10-2. (*a*) In Sec. 10-7, if p_i designates the phone-line timing pulse train and δ_i the range-mark pulse train (6 μsec $\equiv \frac{1}{2}$ nautical mile between pulses), develop the detailed input equations (including the clocks) to flip-flops E_0 and E, when regarded as RS flip-flops.

(*b*) Discuss the behavior of the synchronizing circuit of part *a* when the two clock trains p_i and δ_i approach the same frequency. That is, if one assumes a delay ϵ for the RS flip-flops E_0 and E, when will the circuit first fail to synchronize train p to train δ.?

10-3. From the transfer equations of Sec. 10-7 and Fig. 10-9 write the detailed flip-flop equations for the FGD output. Assume RS flip-flops.

10-4. Let us suppose an air-search radar of 200-mile range, where the time T between transmitted pulses is somewhat greater than $200(12) = 2,400$ μsec; e.g., T might be 3 msec. $T - 2,400$ is called the *dead time* or unused radar range of the radar. If we assume a parallel access, say core memory, of 512-word storage with an access time of less than 6 μsec, develop the structure of a fixed-program FGD machine. Use the dead time of the radar and the extra storage to prepare a phone-line word for transmittal over a phone line.

10-5. (*a*) Let $A_1, A_2, \ldots$ be a sequence of positive or negative integers which satisfy $A_k < K$ for all k, where K is a fixed positive integer. Let U be a register $(U_0, U_1, U_2, \ldots, U_n)$ such that $2^n > \max |A_k| + K$ and where U_0 holds the sign $S_k(= \pm 1)$ of a binary number u_k in U at time k. Now let u_k be related to u_{k-1}, A_k, and S_{k-1} by the following difference equation:

$$u_k = A_k + u_{k-1} - S_{k-1}K$$

where initially $u = u_0$ with sign S_0. Under these conditions show that

$$\sum_{k=1}^{n} S_k = \frac{1}{K}\left\{ \sum_{k=1}^{n} A_k + (u_0 - S_0K) - (u_k - S_kK)\right\}$$

(*b*) If initially $u_0 - S_0K = 0$, show by induction that $|u_k| < 2K$ for all k.

(*c*) With part *b* show that

$$\left| \sum_{k=1}^{n} S_k - \frac{1}{K}\sum_{1}^{n} A_k \right| < 1$$

for all n if $u_0 - S_0K = 0$.

(*d*) If $\Delta x_k (= \pm 1)$, y_k and $\Delta_k z$ are defined as in Sec. 10-8, and if we let $\Delta z_k = S_k$ and $A_k = y_k \Delta x_k$, show that

$$K \sum_{k=1}^{n} \Delta z_k = \sum_{k=1}^{n} y_k \Delta x_k - (u_k - S_kK)$$

This has been called the three-line digital integrator algorithm.

(*e*) Show that the three-line integrator, when interpreted as an integrator, must satisfy the scale-factor relation

$$S_z = \frac{1}{K} S_x S_y$$

(*f*) Show that r_k, the content of the R register of the MADDIDA algorithm, corresponds to the number $u_k - S_kK + K$ of the three-line integrator when $K = 2^n$.

10-6. Design a DDA which uses the three-line integrator of Prob. 5d along the lines of the machine developed in Sec. 10-10.

10-7. (a) Interpret the three-line integrator as an elementary feedback loop (servo loop).

(b) Suppose a variable of some instrument, e.g., the voltage output of a strain gage, is converted continuously to a digital number $x(t)$ in register X at time t and that initially $x(t_0)$ is a known constant. Use a three-line integrator to produce a scaled sequence of changes $\Delta z_k (= \pm 1)$ whose sum is proportional to $x(t)$. Discuss the effect of the sampling or cycling time of the integrator as well as the scaling constant K on accuracy.

10-8. Interconnect a set of integrators to solve the following differential equations:

(a) Bessel's equation

$$\frac{d^2u}{dt^2} + \frac{1}{t}\frac{du}{dt} + \left(1 - \frac{nN^2}{t^2}\right)u = 0$$

(b) Legendre's equation

$$\frac{d}{dt}\left[(1 - t^2)\frac{du}{dt}\right] + \left[n(n + 1) - \frac{m^2}{1 - t^2}\right]u = 0$$

(c) Mathieu's equation

$$\frac{d^2u}{dt^2} + (\lambda - 2n^2 \cos 2t)u = 0$$

(d) Einstein's equation for the perihelion of Mercury

$$\frac{d^2u}{d\partial^2} + u = \frac{km}{h^2}(1 + 3h^2u^2)$$

where $u = 1/r$ and r is the distance from Mercury to the center of sun.

10-9. (a) Program and scale the differential equation

$$\frac{dw}{dt} + w = 0$$

where initially $w(t_0) = 0$.

(b) Show that, when an integrator is so connected as to solve this equation, w satisfies the difference equation

$$\Delta w_i + \frac{1}{2^N}w_{i-1} = \frac{R_i - R_{i-1}}{2^N S_z} \qquad \text{for } i = 1, 2, \ldots$$

where $w_0 = w(t_0)$, $\Delta w_i = w_i - w_{i-1}$, S_z is the scale of the dz output, and $t_n - t_{n-1} = 1/2^N$.

(c) Show that the solution of the difference equation in part b is

$$w_n = \sum \left(\frac{R_i - R_{i-1}}{2^N S_z}\right)\left(1 - \frac{1}{2^N}\right)^{n-i} + \left(1 - \frac{1}{2^N}\right)^n w_0$$

10-10. Error-correcting codes are used to provide reliable communications over a noisy channel. A binary error-correcting code is one in which a k-bit message is mapped into an n-bit transmitted message. The received sequence, corrupted by noise, is mapped into the k-bit message by the decoder. Consider the following example of a Hamming code for $k = 4$, $n = 7$. Let m_1, m_2, m_3, and m_4 designate the three message bits. The transmitted message x_1, x_2, $\ldots$, x_7 is formed by the encoder as follows:

$$x_1 = m_1 \oplus m_2 \oplus m_4$$
$$x_2 = m_1 \oplus m_3 \oplus m_4$$
$$x_3 = m_1$$
$$x_4 = m_2 \oplus m_3 \oplus m_4$$
$$x_5 = m_2$$
$$x_6 = m_3$$
$$x_7 = m_4$$

Let the received message be $y_1, y_2, \ldots, y_7$. The decoder computes the triple (K_1, K_2, K_3), where

$$K_1 = y_4 \oplus y_5 \oplus y_6 \oplus y_7$$
$$K_2 = y_2 \oplus y_3 \oplus y_6 \oplus y_7$$
$$K_3 = y_1 \oplus y_3 \oplus y_5 \oplus y_7$$

The code has the property that if one and only one digit is received incorrectly, say $y_j \neq x_j$, then the value of the triple $(K_1, K_2, K_3) = j$, the location of the bit in error. If no errors are made, $(K_1, K_2, K_3) = 0$. If more than one error is made, the decoding is in general incorrect. Design an encoder and decoder to implement this code.

REFERENCES

1. Alonzo, R. L.: "Design and Operation of Digital Calculating Machinery," Ph.D. Thesis, Harvard University Computation Laboratory, Cambridge, Mass., May, 1957.
2. Bivans, E. W.: Synchronizing Magnetic Drum Storage Speed, *Electronics*, vol. 28, p. 140 (1955).
3. Braun, Edward L.: Design Features of Current Digital Differential Analyzer, *IRE Convention Record*, part 4, 1954 National Convention, pp. 87–97.
4. Bush, Vannevar: The Differential Analyzer, A New Machine for Solving Differential Equations, *J. Franklin Inst.*, vol. 212, no. 4, pp. 447–488, October, 1931.
5. Bush, Vannevar, and S. H. Caldwell: A New Type of Differential Analyzer, *J. Franklin Inst.*, vol. 240, no. 4, pp. 255–326, October, 1945.
6. Dinneen, G. P., and I. S. Reed: An Analysis of Signal Detection and Location by Digital Methods, *IRE Trans. on Inform. Theory*, vol. IT-2, no. 1, March, 1956.
7. Donan, John F.: The Serial-memory Digital Differential Analyzer, *Mathematical Tables and Other Aids to Computation*, vol. VI, no. 38, pp. 102–112, April 1952.
8. Githens, J. A.: "Digital Differential Analyzers," Tradic Final Report for the 1st, 2nd, 3rd, and 4th Quarters, Bell Telephone Laboratories, New York, 1955.
9. Harrington, J. V.: An Analysis of the Detection of Repeated Signals in Noise by Binary Integration, *IRE Trans. on Inform. Theory*, vol. IT-1, no. 1, 1955.
10. Kelley, E. J., I. S. Reed, and W. L. Root: The Detection of Radar Echoes in Noise, *J. Soc. Ind. and Appl. Math*, part I, vol. 8, no. 2, June, 1960; part II, vol. 8, no. 3, September, 1960.
11. Marcum, J. I.: A Statistical Theory of Target Detection by Pulsed Radar, *IRE Trans. on Inform. Theory*, vol. IT-6, no. 2, April, 1960.
12. Meisner, L. P.: "Real-time Digital Differential Analyzer (DART), Preliminary Analysis," *National Bureau of Standards Report* 1231, May, 1953.
13. Mendelson, Myron J.: The Decimal Digital Differential Analyzer, *Aeronaut. Eng. Rev.*, vol. 13, no. 2, February, 1954.

14. Mendelson, Myron J.: "Scaling Techniques for MADDIDA," Northrop Aircraft, Inc., August, 1950.
15. Miller, K. S., and F. J. Murray: Mathematical Basis for Error Analysis of Differential Analyzers, *J. Math. and Phys.*, vol. 32, no. 2-3, pp. 136–163, July-October, 1953.
16. Neumann, H. D.: "A Comparison between Numerical and Differential Analyzer Techniques of Solving Differential Equations," M.S. Thesis, MIT, Cambridge, Mass., May, 1954.
17. Palevsky, M.: The Design of the Bendix Differential Analyzer, *Proc. IRE*, vol. 41, no. 10, pp. 1352–1356, October, 1953.
18. Schneider, M. I.: "Logical Design of Integrators for Digital Differential Analyzers," M.S. Thesis, MIT Instrumentation Laboratory, Cambridge, Mass., Rept. T154, 1957.
19. Shackell, S. M., and J. A. Tryon: The Relative Merits of Incremental and Conventional Digital Computers in Airborne Real-time Control, AIEE Conference Paper 60-120, Presented at AIEE Winter Meeting, New York, 1959.
20. Shannon, Claude E.: Mathematical Theory of the Differential Analyzer, *J. Math. and Phys.*, vol. 20, no. 4, December, 1941.
21. Sprague, R. E.: Fundamental Concepts of Digital Differential Analyzer Method of Computation, *Mathematical Tables and Other Aids to Computation*, vol. VI, no. 37, pp. 41–49, January, 1952.
22. Sprague, R. E.: CRC 105 Computer, *Aero Digest*, vol 67, no. 2, pp. 45–55, August, 1953.
23. Swerling, P.: Maximum Angular Accuracy of a Pulsed Search Radar, *Proc. IRE*, vol. 44, no. 9, September, 1956.
24. Walter, R. W., and F. E. Brinkerhoff: A Comparison of Whole Value and Incremental Digital Techniques by the Use of Patch Panel Logic, AIEE Conference Paper CP60-121, Presented at AIEE Winter Meeting, 1960.
25. Weiss, Eric: Applications of CRC-105, Decimal Digital Differential Analyzer, *Trans. IRE*, PGEC-1, pp. 19–24, December, 1952.

11

Sequential Machines

11-1. Initial Considerations. Chapter 7 introduced some of the basic concepts of networks constructed of both gating and memory elements, and the following chapters have described design procedures in which the concepts were useful, especially in the design of the control sections of digital machines. In the last 30 years a great deal of research has been done in the sequential machine–sequential circuit area and a significant body of theory, both abstract and applied, has been developed. An important fact will be discovered in any attempt to survey the literature in this field: the diverse viewpoints of research workers have led to the reformulation of the same basic problems in a great variety of ways. Perhaps this is all to the good. The abstract, sometimes unnecessarily symbolically oriented viewpoint of the mathematician or logician may serve to counterbalance the engineer's preoccupation with the immediate physical characteristics of digital machines. At the same time the very real problems encountered in designing equipment for specific purposes often tend to illuminate theoretical problems of interest which a more abstract viewpoint might overlook. Nevertheless, one might wish, with Leibnitz, for a more universal language, which would allow for easier communication between members of the different groups studying digital machines, and a subsequent integration and broadening of the over-all theory.

The first problem attacked in this chapter will be that of minimizing the number of states in a sequential circuit for which the input-output relations have been specified. The second problem consists of assigning the actual binary values to the memory cells which are used to construct a given circuit, so that the total circuitry required is reduced. A short, heuristic discussion of Turing machines follows this, along with some of the more abstract results from sequential-machine theory. The presentation of Turing-machine theory is in no sense formal; a more rigorous treatment requires groundwork in modern set theory and/or logic, and Refs. 11 and 22 present such treatments.

Sections 11-2 through 11-5 present a basic procedure for the design of a given sequential circuit, consisting of four steps: (1) a word statement or some listing of the circuit characteristics is converted into a formal representation such as a state table; (2) the number of states which must be taken by the internal memory elements of the circuit is minimized, thus reducing the number of memory cells required to realize the circuit physically and aiding the reduction of the complexity of the overall circuitry; (3) binary values are assigned to the memory cells with distinct assignments of the values made to each state; and (4) the necessary combinational circuitry is designed. In the third step an attempt is made to assign values to the memory cells so that circuit complexity and cost are in some sense minimal. This is a very difficult problem, and no satisfactory technique (short of enumeration) is now available, although it is possible to assign values in a systematic way so that any state table can be realized. Also, an examination of this problem will indicate some steps which can be taken toward reducing the over-all complexity of a particular circuit.

The treatment of the procedures for reducing the number of states in a given sequential circuit proceeds with some rigor, in the belief that this is justified in the later sections of the chapter.

11-2. The Sequential Machine. The analysis and design of certain classes of sequential circuits were described in Chap. 7. The sequential network will now be described from a more abstract viewpoint, and some of the properties of sequential circuits in general will be examined.

Let us therefore use the term sequential machine instead of sequential circuit or network, realizing that the terms are virtually synonymous. Denote the input states to the machine by the symbols $x_1, x_2, \ldots, x_m$, the internal states of the machine by $q_1, q_2, \ldots, q_n$, and the output states or symbols by $z_1, z_2, \ldots, z_p$. Again let time pass in discrete steps, so the machine is synchronous. Let us confine our attention temporarily to a machine A for which the *total state* of the machine is, at any time t, completely defined by the input $X(t)$ and the internal state $Q(t)$. Further let the output Z at t be determined by the relation $Z(t) = \zeta_A[X(t),Q(t)]$, and the next internal state by the relation

$$Q(t + 1) = \nu_A[X(t),Q(t)]$$

A sequential machine with a finite number of input symbols, internal states, and output symbols can be described by either a state diagram or a state table. To facilitate description let $Z(r) = \zeta_A[q_i/X(0), \ldots, X(r)]$ be the final output value obtained when machine A is started in state q_i and the input sequence $X(0), \ldots, X(r)$ is applied; this can be calculated using recursion.

$$Z(r) = \zeta_A\{X(r),\nu_A[X(r-1),\nu_A[X(r-2), \ldots ,\nu_A[X(0),q_i]]]\}$$

This can be written in shorter form. Let $X^r = X(0), X(1), \ldots, X(r)$ be the input sequence and $Z(r)$ the final output state; then $Z(r) = \zeta_A(q_i/X^r)$. Also, let $\nu_A(q_i/X^r)$ be the terminal state q_j which machine A assumes when the input sequence $X^r = X(0), \ldots, X(r)$ is applied to machine A. If $Z^r = Z(0), Z(1), \ldots, Z(r)$ is the output sequence corresponding to the input sequence $X^r = X(0), X(1), \ldots, X(r)$ when machine A is started in state q_i, we write

$$Z^r = \bar{\zeta}_A(q_i/X^r)$$

The r indicates the length of the input and output sequences in these definitions.

When the state diagram for machine A can be obtained from the state diagram for machine B by simply relabeling the states, that is, by permuting the state designations, then machine A is *isomorphic* to machine B. The intent here should be clear: a machine A which is isomorphic to a machine B is really the same machine, but with the states given different names. Let us then define the following binary relations:

A state q_i of a machine A is *equivalent* to a state q_j, also in A, if *any* sequence of input symbols will produce the same sequence of output symbols when the machine is started in either q_i or q_j. If (q_i,q_j) are equivalent states, then†

$$\bar{\zeta}_A(q_i/X^r) = \bar{\zeta}_A(q_j/X^r)$$

for all input sequences X^r; and we write $q_i \cong q_j$.

A machine A^* *covers* a machine A if, for every state q_i of A, there is a state q_j^* of A^* such that if q_i and q_j^* are taken as initial states for A and A^*, a given sequence of input symbols will generate the same sequence of output symbols when applied to either A or A^*. More formally, we write $A^* \geq A$ if and only if there exists a q_j^* in A^* for each q_i in A such that $\zeta_{A^*}(q_j^*/X^r) = \zeta_A(q_i/X^r)$ for every input sequence X^r. If $A^* \geq A$ and if $A \geq A^*$ then we say that machines A and A^* are equivalent and write $A \cong A^*$.

The situation here is similar to that for combinational networks. Since we are fundamentally interested in input-output relations, if a machine A will produce the same transformations as A^*, and A is simpler than A^*, we will be sorely tempted to construct machine A.

---

† A *binary relation* is a set R in which each member is an ordered pair. A relation R of a set S is a mapping of $S \times S$ into the set consisting of two values, say true and false, and we write Q_i R Q_j if the image of (Q_i,Q_j) is *true* and Q_i Ʀ Q_j if the image is *false*. A relation R of a set S is an *equivalence relation* if and only if (1) Q_i R Q_i for every Q_i in S, and we then say that R is reflexive; (2) Q_i R Q_j implies Q_j R Q_i, and we then say the relation R is symmetric; and (3) Q_i R Q_j and Q_j R Q_k imply Q_i R Q_k, in which case the relation is *transitive*.

A first consideration in the design would seem to consist of minimizing the number of states when this is possible.

If two states q_i and q_j of machine A are *equivalent*, then q_i can be replaced everywhere in our description of the machine by q_j, thereby reducing the total number of different states. In actual practice there may be sets of states which are equivalent, and for a long time it was thought that the problem of reducing the number of states in a given machine consisted of *partitioning*† the states into *equivalence classes* so that the number of equivalence classes was minimal. This has proved to be the case for machines which are *completely specified*, that is, for machines for which all input sequences are allowable and no output symbols or next states are unspecified (don't care). Accordingly, the first technique which will be described for minimizing the number of states in a given machine will be the technique for completely specified machines, sometimes referred to as the *original* technique, primarily the work of Huffman,[20] Moore,[28] and Mealy.[26]

A completely specified machine may be easily recognized by an examination of the state table for the machine. If every next-state entry and every output entry in the table is filled in, then the machine is completely specified. If some next-state or output values are not specified, then the machine is *partially specified*.

One other classification of machines will be useful: a machine is *strongly connected* if for every pair of states (q_i, q_j) there exists some input sequence (not necessarily a single input) which will cause the internal state of the machine to go from q_i to q_j. A machine which is not strongly connected has states which are inaccessible from other states. If the directed line labeled x_1/z_2 were removed from the machine diagram in Fig. 7-11a, the state q_3 would not be accessible from either of the other states, and the machine would therefore not be strongly connected. Also, it would be partially specified.

11-3. The Minimization Technique for Completely Specified Machines. Often, the first step in the design procedure for a machine (or network) consists of specifying the performance of the machine in some formal representation system such as the state diagram or state table. The second step then consists in attempting to reduce the number of states appearing in the formal representation, thus simplifying the machine

---

† A partition of a set S is a collection $\{P_1, P_2, \ldots P_n\}$ of subsets of S such that the subsets P_i are pairwise disjoint and the union of the subsets P_i is S. If σ is an equivalence relation in S, then the *equivalence class* of an element q_i in S is the set of all elements q_j such that $q_i \, \sigma \, q_j$. Also, a partition of a set S defines an equivalence relation β in S by $q_i \, \beta \, q_j$ if and only if q_i and q_j are in the same subset of the partition. We then call the subsets of the partition equivalence classes and say that the relation β is *induced* by the partition P.

and possibly reducing the number of memory elements required. For completely specified machines, this step basically consists in finding some machine A^* which is equivalent to A, but which has fewest states.

We can define a relation $\cong$ from A to A^* of the states of machine A and machine A^* such that (q_i, q_j^*) is a member of $\cong$ if every sequence of inputs will produce an identical sequence of outputs if applied to both machine A and machine A^* when started in q_i and q_j^* respectively.† Now if machine A^* has a minimal number of states, no two states of A^* can be equivalent to each other, for then these two states could be combined into a single state, thereby reducing the total number of states. Therefore, if (q_i, q_j^*) is a member of $\cong$, (q_i, q_k^*) cannot also be a member, for this would imply q_j^* and q_k^* were equivalent states of A^*. The relation $\cong$, therefore, defines a *partition* of A into *equivalence classes* such that all the states of A which are paired with a given state of A^* in the relation $\cong$ are equivalent to each other.

Notice also that if any other machine A^{**} is equivalent to A^* and has the same number of states, we can define a mapping from A^* to A^{**} such that the states of A^* and A^{**} will be in a one-to-one correspondence and the machines can be seen to be isomorphic.

The fact that the minimal A^* defines a partition of the states of A into equivalence classes can be used to construct a procedure for minimizing the number of states of A. This procedure could well be: start machine A in state q_i; apply all possible input sequences;‡ do this for each state q_i in A. If a given state q_j produces the same output sequence as state q_k for each input sequence, that is, if $\bar{\zeta}(q_j/X^r) = \bar{\zeta}(q_k/X^r)$ for all X^r, combine these into a single state.

The number of steps required can be significantly reduced by using the following procedure.

1. Form a state table for the machine to be designed. Figure 11-1a shows a table for a sample problem.

2. Merge any rows q_i and q_j in the table in which all entries in corresponding columns x_i are identical. To merge two states q_i and q_j, change every q_j entry in the table to read q_i and then delete the row for

† A binary relation $\cong$ between the states Q of machine A and the states Q^* of machine A^* is a subset $\cong$ of the set of ordered pairs (q_i, q_j^*), q_i in A and q_j^* in A^*. If q_i is in A and q_j^* is in A^*, then $q_i \cong q_j^*$ is *true* or *holds* if and only if (q_i, q_j^*) is in $\cong$. The binary relation $\cong$ can also be defined as a mapping of the set of ordered pairs (q_i, q_j^*) from the product set $Q \times Q^*$ into the values *true* and *false;* a given (q_i, q_j^*) is a member of the relation if and only if (q_i, q_j^*) maps onto *true*.

‡ Only input sequences of length $\leq n - 1$, where n is the number of states, need be applied, as has been shown by Moore.[28] Cadden[7] has provided a (sometimes) lower bound of $< n - m + 1$ where m is the number of states, which differ in output values. That is, m is the number of subsets in the original partition.

(a) MACHINE A

	Next-state section				Output section			
	x_1	x_2	x_3	x_4	x_1	x_2	x_3	x_4
q_1	q_1	q_4	q_6	q_6	z_4	z_1	z_3	z_2
q_2	q_6	q_4	q_5	q_7	z_3	z_3	z_1	z_4
q_3	q_3	q_6	q_4	q_8	z_3	z_2	z_4	z_1
q_4	q_6	q_4	q_1	q_3	z_3	z_3	z_1	z_4
q_5	q_1	q_4	q_6	q_6	z_4	z_1	z_3	z_2
q_6	q_3	q_5	q_6	q_3	z_1	z_3	z_2	z_3
q_7	q_7	q_6	q_2	q_8	z_3	z_2	z_4	z_1
q_8	q_1	q_8	q_4	q_6	z_3	z_2	z_4	z_1

(b) T^1: PARTITIONED TABLE FOR MACHINE A

		x_1	x_2	x_3	x_4	x_1	x_2	x_3	x_4
S_1	q_1	1	4	6	6	4	1	3	2
S_2	q_2	$6,S_4$	$4,S_2$	$1,S_1$	$7,S_3$	3	3	1	4
	q_4	$6,S_4$	$4,S_2$	$1,S_1$	$3,S_3$	3	3	1	4
S_3	q_3	$3,S_3$	$6,S_4$	$4,S_2$	$8,S_3$	3	2	4	1
	q_7	$7,S_3$	$6,S_4$	$2,S_2$	$8,S_3$	3	2	4	1
	q_8	$1,S_1$	$8,S_3$	$4,S_2$	$6,S_4$	3	2	4	1
S_4	q_6	3	1	6	3	1	3	2	3

(c) T^2: SECOND PARTITION OF TABLE 1

		x_1	x_2	x_3	x_4	x_1	x_2	x_3	x_4
$q_1^* = S_1$	q_1	1	4	6	6	4	1	3	2
$q_2^* = S_2$	q_2	6	4	1	7	3	3	1	4
	q_4	6	4	1	3	3	3	1	4
$q_3^* = S_3$	q_3	3	6	4	8	3	2	4	1
	q_7	7	6	2	8	3	2	4	1
$q_4^* = S_4$	q_6	3	1	6	3	1	3	2	3
$q_5^* = S_5$	q_8	1	8	4	6	3	2	4	1

(d) T^3: MACHINE A^*

	x_1	x_2	x_3	x_4	x_1	x_2	x_3	x_4
q_1^*	q_1^*	q_2^*	q_4^*	q_4^*	z_4	z_1	z_3	z_2
q_2^*	q_4^*	q_4^*	q_1^*	q_3^*	z_3	z_3	z_1	z_4
q_3^*	q_3^*	q_4^*	q_2^*	q_5^*	z_3	z_2	z_4	z_1
q_4^*	q_3^*	q_1^*	q_4^*	q_3^*	z_1	z_3	z_2	z_3
q_5^*	q_1^*	q_5^*	q_2^*	q_4^*	z_3	z_2	z_4	z_1

FIG. 11-1. Minimal-state machine synthesis.

state q_j. In the table in Fig. 11-1a, state q_5 can be merged with state q_1, for it is obvious that the two states are equivalent.

3. Form a table T^1 partitioned into subsets so that all states with identical output values listed in corresponding columns are placed in the same section of the table. Designate the sections of the table (subsets of the partition) by the symbols S_1, S_2, . . . , S_n.

4. "Operate" the machine by applying each input state x_i (all experiments of length 1†) to all states in subsets containing more than one state. This consists of placing beside each next-state symbol q_i the symbol used to denote the subset S_j of the partition in which q_i has been placed. *If some input signal causes a transition from two states q_j and q_k which are in the same subset S_m to two states in different subsets S_r and S_t, then q_j and q_k are not equivalent.* Therefore, form a new table T^2, placing the states which have been shown to be not equivalent in a new subset S_{n+1}. For instance, if input x_i takes the machine from two states q_i and q_j which are in the same subset S_k into different subsets S_r and S_t respectively, then place q_j and any other states in S_k which are taken into S_t by x_i in a new subset S_{n+1} and call the resulting table T^2. Again operate the machine; if two states in the same partition are again shown to be not equivalent, form a new table T^3 with a new subset S_{n+2} and continue this process until every state in each subset is taken into the same subset by each input value. The state table will now be partitioned into sets of states which comprise equivalence classes. Each set of equivalent states can now be merged into a single state, so that the number of states will be equal to the number of equivalence classes.

Let us examine the use of this procedure on the table in Fig. 11-1a. First, the table in Fig. 11-1b is formed, in which state q_5 has been combined with state q_1. In this table only the subscripts of the next-state symbols and output symbols have been copied. For instance q_1 becomes 1, z_2 becomes 2, etc. The table has also been partitioned into sets of states having equivalent output values for each input value, and then the symbols S_1, S_2, S_3, and S_4 have been assigned to the four subsets of the partition. Input sequences of length 1 have then been applied in the sense that beside each next-state entry the symbol denoting the subset in which the next state lies has been added. For instance, an x_1 input takes state q_2 into state q_6, and since q_6 lies in partition S_4, an S_4 is placed in the table at the q_2,x_1 intersection point.

An examination of the table indicates that states q_2 and q_4 are taken into the same subset by each input signal, and that states q_3 and q_7 are taken into the same subset by any given input. State q_8, however, is

---

† Moore has defined the process of applying an ordered sequence of input signals of length n to a sequential machine as an *experiment of length n*, and this terminology is often used.

not equivalent to either state q_3 or q_7, for an x_1 input takes q_8 into partition S_1 while taking q_3 and q_7 into S_3; an x_2 takes q_8 into S_3, and q_3 and q_7 into S_4; and an x_4 input takes q_8 into S_4, and q_3 and q_7 into S_3.

A new table is therefore formed, with q_8 separated from q_3 and q_7, and the new partition is given the letter designation S_5. Input sequences of length 1 are again applied; this time all the states within partitions containing more than one state prove to be equivalent. (Notice that it is not necessary to copy the output values after the first partitioning.) The table has now been partitioned into five subsets; thus there will be five states in the minimal-state machine. A new table is then drawn up, and state numbers are assigned according to the results of Fig. 11-1c. In this table, state q_1^* corresponds to the partition S_1, state q_2^* to S_2, state q_3^* to S_3, etc. The next-state and output values are then placed in the new table so that the output values in row q_i^* of the new table correspond to the output values in each row of the q_i^* of the partitioned table, and the next-state values in row q_i^*, column x_i correspond to the section of the partitioned table in which fall the next-state entries in section q_i^*, row x_i of the partitioned table. Figure 11-1d shows the new table. If the machine in Fig. 11-1d is referred to as Machine A^*, and the machine in Fig. 11-1a as machine A, then the following equivalence relations have been formed:

$$
\begin{array}{ccc}
Machine\ A & & Machine\ A^* \\[4pt]
\left.\begin{array}{c} q_1 \\ q_5 \end{array}\right\} & \cong & q_1^* \\[10pt]
\left.\begin{array}{c} q_2 \\ q_4 \end{array}\right\} & \cong & q_2^* \\[10pt]
\left.\begin{array}{c} q_3 \\ q_7 \end{array}\right\} & \cong & q_3^* \\[10pt]
q_6 & \cong & q_4^* \\[4pt]
q_8 & \cong & q_5^*
\end{array}
$$

The relation $\cong$ from A to A^* of the states of machines A and A^* contains the following ordered pairs: (q_1,q_1^*), (q_5,q_1^*), (q_2,q_2^*), (q_4,q_2^*), (q_3,q_3^*), (q_7,q_3^*), (q_6,q_4^*), and (q_8,q_5^*); these define the partition promised.

11-4. The Partially Specified Machine. The state tables and state diagrams in the preceding description have all been completely specified; that is, every row of the state table for the machine contains an entry in every column. Often, a state table contains unspecified or don't care entries in some of the next-state or output positions. These unspecified conditions may arise due to the manner in which the state table is derived during the preliminary design stages or may relate to actual characteristics of the machine. For instance, some input sequences may never occur (especially in asynchronous machines), or, when the machine is in some state and a particular input is applied, the memory cells may always be set externally to some new state before another input is applied. Also, output signals which occur when the machine is in some state and a particular input is applied may not be used. In any case it may be seen that unspecified entries may occur in a given table, and it would be desirable to use this fact in minimizing the number of states, thus deriving a minimal-state machine.

The amount of reduction possible in the number of states will sometimes be dependent on the interpretation of what an unspecified position in a table means. The earlier literature on sequential machines assumed that the unspecified positions in the table were to be filled in with specific entries, but that any state or output could be used, and that the entries should be chosen so that the number of states in the resulting machine after reduction would be minimal. Huffman[20] described a scheme which efficiently accomplished the filling in of the blanks in an optimum way and the subsequent minimizing, and a modified version of this technique, suggested by Mealy,[26] was standard for several years. Later, however, Ginsburg[16-18] found cases where filling in the blank positions of certain tables in an optimum way and minimizing did not lead to a machine with as few states as might be obtained if the unspecified positions were assumed to be completely don't care, in that, effectively, the entries in the unspecified positions may change value as the operating point moves around the table. More will be said about this later. The technique for reduction described in this section will reduce all partially specified machines to minimal-state machines subject to the Ginsburg interpretation of unspecified entries, and this appears to be a more realistic interpretation. Some enumeration is sometimes required in the procedure which is largely due to Unger and Paull.[47] This technique will be preceded by a brief description of the original Huffman-Mealy technique.

The principle behind the original Huffman-Mealy procedure is, at face value, the same as that behind the technique for handling don't cares in combinational circuits. Consider the state table in Fig. 11-2. There are six blank positions in the next-state section of the table and also six blanks in the section listing the output values. Since the machine has six states, we could fill in the next-state section of the table with 6^6 different combinations of next-state values, and for each of these the output table could be filled in with 2^6 different values, so there are $6^6 \times 2^6$ different ways to fill in the table. If each table were to be partitioned into equivalence classes (subsets containing equivalent states) as in the preceding section, then the table(s) with the fewest classes (subsets) would define, in this case, minimal-state machines, each of which would do everything the original machine would.

We can shorten this procedure in a direct way, using the following rule:

Partition and Merger Rule. To reduce the number of states in a given machine A, partition the rows of the state table for machine A so that if two states are in the same partition, none of the *specified* output values differ. (Unspecified output values are assumed not to differ from specified values.) Call this partition P_1 and the subsets of the partition $S_1, S_2, \ldots, S_n$. Apply all input sequences of length 1 to the resulting table (perform all experiments of length 1) to see if any two states q_i and q_j in a given subset S_k of the partition have next-state entries such that some input state will cause the operating point to move into differing subsets S_s and S_t of the partition. If two such states are found, form a new partition P_2 with an additional subset S_{n+1} containing one of these states, say q_j, and also all other states from S_k which have specified next-state entries falling in the same subsets as q_j. Continue this process until no new partitions need be formed. Then form a new machine A^* by merging the states in each subset of the final partition into a single state.

To merge a partitioned state table for machine A, first assign a set of labels $q_1^*, q_2^*, \ldots, q_n^*$ to the n different subsets of the state table for A. Then form a new state table for a machine which we shall call A^*, using the n different q_i^*'s assigned to the subsets of A as state designations.

	Next-state section			Output section		
	x_1	x_2	x_3	x_1	x_2	x_3
q_1			q_6			0
q_2		q_1	q_3	1	0	
q_3	q_5	q_2	q_3	0	0	0
q_4	q_2	q_3		1	0	
q_5	q_1	q_4	q_6	1	0	0
q_6	q_2			0		

(a)

		x_1	x_2	x_3	x_1	x_2	x_3
S_1	q_1			$6\,S_1$			0
	q_3	$5\,S_2$	$2\,S_2$	$3\,S_1$	0	0	0
	q_6	$2\,S_2$			0		
S_2	q_2		$1\,S_1$	$3\,S_1$	1	0	
	q_4	$2\,S_2$	$3\,S_1$		1	0	
	q_5	$1\,S_1$	$4\,S_2$	$6\,S_1$	1	0	0

(b)

FIG. 11-2. Partition and merger of unspecified tables.

		x_1	x_2	x_3
S_1	q_1			$6\,S_1$
	q_3	$5\,S_3$	$2\,S_3$	$3\,S_1$
	q_6	$2\,S_2$		
S_2	q_2		$1\,S_1$	$3\,S_1$
	q_4	$2\,S_2$	$3\,S_1$	
S_3	q_5	1	4	6

(c)

		x_1	x_2	x_3
S_1	q_1			$6\,S_1$
	q_3	5	2	$3\,S_4$
S_2	q_2		$1\,S_1$	3
	q_4	2	$3\,S_1$	
S_3	q_5	1	4	6
S_4	q_6	2		

(d)

		x_1	x_2	x_3
S_1	q_1			6
	q_6	2		
S_2	q_2	1	$1\,S_1$	3
	q_4	2	$3\,S_5$	
S_3	q_5	1	4	6
S_4	q_3	5	2	3

(e)

		Next-state section			Output section		
		x_1	x_2	x_3	x_0	x_1	x_2
S_1	q_1			$6\,S_2$			0
	q_2		$1\,S_1$	$3\,S_2$	1	0	
	q_4	$2\,S_1$	$3\,S_2$		1	0	
	q_5	$1\,S_1$	$4\,S_1$	6	1	0	0
S_2	q_3	$5\,S_1$	$2\,S_1$	$3\,S_2$	0	0	0
	q_6	$2\,S_1$			0		

(f)

		x_1	x_2	x_3
S_1	q_1			$6\,S_2$
	q_2		$1\,S_1$	$3\,S_2$
	q_5	1	$4\,S_3$	$6\,S_2$
S_2	q_3	$5\,S_1$	2	
	q_6	$2\,S_1$		
S_3	q_4	2	3	

(g)

		x_1	x_2	x_3
S_1	q_1			$6\,S_3$
	q_4	$2\,S_2$	$3\,S_3$	
S_2	q_2		$1\,S_1$	$3\,S_3$
	q_5	$1\,S_1$	$4\,S_1$	$6\,S_3$
S_3	q_3	$5\,S_2$	$2\,S_2$	$3\,S_3$
	q_6	$2\,S_2$		

(h)

	Next-state section			Output section		
	x_1	x_2	x_3	x_1	x_2	x_3
q_1^*	2	3	3	1	0	0
q_2^*	1	1	3	1	0	0
q_3^*	2	2	3	0	0	0

(i)

FIG. 11-2. (Continued)

There will therefore be n rows in the state table for A^*. Row q_i^* of the table for A^* is filled in as follows:

The entry at the intersection of row q_i^* and column x_j in the output section of A^* is left unspecified if each state in subset q_i^* of table A has an unspecified entry in column x_j of the output section of table A. If some state in subset q_i^* of A has a specified entry z_k in column x_j, then z_k is copied into the intersection of row q_i^* and column x_j of table A^*.

The next-state entry at the intersection of row q_i^* and column x_j in table A^* is left unspecified if every state in subset q_i^* of table A is unspecified in column x_j of the next-state section of the table. If some state in subset q_i^* of A has a specified next-state entry, say q_k, in column x_j, then the subset q_m^* in which this next state q_k lies is copied in table A^* at the intersection of row q_i^* and column x_j.

The machine in Fig. 11-2a, which we will designate machine A, has a state table containing six states. The original partition of this table may be formed in several ways. One of these is shown in Fig. 11-2b, where states q_1, q_3, and q_6 are placed in subset S_1 and q_2, q_4, and q_5 are placed in subset S_2. When all input sequences of length 1 are applied (by writing the designation for the subset in which a given next state lies beside the next state in the table), it will be found that an x_1 input takes states q_4 and q_5 into subsets S_2 and S_1 respectively, and so a new partition is formed with q_5 in a new subset S_3. This time an x_1 input takes q_3 and q_6 into differing subsets. Now q_1 may be placed in the same subset with q_3 or q_6, so two tables are now formed, one with q_1 and q_3 in subset S_1 (Fig. 11-2d) and the other with q_1 and q_6 in subset S_1 (Fig. 11-2e). Both these tables will need further partitioning, for an x_3 input takes q_1 and q_3 into differing subsets in Fig. 11-2d, and an x_2 input takes q_2 and q_4 into differing subsets in Fig. 11-2e. The next partitions of these tables will contain five subsets each and are not shown, but these will terminate the process for this particular initial partition.

Another initial partition is shown in Fig. 11-2f. States q_2 and q_5 are taken into subsets S_1 by x_2, and q_4 is taken into S_2, so the states must be separated, but q_1 can be placed in the next partition with either q_2 and q_5 or with q_4. The resulting tables are shown in Fig. 11-2g and h. The table in Fig. 11-2h needs no further partitioning and shows an optimal partition for Fig. 11-2a. The rows of Fig. 11-2h are therefore merged, subset S_1 becomes state q_1^*, S_2 becomes q_2^*, S_3 becomes q_3^*, and the new machine A^* is shown in Fig. 11-2i.

Several techniques have been described for reducing the number of steps required in the above process, although with a little practice and insight a great deal of the labor involved will be reduced. Interested readers may find many details and examples in Caldwell.[8] Notice that even the relation between states which is used to form the initial partition in the merging procedure does not define a unique partition of a state table. Let us assume that the fact that two states do not differ in specified output entries defines a binary relation α between the two states; we shall call two states in this relation *output-compatible*. The fact that α does not define an equivalence relation may be seen by means of the following state table:

	x_0	x_1	x_0	x_1
q_1	q_1	q_3	z_0	z_1
q_2	q_2	q_3		z_1
q_3	q_3	q_1	z_1	z_1

Now α is reflexive, for $q_i \, \alpha \, q_i$ is always true; α is also symmetric, for $q_i \, \alpha \, q_j$ implies $q_j \, \alpha \, q_i$, but it is not transitive, for (referring to the table above) $q_1 \, \alpha \, q_2$ and $q_2 \, \alpha \, q_3$, but $q_1 \not\alpha \, q_3$. The three states in the machine above may therefore be partitioned in two ways, but only one of the partitions will yield the true minimal-state machine which follows:

Machine A *

	x_0	x_1	x_0	x_1
q_1^*	q_1^*	q_2^*	z_0	z_1
q_2^*	q_2^*	q_1^*	z_1	z_1

Now let us examine a more complicated example. The following five-state machine contains two unspecified entries:

	x_0	x_1	x_0	x_1
q_1		q_3	z_1	z_2
q_2	q_4		z_1	z_2
q_3	q_4	q_5	z_1	z_2
q_4	q_1	q_2	z_2	z_2
q_5	q_3	q_2	z_1	z_1

We can partition this table in two ways, attempting to merge either q_1 and q_2 or q_2 and q_3. Either merger will work, yielding the following two machines:

Machine A^*

	x_0	x_1	x_0	x_1
q_1^*	q_3^*	q_2^*	z_1	z_2
q_2^*	q_3^*	q_4^*	z_1	z_2
q_3^*	q_1^*	q_2^*	z_2	z_2
q_4^*	q_2^*	q_1^*	z_1	z_1

Machine $\bar{A}$

	x_0	x_1	x_0	x_1
$\bar{q}_1$		$\bar{q}_2$	z_1	z_2
$\bar{q}_2$	$\bar{q}_3$	$\bar{q}_4$	z_1	z_2
$\bar{q}_3$	$\bar{q}_1$	$\bar{q}_2$	z_2	z_2
$\bar{q}_4$	$\bar{q}_2$	$\bar{q}_2$	z_1	z_1

This simple example shows that the partition and merger technique does not always lead to a unique machine; nor does it necessarily lead to equivalent machines, for, if the blank in $\bar{A}$ is filled with anything, even q_3, then $\bar{A} \neq A^*$. We might, indeed, ask what are the relations among machines A, A^*, and $\bar{A}$. Define a sequence of inputs X^r to a machine A in state q_i as *applicable* if the next state is specified for each input except possibly the last. That is, $\nu_A(q_i/X^j)$ is specified for every $j < r$.

An applicable input sequence X^r will generate a sequence of outputs Z^r, some terms of which may not be specified. Given two output sequences Z^r and $\bar{Z}^r$, we say that $\bar{Z}^r$ *covers* Z^r if every *specified term* $Z(i)$ in Z^r is equal to the corresponding term $\bar{Z}(i)$ in $\bar{Z}^r$, and we then write $\bar{Z}^r \geq Z^r$. For instance $\bar{Z}^3 = \bar{Z}(0), \bar{Z}(1), \bar{Z}(2) = z_1, z_3, z_4$ covers $Z^3 = Z(0),$ $Z(1), Z(2) = z_1, z_3, z_4$ and $\bar{Z}^3 = z_4, z_3, z_2$ covers $Z^3 = z_4, -, z_2$ where the $-$ indicates an unspecified output value. Let us now extend the definition of cover to include states and partially specified machines.

A machine A^* *covers* a machine A (written $A^* \geq A$) if and only if for every state q_i in A there is a state q_j^* in A^* such that $\zeta_A(q_i/X^r) \leq$ $\zeta_{A^*}(q_j^*/X^r)$ for every input sequence X^r which is applicable to A with initial state q_i. We then say that state q_j^* covers state q_i and write $q_j^* \geq q_i$.

In order to explain more clearly the minimization procedure it will be expedient to define the quality of output sequences of not "differing in any specified entry" as follows: an output sequence $Z(0), Z(1), \ldots ,$ $Z(r) = Z^r$ is *compatible* with an output sequence $\bar{Z}(0), \bar{Z}(1), \ldots ,$ $\bar{Z}(r) = \bar{Z}^r$, if, for all $i \leq r$, $Z(i) = \bar{Z}(i)$ when both are specified.

Further, two states q_i and q_j of a machine A are *compatible* if

$$Z^r = \zeta_A(q_i/X^r)$$

is compatible with $\bar{Z}^r = \zeta_A(q_j/X^r)$ for all sequences X^r which are applicable to machine A when started in both q_i and q_j (that is, only the X^r in the intersection of the two sets of sequences which are applicable to machine A when started in q_i and q_j respectively are to be considered). If two states are compatible, we write $q_i \sigma q_j$, which is a reflexive, symmetric, not necessarily transitive, binary relation.   Two states which are *not* compatible are *incompatible*, and we write $q_i \not\sigma q_j$.

If two states are compatible they can be merged, and the new state thus formed will cover both of the original states.   Thus, if two states q_i and q_j in machine A are merged to form state q_k^* in a new machine A^*, then

$$\zeta_{A^*}(q_k^*/X^r) \geq \zeta_A(q_i/X^r) \cup \zeta_A(q_j/X^r)$$

where $\cup$ indicates a term-by-term union of the terms of the sequences Z^r according to the rules $z_i \cup z_i = z_i$, $z_i \cup - = z_i$, $- \cup z_i = z_i$, and $- \cup - = -$. Thus $(z_1,-,z_3,-) \cup (z_1,z_4,-,-,) = (z_1,z_4,z_3,-)$.   So q_k^* covers both q_i and q_j.

Compatibility is a key concept in reducing partially specified tables, so we need a test for compatibility which can be applied to the states of a table.   Now compatibility is a reflexive, symmetric relation, so we can handle pairs of states which are in the relation σ as *unordered pairs*, thus cutting the number of pairs in σ in half.   Also, since $q_i \sigma q_i$ always holds, we will omit the pairs (q_i,q_i).   Notice that the *output-compatible* relation α is also reflexive and symmetric, so we can perform calculations using members of α as unordered pairs.   Also, notice that the pairs of states which are *output-incompatible* form a subset of the pairs of states which are incompatible.

A technique for determining which states in a machine A are compatible will be described using machine A in Fig. 11-3a.   First list the unordered pairs of states in A which are output-incompatible; states q_i and q_j are an output-incompatible pair if both outputs are specified for some input x_k and

$$\zeta_A(q_i,x_k) \neq \zeta_A(q_j,x_k)$$

In Fig. 11-3a the pairs (1,2), (1,4), (2,3), (2,6), (3,4), (4,5), and (4,6) are output-incompatible and are listed in the first and second rows of Fig. 11-3b as members of $\not\alpha$.   A table is then formed containing the remaining unordered pairs (each of which is output-compatible) along the ordinate and the input values for the machine along the abscissa. The intersection of column x_k and row (q_i,q_j) is then filled in with the value(s) of the next-state functions $\nu(q_i/x_k)$ and $\nu(q_j/x_k)$.   If these values are both unspecified, the entry is left blank; if only one value is specified, the entry may be left blank, although these single values are shown in Fig. 11-3b for illustrative purposes.

If any entry in the resulting table contains an output-incompatible pair, then the pair of states on the ordinate is incompatible and is added to the list of incompatible pairs.   In Fig. 11-3b the pairs (1,5), (3,5), and (5,6) are incompatible, for with an x_3 input (1,5) goes to (2,6), an output incompatible pair; (3,5) goes to (2,6) under x_3; (5,6) goes to (1,2) under x_3.   The pairs (1,5), (3,5), and (5,6) are added to the list of incompatibles, and the process is repeated. This time (2,5) is found to be incompatible, for an x_1 input takes the pair (2,5) into (3,5), an incompatible. In Fig. 11-3b the incompatibles are checked, and the remaining pairs [(1,3), (1,6), (2,4), and (3,6)] along the ordinate comprise the compatible pairs.

	Next-state section			Output section		
	x_1	x_2	x_3	x_1	x_2	x_3
q_1	6	5	6	z_1	z_2	z_1
q_2	3	5	4	z_2	z_1	
q_3	3	5	6	z_1		z_1
q_4	6		4	z_2		z_2
q_5	5	1	2			z_1
q_6	3		1		z_2	z_1

(a)

	x_1	x_2	x_3
(1,3)	(3,6)	(5)	(6)
✓(1,5)	(5,6)	(1,5)	(2,6)
(1,6)	(3,6)	(5)	(1,6)
(2,4)	(3,6)	(5)	(4)
✓(2,5)	(3,5)	(1,5)	(2,4)
✓(3,5)	(3,5)	(1,5)	(2,6)
(3,6)	(3)	(5)	(1,6)
✓(5,6)	(3,5)	(1)	(1,2)

Incompatible pairs ∮

(1,2) (1,4) (2,3) (2,6) ⎫ members of
(3,4) (4,5) (4,6)　　　 ⎬ ∮ and ∮
(1,5) (3,5) (5,6) ⎫
(2,5)　　　　　 ⎬ ∮ only

Compatible pairs (members of σ)
　　(1,3), (1,6), (2,4), (3,6)

(b)

FIG. 11-3. Determination of compatible pairs from state tables. (a) Partially specified state table; (b) derivation of compatible pairs.

These results indicate that the pairs q_1, q_3, and q_6 are mutually compatible, for (q_1,q_3), (q_3, q_6), and (q_1,q_6) are all members of σ. We might therefore merge all three states into a single state. In order to express this possibility we define a ternary relation $\overset{3}{\sigma}$, a quaternary relation $\overset{4}{\sigma}$, and in general an n-ary relation $\overset{n}{\sigma}$ so that (q_i,q_j,q_k) is a member of $\overset{3}{\sigma}$ if and only if (q_i,q_j), (q_j,q_k), and (q_i,q_k) are members of σ, and, in general, an element $(q_i,q_j, \ldots ,q_m)$ containing n states is a member of the n-ary relation $\overset{n}{\sigma}$ if and only if (q_i,q_j) is a member of σ for all $i, j, \ldots , m$. The states in a given $\overset{n}{\sigma}$ are said to be *mutually compatible* [$n(n - 1)/2$ pairs are required to form an n-ary relation]. It will be convenient to form such relations where n is maximal; we call these *maximal compatibles*, after Unger and Paull,[47] and define a maximal compatible as a set of states which are mutually compatible and which form a proper subset of no other set of states which are mutually compatible.

Now consider machine A in Fig. 11-4a. The compatible pairs of states are (q_0,q_1) and (q_1,q_2). An attempt to partition the table will fail, however, for if q_0 and q_1 are merged, then the resulting machine will not cover machine A; nor will the machine obtained by merging q_1 and q_2. If, however, we form two subsets, not disjoint, consisting of q_0,q_1 and q_1,q_2,

(a) MACHINE A

	x_0	x_1	x_0	x_1
q_0	q_2	q_1	z_1	z_2
q_1	q_1	q_2		z_2
q_2	q_0	q_1	z_0	z_2

(b)

		x_0	x_1	x_0	x_1
q_0^*	q_0	q_2	q_1	z_1	z_2
	q_1	q_1	q_2		z_2
q_1^*	q_1	q_1	q_2		z_2
	q_2	q_0	q_1	z_0	z_2

(c) MACHINE A^*

	x_0	x_1	x_0	x_1
q_0^*	q_1	q_1	z_1	z_2
q_1^*	q_0	q_1	z_0	z_2

(d) MACHINE B

	x_0	x_1	x_2	x_0	x_1	x_2
q_0	q_0	q_1	q_2	z_1	z_2	z_0
q_1	q_2	q_0	q_1	z_1	z_2	z_1
q_2	q_0	q_2		z_1	z_2	

(e)

		x_0	x_1	x_2	x_0	x_1	x_2
q_0^*	q_0	q_0	q_1	q_2	z_1	z_2	z_0
	q_2	q_0	q_2		z_1	z_2	
q_1^*	q_1	q_2	q_0	q_1	z_1	z_2	z_1
	q_2	q_0	q_2		z_1	z_2	

(f) MACHINE B^*

	x_0	x_1	x_2	x_0	x_1	x_2
q_0^*	0	1	1	z_1	z_2	z_0
q_1^*	0	0	1	z_1	z_2	z_1

FIG. 11-4. Partially specified machines.

the machine A^* (Fig. 11-4b and c) formed by merging *both* pairs will cover machine A and will be a minimal state machine.

The same point is illustrated by machine B in Fig. 11-4d, where both next state and output are unspecified for (q_2,x_2). In machine B, (q_0,q_2) and (q_1,q_2) are compatible pairs, but in order to reduce the number of states, two overlapping subsets must be formed, one containing (q_0,q_2) and the other containing (q_1,q_2).

When the states in each subset of the machine in Fig. 11-4e are merged machine B^* in Fig. 11-4f is formed; this is the minimal-state machine.

A procedure for deriving a minimal-state machine which covers a given machine A may now be described.

1. Determine which sets of states of A are compatible.

2. Place the states of A into m subsets S_1, S_2, . . . , S_m, where m is minimal, such that

   a. Each S_i contains only states which are mutually compatible.

   b. Each state q_j of A is in at least one subset S_i.

   c. For each input x_i and each subset S_j a subset S_p can be found such that x_i takes all the states q_k, q_m, . . . in S_j into S_p. [Given x_i and S_j, a subset S_p can be found such that $\nu(q_k,x_i)$ is a member of S_p for each q_k in S_j.]

3. Form a new table called A^* with states q_1^*, . . . , q_m^* corresponding to S_1, . . . , S_m in original machine A, and such that the intersection of q_i^*,x_k in the output section of the table contains $\zeta(q_j,x_k)$ if the value is specified for all q_j in S_i, otherwise the entry is left unspecified, and the intersection q_i^*,x_k is left blank if $\nu(q_i,x_k)$ is unspecified for all q_j in S_i; otherwise a set S_r to which all q_j in S_i are taken by x_k can be found, and so q_r^* is entered in the table.

The problem is to form a table consisting of m subsets S_1, S_2, . . . , S_m of the states of A such that m is minimal and the three conditions 2a to 2c are satisfied. Obviously the minimum m which might be realized would equal the smallest number l of subsets formed of mutually compatible states such that each state was in some subset. Also the maximum m which must be considered is the number of states in the original machine A. A reasonable approach therefore consists of first trying groupings of the states of A into l subsets, then trying $l + 1$ subsets, then $l + 2$, . . . , until a table satisfying conditions 2a and 2c is formed.

Consider the machine in Fig. 11-5a. The state pairs (1,4), (1,5), (1,6), (3,6), (4,5), (4,6), and (5,6) of this machine are compatible as shown in Fig. 11-5b, and so (3,6) and (1,4,5,6) are the maximal compatibles. The table in Fig. 11-5c is formed with $m = 3$, and this table is consistent with rules 2a to 2c, so the states in each subset S_i of the table in Fig. 11-5c are merged to form state q_i^* of the machine in Fig. 11-5c. Examples, further observations, and other procedures for deriving minimal-state

	x_1	x_2	x_3	x_4	x_1	x_2	x_3	x_4
q_1		5	4	3		0	1	1
q_2	3	2	4	1	1	1	1	0
q_3	6	5	1	3	1	1	1	0
q_4		6		6		0		1
q_5	5	1		3	1	0		1
q_6	5		4		1		1	

(a)

Compatible pairs

Incompatible pairs

(1,2), (1,3), (2,4), (2,5), (3,4), (3,5), (2,3), (2,6)

	x_1	x_2	x_3	x_4
(1,4)		(5,6)	(4)	(3,6)
(1,5)	(5)	(5,1)	(4)	(3)
(1,6)	(5)	(5)	(4)	(3)
✓(2,3)	(3,6)	(2,5)		
✓(2,6)	(3,5)			
(3,6)	(5,6)	(5)	(1,4)	(3)
(4,5)	(5)	(1,6)		(3,6)
(4,6)	(5)	(6)	(4)	(6)
(5,6)	(5)	(1)	(4)	(3)

n-ary relations require $\dfrac{n(n-1)}{2}$ binary relations

(b)

	x_1	x_2	x_3	x_4	x_1	x_2	x_3	x_4
$S_1 = q_1^*$ q_1		5	4	3		0	1	1
q_4		6		6		0		1
q_5	5	1		3	1	0		1
q_6	5		4		1		1	
$S_1 = q_2^*$ q_2	3	2	4	1	1	1	1	0
$S_1 = q_3^*$ q_3	6	5	1	3	1	1	1	0
q_6	5		4		1		1	

(c)

	x_1	x_2	x_3	x_4	x_1	x_2	x_3	x_4
q_1^*	1	1	1	3	1	0	1	1
q_2^*	3	2	1	1	1	1	1	0
q_3^*	1	1	1	3	1	1	1	0

(d)

Fig. 11-5. Derivation of minimal-state machines.

machines may be found in Unger and Paull.[47] Ginsburg also presents some examples and further theory in Refs. 16 to 18.

When a machine is completely specified, the determination of the states which are compatible will completely define the minimal-state machine, for the compatible relation for completely specified machines is an equivalence relation and defines a unique partition of the states. The test for compatibility given in this section therefore comprises an alternate technique for minimizing completely specified machines.

11-5. Assignment of Values to Memory Cells. The final steps in the design of sequential machines or circuits remain to be described. The preceding sections have introduced several formal representation schemes and given a technique for the reduction of the number of states in a given machine, and hence the number of memory cells. The states of the machine have, however, been represented by symbols, i.e., q_1, q_2, $\ldots$, q_m. Now if the number of states has been reduced to a minimum of m, we know that at least $c \geq \log_2 m$ binary memory cells, where c is an integer, will be required in the actual register Q used to construct the machine. The next problem involves assigning actual state values to the register and cells of the register and assigning these values so the cost of the final network is minimal in some sense. Here, however, the problem becomes quite complicated. Often by using more memory cells the cost of the combinational networks required may be reduced. For instance, in Sec. 10-8 we noted that a ring counter required more memory cells than a binary counter plus decoder, but less combinational elements. The choice of circuitry in this case would center around the period of the counter, type of gates and memory cells available, etc. Also, constant weight counters were shown which required a number of memory cells and gates lying between the extreme of the binary counter and ring counter. The decision here, and hence the design procedure, must, to some extent, remain unspecified, for each instance must almost certainly be treated as a special case.

To complicate the problem further, given a specific set of gates and memory cells and also a specific state table or flow table, there is, at present, no systematic way to assign values to a register of given length, short of enumeration of all ways, such that the resulting combinational network will be minimal. The situation is not quite that bad, however, for it is possible to implement a given table in a straightforward manner, and by using some judgement it is generally possible to implement the circuit in a relatively efficient way, so that the absolute minimal-cost circuit will not be too different in cost from the circuit designed.

In order to describe a procedure for assigning values to the memory cells, let us design a *sequence detector*. Consider a machine with two input lines X_1 and X_2 which will contain binary signals at time $t = 0$,

$1, 2, \ldots, n$, and two output lines Z_1 and Z_2 which will also carry binary signals.

We assign a value x_i to each input state $X(t)$ as follows:

$X_1(t)$	$X_2(t)$	$X(t)$
0	0	x_0
0	1	x_1
1	0	x_2
1	1	x_3

We assign values to $Z(t)$ similarly, i.e., z_2 represents the output state $Z(t)$ when $Z_1(t) = 1$ and $Z_2(t) = 0$.

The sequence detector is to detect the two input sequences $\{x_1, x_3, x_2, x_3\}$ and $\{x_2, x_3, x_1, x_3\}$. If either of these input sequences occur, than Z_2 is to carry a 1 and Z_1 a 0 when the fourth term of the sequence is applied. If any other sequence except the null sequence $x_0, x_0, x_0, \ldots$ is applied, output line Z_1 is to carry a 1 and Z_2 a 0. When the null sequence $x_0, x_0, x_0, \ldots$ is applied the output is to be z_0 or $(Z_1 = 0, Z_2 = 0)$. The sequence detector may be thought of as a combination lock which either of the sequences x_1, x_3, x_2, x_3 or x_2, x_3, x_1, x_3 will "unlock" (e.g., produce a 1 output on Z_2), and any erroneous attempt to open the lock will result in a warning signal on line Z_1.

Figure 11-6a shows a flow table for this sequence detector. If either of the correct sequences is applied a z_1 output will result, and as long as the fourth term x_3 of either sequence is applied after the sequence, the output will remain z_1. If the next input is x_0, than a z_0 output will result. Any but a correct input sequence will result in a z_2 output.

The table for the machine in Fig. 11-6a may be reduced from nine to six states, and the resulting table is shown in Fig. 11-6c, where the states q_i of the new machine are no longer called q_i^*, in order to simplify the notation in future steps.

Having derived the minimal-state circuit described by Fig. 11-6c, we now assign values to the states q_i. Let us first rewrite the description of the circuit in Fig. 11-6c as a set of 4-tuples as follows. Each 4-tuple is of the form $x_i q_j q_k z_m$, where x_i is the input state, q_j the internal state, q_k the next state or the value $\nu(q_j, x_i)$, and z_m the output value $\zeta(q_j, x_i)$. Since there are four input values x_i, x_2, x_3, and x_4, six internal states $q_1, \ldots, q_6$, and the circuit is completely specified, there will be twenty-four 4-tuples; these are listed in Fig. 11-7a.

Since the binary values for the x_i's and the z_j's are known, we have only to assign values to the q_k's. First we will use the obvious assignment q_0 is $Q_1 = 0$, $Q_2 = 0$, $Q_3 = 0$; q_1 is $Q_1 = 0$, $Q_2 = 0$, $Q_3 = 1$; and, for instance, q_6 is $Q_1 = 1$, $Q_2 = 1$, $Q_3 = 0$. The list of 4-tuples in Fig. 11-7a can be used directly to design the combinational network for a class A sequence detector (refer to Fig. 7-2). Assume that delay lines are used as the memory cells, so the state of a cell at time t is equal to the input value to the cell at time $t - 1$. Then form a table of combinations for the combinational network by letting the input x_i and the present

	x_0	x_1	x_2	x_3	x_0	x_1	x_2	x_3
q_0	0	1	5	0	z_0	z_0	z_0	z_2
q_1	0	0	0	2	z_2	z_2	z_2	z_0
q_2	0	0	3	0	z_2	z_2	z_0	z_2
q_3	0	0	0	4	z_2	z_2	z_2	z_1
q_4	0	0	0	4	z_0	z_2	z_2	z_1
q_5	0	0	0	6	z_2	z_2	z_2	z_0
q_6	0	7	0	0	z_2	z_0	z_2	z_2
q_7	0	0	0	8	z_2	z_2	z_2	z_1
q_8	0	0	0	8	z_0	z_2	z_2	z_1

(a)

		x_0	x_1	x_2	x_3	x_0	x_1	x_2	x_3
a	q_0	0	1	5	0	0	0	0	2
b	q_1	0a	0a	0a	zc	2	2	2	0
	q_5	0a	0a	0a	6f	2	2	2	0
c	q_2	0	0	3	0	2	2	0	2
d	q_3	0a	0a	0a	4e	2	2	2	1
	q_7	0a	0a	0a	8e	2	2	2	1
e	q_4	0a	0a	0a	4e	0	2	2	1
	q_8	0a	0a	0a	8e	0	2	2	1
f	q_6	0	7	0	0	2	0	2	2

(b)

	x_0	x_1	x_2	x_3	x_0	x_1	x_2	x_3
q_0	q_0	q_1	q_5	q_0	z_0	z_0	z_0	z_2
q_1	q_0	q_0	q_0	q_2	z_2	z_2	z_2	z_0
q_2	q_0	q_0	q_3	q_0	z_2	z_2	z_0	z_2
q_3	q_0	q_0	q_0	q_4	z_2	z_2	z_2	z_1
q_4	q_0	q_0	q_0	q_4	z_0	z_2	z_2	z_1
q_5	q_0	q_0	q_0	q_6	z_2	z_2	z_2	z_0
q_6	q_0	q_3	q_0	q_0	z_2	z_0	z_2	z_2

(c)

$q_0 \to q_0$ $(q_5, q_1) \to q_1$ $q_2 \to q_2$ $(q_3, q_7) \to q_3$

$(q_4, q_8) \to q_4$ $q_6 \to q_6$

FIG. 11-6. State table for sequence detector.

				Input					Output				
				$X(t)$		$Q(t)$			$Q(t+1)$			$Z(t)$	
				X_1	X_2	Q_1	Q_2	Q_3	$\bar{Q}_1$	$\bar{Q}_2$	$\bar{Q}_3$	Z_1	Z_2
x_0	q_0	q_0	z_0	0	0	0	0	0	0	0	0	0	0
x_1	q_0	q_1	z_0	0	1	0	0	0	0	0	1	0	0
x_2	q_0	q_5	z_0	1	0	0	0	0	1	0	1	0	0
x_3	q_0	q_0	z_2	1	1	0	0	0	0	0	0	1	0
x_0	q_1	q_0	z_2	0	0	0	0	1	0	0	0	1	0
x_1	q_1	q_0	z_2	0	1	0	0	1	0	0	0	1	0
x_2	q_1	q_0	z_2	1	0	0	0	1	0	0	0	1	0
x_3	q_1	q_2	z_0	1	1	0	0	1	0	1	0	0	0
x_0	q_2	q_0	z_2	0	0	0	1	0	0	0	0	1	0
x_1	q_2	q_0	z_2	0	1	0	1	0	0	0	0	1	0
x_2	q_2	q_3	z_0	1	0	0	1	0	0	1	1	0	0
x_3	q_2	q_0	z_2	1	1	0	1	0	0	0	0	1	0
x_0	q_3	q_0	z_2	0	0	0	1	1	0	0	0	1	0
x_1	q_3	q_0	z_2	0	1	0	1	1	0	0	0	1	0
x_2	q_3	q_0	z_2	1	0	0	1	1	0	0	0	1	0
x_3	q_3	q_4	z_1	1	1	0	1	1	1	0	0	0	1
x_0	q_4	q_0	z_0	0	0	1	0	0	0	0	0	0	0
x_1	q_4	q_0	z_2	0	1	1	0	0	0	0	0	1	0
x_2	q_4	q_0	z_2	1	0	1	0	0	0	0	0	1	0
x_3	q_4	q_4	z_1	1	1	1	0	0	1	0	0	0	1
x_0	q_5	q_0	z_2	0	0	1	0	1	0	0	0	1	0
x_1	q_5	q_0	z_2	0	1	1	0	1	0	0	0	1	0
x_2	q_5	q_0	z_2	1	0	1	0	1	0	0	0	1	0
x_3	q_5	q_6	z_0	1	1	1	0	1	1	1	0	0	0
x_0	q_6	q_0	z_2	0	0	1	1	0	0	0	0	1	0
x_1	q_6	q_3	z_0	0	1	1	1	1	0	1	1	0	0
x_2	q_6	q_0	z_2	1	0	1	1	0	0	0	0	1	0
x_3	q_6	q_0	z_2	1	1	1	1	0	0	0	0	1	0

(a)

Redundancy (don't cares)

0	0	1	1	1
0	1	1	1	1
1	0	1	1	1
1	1	1	1	1

(b)

Fig. 11-7. Assignment of state values.

outputs from the memory cells q_j form the inputs to the network; the outputs are then the two-digit values of y_k and the three-digit values of the next state for Q, which we designate $\bar{Q} = \bar{Q}_1\bar{Q}_2\bar{Q}_3$ as in Fig. 11-7b. The redundancy or don't care inputs consist of the four unused input states and are also shown. Figure 11-7b may be used as a table of combinations describing a Boolean function, and when simplified yields the expression:

$$\bar{Q}_1 = Q_1Q_2Q_3X_1'- + Q_1Q_2'Q_3'X_1'X_2' + Q_1Q_2Q_3'X_1X_2$$
$$\bar{Q}_2 = Q_1'Q_2Q_3X_1X_2 + Q_1Q_2-X_1'X_2 + Q_1Q_2'Q_3'X_1X_2'$$
$$\bar{Q}_3 = Q_1'Q_2Q_3'X_1'X_2' + Q_1Q_2'Q_3'X_1'X_2' + Q_1Q_2'Q_3'X_1X_2' + Q_1'Q_2Q_3X_1X_2$$
$$\bar{Q}_4 = -Q_2'Q_3X_1X_2' + Q_1Q_2'Q_3X_1'- + Q_1'--X_1'X_2 + Q_1-Q_3X_1X_2'$$
$$\qquad\qquad + Q_1'-Q_3'X_1- + Q_1Q_2Q_3'-X_2' + -Q_2'Q_3'-X_2 + Q_1'Q_2Q_3X_1'-$$
$$\bar{Q}_5 = Q_1Q_2Q_3X_1'X_2' + Q_1Q_2Q_3'X_1X_2$$

If the circuit is to be designed using complementing flip-flops (refer to Sec. 7-5), then the binary input-output relations for the combinational network may be derived from Fig. 11-7a by simply noting the changes in Q_1, Q_2, and Q_3 corresponding to each input. If $C(t) = C_1(t)$, $C_2(t)$, $C_3(t)$ is the C input to flip-flops Q_1, Q_2, and Q_3, the first rows of the table of combinations for the complementing type flip-flop are:

$X(t)$		$Q(t)$			$C(t)$			$Z(t)$	
X_1	X_2	Q_1	Q_2	Q_3	C_1	C_2	C_3	Z_1	Z_2
0	0	0	0	0	0	0	0	0	0
0	1	0	0	0	0	0	1	0	0
1	0	0	0	0	1	0	1	0	0
. . . . . .		. . . .			. . . .			. . . . . .	

Note that row 8 would be

$$1\,1 \quad 0\,0\,1 \quad 0\,1\,1 \quad 0\,0$$

For the RS flip-flops described in Sec. 7-5 we need six output lines (one to the R and S input of each flip-flop and two output lines Z_1 and Z_2), so the net has five input lines and eight output lines and the first few rows of the table will be:

Input					Output								
$X(t)$		$Q(t)$			$R(t)$ and $S(t)$						$Z(t)$	
X_1	X_2	Q_1	Q_2	Q_3	R_1	S_1	R_2	S_2	R_3	S_3	Z_1	Z_2
0	0	0	0	0	d	0	d	0	d	0	0	0
0	1	0	0	0	d	0	d	0	0	1	0	0
1	0	0	0	0	0	1	0	1	d	0	0	0
. . . . . .		. . . .			. . . .							. . . .	

where the d's indicate don't care outputs.

The choice of the type of memory elements used and the assignment of values to the memory elements are very difficult and no completely satisfactory technique is now available.   References 13 and 29 discuss this problem in some detail, and the chapter by Phister in Ref. 23 gives some practical advice and further examples.   We could, of course, try assigning all possible combinations of values to the states of the memory cells, but for all but the smallest of tables this would be prohibitive.   McCluskey and Unger describe this problem in Ref. 24, giving formulas for determining the number of assignments which affect circuit size, given number of memory elements and number of states.

Here is an assignment of values to the q_i for the problem in Fig. 11-7a which yields the minimal combinational circuit we have discovered to date,[†] assuming delay elements and two-level logic:

$$q_0 = 010; q_1 = 100; q_2 = 111; q_3 = 000; q_4 = 001; q_5 = 101; q_6 = 011$$

The resulting expressions for the $\bar{Q}_i$ are:

$$\bar{Q}_1 = Q_1 \text{-} Q_3' X_1 X_2 + -Q_2 Q_3' X_1' X_2 + \underline{-Q_2 Q_3' X_1 X_2'}$$
$$\bar{Q}_2 = ---X_1' X_2' + \underline{Q_1' Q_2 Q_3 \text{-} X_2'} + \underline{-Q_2' - X_1 X_2'} + Q_1 ---X_2$$
$$\qquad\qquad\qquad\qquad + \underline{-Q_2' \text{-} X_1' X_2} + \underline{-Q_2 \text{-} X_1 X_2}$$
$$\bar{Q}_3 = -Q_2' \text{-} X_1 X_2 + \underline{-Q_2 Q_3' X_1 X_2'}$$
$$Z_1 = -Q_2' Q_3' \text{-} X_2' + \overline{Q_1 ---X_1' \text{-}} + \underline{-Q_2' \text{-} X_1 X_2'} + \underline{-Q_2' \text{-} X_1' X_2}$$
$$\qquad\qquad\qquad\qquad + \underline{-Q_2 \text{-} X_1 X_2} + \overline{Q_1' Q_2 Q_3 \text{-} X_2'}$$
$$Z_2 = Q_1' Q_2' \text{-} X_1 X_2$$

The underlined terms appear in more than one expression.

11-6. Turing Machines and Other Types of Automata.   The machines studied in previous sections were all *finite-state* machines in which the number of internal states of the machine and the number of input and output values, etc., was fixed.   The finite-state machine is the only type of digital machine which we now design and construct.   It is possible to formulate infinite-state machines, but little has been done in this area. On the other hand, a number of important studies have been made of *potentially infinite* machines or potentially infinite automata.   A potentially infinite machine contains a finite number of elements (components) at any given time, but is able to grow or to expand indefinitely.   The most studied example of a potentially infinite machine is the *Turing* machine, named in honor of its inventor, A. M. Turing.[43-46]   Turing machines have been formulated in various ways, but basically the machine consists of a *computing element* and a *tape* which is assumed to be potentially infinite in length.   The computing element appears in

[†] This particular assignment was derived by M. Schneider of Lincoln Laboratory, MIT, during a study of the assignment problem.

several different forms in the literature, but generally the computing element is assumed to be capable of reading any symbols written on the tape, moving the tape either right or left, and printing any of a given set of symbols on the tape (including overprinting a symbol already written on the tape). Each symbol written on the tape occupies some finite area along the tape, and the set of symbols which can be read and which can be written may be denoted $S_1, S_2, \ldots, S_m$, where m is an integer.

When a computation is to be performed, a program, or initial set of symbols, is written on the tape (the total number of symbols on the tape is assumed to be finite) and the starting position for the computing element is indicated. Since the total length or amount of tape available is assumed to be potentially infinite, computation may continue indefinitely. (Potentially infinite can be taken to mean that while the tape will be finite in length at any given time, some machine operator will always be present to add tape when required.)

Several different representation schemes for Turing machines have been used. Let the internal states of the computing element be designated $q_1, q_2, \ldots, q_n$, and the tape symbols or machine alphabet $S_1, S_2, \ldots, S_m$. Also let L designate a left movement of the tape the distance occupied by one tape symbol, and R a corresponding right movement of the tape. If it is assumed that the tape passes through the machine one square at a time, the operation of the computing element can be described by means of a finite set of 4-tuples of the following kinds:

$$q_i S_j S_k q_l$$
$$q_i S_j R q_k$$
$$q_i S_j L q_k$$

The first two symbols of each 4-tuple designate the present state of the computing element and the symbol being read on the tape; the last two symbols designate the action taken by the computing element and the next stage of the computing element. For instance, the 4-tuple $q_2 S_3 S_4 q_3$ indicates that if the computing element is in state q_2 when the symbol S_3 is read, the computing element will print the symbol S_4 over the symbol S_3 and then go to state q_3. The 4-tuple $q_1 S_4 L q_5$ indicates that if the computing element is in state q_1 and an S_4 is read, the computing element will move the tape left one square and go to state q_5.

The question of the *total state* of the machine (called "tape versus machine configuration" by Kleene, "complete configuration" by Turing, and "instantaneous description" by Davis) now arises. By assuming the tape to be part of the machine, and by listing the squares on the tape which contain symbols, the position of the computing element along the tape, and the present state of the computing element, we may realize a complete description of the machine; this is defined as the *total state*.

Most of the studies of Turing machines have been concerned with which numbers and functions are computable, where a number (function) is called computable if it can be realized as the result of some Turing-machine computation. Turing, in his original paper, showed that most of the numbers and functions which we generally think of as being "effectively calculable" can indeed be computed by a Turing machine. The term effectively calculable is taken to mean that a set of rules (an algorithm) can be written down so that a given number or value of a function can be calculated to n places, where n is finite, by following these rules and using some finite number of calculations. The thesis that all numbers (functions) which are effectively calculable are also computable by means of a Turing machine was the subject of several of the earlier papers in this area. Although this thesis can be shown to be unprovable (in that effectively calculable is not precisely definable), it is commonly accepted. Most users of modern computers would certainly acquiesce.

Let us examine a simple Turing machine which adds two positive integers m and n to yield the function $f(m,n) = m + n$. Let the tape symbols be the binary symbols 0 and 1 and let the integer m be represented on the tape by a sequence of $m + 1$ ones preceded and followed by a 0. Let n also be represented by $n + 1$ ones and let the tape be prepared so that a single 0 separates representations for m and n.† Thus, the sequence of symbols $\cdots$ 0011101111000 $\cdots$ represents the positive integer $m = 2$ followed by the positive integer $n = 3$. A simple Turing machine which will compute $m + n$ is given by the following table:

Tape-internal-state configuration		Behavior	
Present state	Tape symbol	Operation	Next state
q_1	0	move right	q_1
q_1	1	write 0	q_2
q_2	0	move right	q_2
q_2	1	write 0	q_3
q_3	0	move right	q_3
q_3	1	move right	q_4
q_4	0	write 1	q_5
q_4	1	move right	q_4
q_5	0	stop	
q_5	1	move right	q_5

The first rows of this table can be represented by the 4-tuples $q_1 0 R q_1$, $q_1 1 0 q_2$, $q_2 0 R q_2$, and $q_2 1 0 q_3$. It is not difficult to make up machines which

† It is common practice to let the integer n be represented by a sequence of $n + 1$ ones. (The use of an additional 1 arises in the representation of 0.) In his original paper, Turing used a single tape and left alternate squares (cells) of the tape blank to

will perform all the arithmetic operations and then to invent machines which will perform a given sequence of arithmetic operations. In this way we are moved to the conclusion, with Turing, that all effectively calculable functions are indeed Turing-machine computable. Please note that Turing's basic paper was written in 1936, predating even the first of the modern digital machines.

In his original paper Turing also showed, by demonstration, that it is possible to construct a *universal* Turing machine which can perform any calculation which can be performed by any other Turing machine. In effect, the universal Turing machine imitates a given machine when supplied with a description of that machine or its input tape. To be a little more precise, Turing set up a system whereby the operation of any machine M was described by means of a sequence of symbols called the *standard description* of the machine. This standard description was simply the set of 4-tuples for the computing element of a machine, with each 4-tuple separated by a semicolon. Turing then showed that a universal machine U could be designed such that if the input tape to U contained the standard description of M, then, given a sequence of symbols S, U would compute the same output sequence on the tape as M. In effect, the standard description was a program and the sequence of symbols S the input data, so that U was a general-purpose computer and M a given digital machine.

The question of the minimal universal Turing machine now arises, but this concept requires a definition of minimality. Shannon[40] has shown that it is possible to construct a universal Turing machine using only either two input states or two internal states. Shannon has further pro-proposed that a criterion for minimality be number of internal states times number of tape symbols. According to this criterion, the record for the minimal machine at this time is held by N. Ikeno of Japan with a machine having 10 internal states and 6 tape symbols, hence a product of 60. (No one has yet shown that a two-symbol, two–internal state universal machine cannot be designed.)

It is possible to formulate the questions concerning computability in terms of recursive function theory, and it can be shown that the functions which are computable are also *recursive functions.*† Further, the set of all

---

be used for partial results (as "scratch paper"). It can be shown that a single-tape machine can compute anything computable by a multitape machine, so there would be no objection to using a tape for the program, another for numbers or other input data, and still another for intermediate results. Such convenient extensions would lead us to the machine in Chap. 9.

† The Appendix contains a definition of addition and multiplication using recursion. Reference 22 contains a most complete treatment of recursive function theory along with many details concerning Turing machines.

functions which are computable is countably infinite, and therefore the computable functions form only a subset of all functions, for there are an uncountably infinite number of functions.

A well-known and interesting result in the theory of Turing machines is that no Turing machine exists which when presented with the input tape and standard description of the computing element of a given Turing machine can decide whether that machine will stop. This is called the *halting* or *stopping* problem and involves quasi- or partial functions, which are functions defined only on a proper subset of the possible input values. Determining whether a given Turing machine will stop when presented with a particular tape is equivalent to determining whether the function represented by the machine includes the value on the tape in its domain, and it can be shown that the answer to this question is not computable in a finite number of operations.

A thorough and organized treatment of Turing-machine theory can be found in Davis,[11] and a number of the basic papers may be found in the list of references.

The machines which have been discussed in previous sections are *deterministic* in that, given the starting state and input sequence of the machine, the behavior of the machine is uniquely defined, and we have assumed that there was no chance of a specified state transition or output value being in error. Studies have also been made of *probabilistic* machines for which, given the present state and input, the next state and/or output are not uniquely determined, but for which probabilities may be assigned to alternate courses of action. In effect, studies of probabilistic machines are concerned with the analysis and design of machines whose behavior is not deterministic. Machines have also been studied where some probability that a given element or set of elements might malfunction was introduced. Sometimes included in this general category are studies of *redundant* machines, or machines containing redundant components so that if components malfunction, the machine will still behave in the described manner.

The paper by Von Neumann[48] is a good basic paper on probabilistic machines. The book by Peterson[31] contains some data on schemes for detecting or correcting errors occurring during operation, and the papers by Brown,[4] Elias,[14] and Garner[15] give various results in this area. A study of whether probabilistic machines can compute functions not computable by deterministic machines may be found in the work of De Leener et al.[12] This paper also introduces some elements of probability theory into finite automaton theory, showing, for instance, that for each stochastic machine there is an equivalent Markoff machine, so that, conversely, underlying each stochastic process there is a Markoff process.

PROBLEMS

11-1. Design a synchronous sequential machine with two input lines and two output lines so that at time $t + 1$ the two output lines contain the same signals which were on the two input lines at time t. Use RS flip-flops, AND gates, and OR gates as required.

11-2. Design a sequential machine with two input lines X_1 and X_2 and one output line Y such that the output line will be 1 only after the sequence $(0,1)$, $(1,0)$, $(1,1)$ appears on the input lines X_1 and X_2. The machine is to be started with $Y = 0$ and with inputs $X_1 = 0$ and $X_2 = 0$. After the sequence $(0,1)$, $(1,0)$, $(1,1)$ appears at the input, the Y output is to remain a 1 until the machine is restarted.

11-3. Reduce the following flow table:

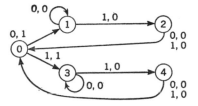

11-4. Design a machine using RS flip-flops, AND gates, and OR gates which realizes the flow table in Prob. 11-3.

11-5. Draw a state table and flow chart and then design a machine with a single input X and eight outputs $y_0, y_1, \ldots, y_7$ such that

(a) The inputs to X consist of a binary sequence occurring at times $t_0, t_1, \ldots t_n$, etc.

(b) The machine will have 0 outputs on $y_0, \ldots, y_7$ except in the interval between every third and fourth input value on X. After each third input value is received, the output line whose subscript corresponds to the binary value of the input triple is to carry a 1 signal. That is, if the sequence of signals on X at time $t_0, t_1, \ldots, t_5$ is 0, 1, 1, 0, 1, 0, then in the interval (t_3,t_4) output y_3 will be a 1, and in the interval (t_6,t_7), y_2 will be a 1. Use C flip-flops or RS flip-flops, AND gates, and OR gates as required.

11-6. Reduce the following partially specified state table to the minimum number of states (Mealy).

Present state	Next state				Output			
	X_0	X_1	X_2	X_3	X_0	X_1	X_2	X_3
1	1	6		2	y_0			y_1
2			3	2				y_1
3		4	3	5	y_2			y_3
4	1	4			y_2			
5	1			5				y_3
6		6	7		y_0			
7		9	7	8				
8	1			8	y_2			y_3
9	1	9			y_0			y_3

11-7. Design a Turing machine which "decodes" an input tape. The input tape is to contain the symbols a, b, c, d, . . . , z and spaces. The code is as follows: a *space* represents an a; a z represents a *space;* a represents b; b represents c; . . . ; y represents z. The machine is capable of reading and writing on the single tape and only one character may be read or written at a time.

11-8. A Turing machine has three input and output symbols, S_1, S_2, and S_3. The input tape to the Turing machine is recorded as follows:

blank	X_1	X_2	X_3	X_4	X_5	blank	X_1	X_2	. . .

where each variable X_i is either S_1, S_2, or S_3. The Turing machine is to rearrange the sextuples as follows:

blank	X_5	X_4	X_3	X_2	X_1	blank	X_5	X_4	X_3	. . .

Draw a state table and flow table for the machine, which is to read and print one symbol X_i at a time. This machine realizes the function f $(X_1, X_2, . . . , X_5) = (X_5, X_4, . . . , X_1)$.

11-9. How many different completely specified state tables can be drawn for a Turing machine with two input symbols X_1 and X_2, two output symbols Z_1 and Z_2, and two internal states Q_1 and Q_2? How many partially specified state tables can be drawn? Do any of the completely specified tables correspond to equivalent machines? Do any of the partially specified tables correspond to equivalent machines? Discuss the relationship between the state table for a Turing machine and the Turing machine. How many different Turing machines with two input symbols, two output symbols. and two internal states are there?

11-10. Reduce the following state tables:

	Next state		Output	
	X_0	X_1	X_0	X_1
1		1	0	0
2	1	3	0	0
3	4	2	0	0
4	4	2	1	0

	Next state			Output		
	X_0	X_1	X_2	X_0	X_1	X_2
1	2		3		0	
2	3	5	2	0	0	0
3	3	4	2		0	
4		1	2			
5		4	1	1		1

	Next state				Output			
	X_0	X_1	X_2	X_3	X_0	X_1	X_2	X_3
1	4		3	2		0		
2			5	3		0	0	
3		3	6	4	1	1		
4	1		3	5		0		1
5			6			1		
6	2	4			1	1	1	

REFERENCES

1. Aufenkamp, D. D.: Analysis of Sequential Machines II, *IRE Trans. on Electronic Computers*, vol. EC-7, pp. 299–306, December, 1958.
2. Aufenkamp, D. D., and F. E. Hohn: Analysis of Sequential Machines, *IRE Trans. on Electronic Computers*, vol. EC-6, pp. 276–285, December, 1957.
3. Aufenkamp, D. D., S. Seshu, and F. E. Hohn: The Theory of Nets, *IRE Trans. on Electronic Computers*, vol. EC-6, pp. 154–162, September, 1957.
4. Brown, D. T.: Error Detecting and Correcting Binary Codes for Arithmetic Operations, *IRE Trans. on Electronic Computers*, vol. EC-9, pp. 333–337, 1960.
5. Burks, A. W.: "The Logic of Fixed and Growing Automata," Willow Run Laboratories, Eng. Res. Inst., University of Michigan, Ypsilanti, Michigan, Rept. no. 2144-231-T, February, 1958.
6. Burks, A. W., and H. Wang: The Logic of Automata, *J. Assoc. for Computing Machinery*, vol. 4, pp. 193–218, April, 1957; pp. 279–297, July, 1957.
7. Cadden, W.: "Sequential Circuit Theory," Ph.D. dissertation, Princeton University, Princeton, N.J., 1956.
8. Caldwell, S. H.: "Switching Circuits and Logical Design," John Wiley & Sons, Inc., New York, 1958.
9. Church, A.: "Introduction to Mathematical Logic," vol. 1, Princeton University Press, Princeton, N.J., 1956.
10. Church, A.: A Note on the Entscheidurgsproblem, *J. Symbolic Logic*, vol. 1, pp. 40–41, 1936.
11. Davis, M.: "Computability and Unsolvability," McGraw-Hill Book Company, Inc., New York, 1958.
12. De Leener, R., E. F. Moore, C. E. Shannon, and N. Shapiro: Computability by Probabilistic Machines, "Automata Studies," Annals of Mathematics Studies, no. 34, Princeton University Press, Princeton, N.J., 1956.
13. Dolotta, T. A.: "The Coding Problem in the Design of Switching Circuits," Ph.D. dissertation, Princeton University, Princeton, N.J., May, 1961.
14. Elias, P.: Computation in the Presence of Noise, *IBM J. Research and Develop.*, vol. 2, pp. 346–353, 1958.
15. Garner, H. L.: Generalized Parity Checking, *IRE Trans. on Electronic Computers*, vol. EC-7, pp. 207–213, 1958.
16. Ginsburg, S.: On the Length of the Smallest Uniform Experiment which Distinguishes the Terminal States of a Machine, *J. Assoc. for Computing Machinery*, vol. 5, pp. 266–280, July, 1958.
17. Ginsburg, S.: A Technique for the Reduction of a Given Machine to a Minimal

State Machine, *IRE Trans. on Electronic Computers*, vol. EC-8, pp. 346–356, September, 1959.

18. Ginsburg, S.: Synthesis of Minimal State Machines, *IRE Trans. on Electronic Computers*, vol. EC-8, pp. 441–449, December, 1959.

19. Huffman, D. A.: "The Design and Use of Hazard-free Switching Networks," *J. Assoc. for Computing Machinery*, vol. 4, pp. 47–62, January, 1957.

20. Huffman, D. A.: The Synthesis of Sequential Switching Circuits, *J. Franklin Inst.*, vol. 257, pp. 161–190, March, 1954; pp. 275–303, April, 1954.

21. Kleene, S. C.: λ-definability and Recursiveness, *Duke Mathematical J.*, vol. 2, pp. 340–353, 1936.

22. Kleene, S. C.: "Introduction to Metamathematics," D. Van Nostrand Company, Inc., Princeton, N.J., 1952.

23. McCluskey, E. J., and T. C. Bartee: "A Survey of Switching Circuit Theory," McGraw-Hill Book Company, Inc., New York, 1962.

24. McCluskey, E. J., and S. H. Unger: A Note on the Number of Internal Variable Assignments for Sequential Switching Circuits, *IRE Trans. on Electronic Computers*, vol. EC-8, no. 4, pp. 439–440, September, 1959.

25. McCulloch, W. S., and W. Pitts: A Logical Calculus of the Ideas Immanent in Nervous Activity, *Bull. Mathematical Biophys.*, vol. 5, pp. 115–133, 1943.

26. Mealy, G. H.: A Method for Synthesizing Sequential Circuits, *Bell System Tech. J.*, vol. 34, pp. 1045–1079, September, 1955. Also Bell Telephone System Monograph 2458.

27. Metze, G., R. E. Miller, and S. Seshu: Transition Matrices of Sequential Machines, *IRE Trans. on Circuit Theory,*, vol. CT-6, pp. 5–11, March, 1959.

28. Moore, E. F.: Gedanken-experiments on Sequential Machines, "Automata Studies," Annals of Mathematics Studies, no. 34, pp. 129–153, Princeton University Press, Princeton, N.J., 1956.

29. Netherwood, D. B.: Minimal Sequential Machines, *IRE Trans. on Electronic Computers*, vol. EC-8, pp. 367–380, September, 1959.

30. Peterson, W., and D. T. Brown: Cyclic Codes for Error Detection, *Proc. IRE*, vol. 49, no. 1, pp. 228–235, January, 1961.

31. Peterson, W.: "Error Correcting Codes," The M.I.T. Press, Cambridge, Mass., and John Wiley & Sons, Inc., New York, 1960.

32. Peterson, W.: On Checking an Adder, *IBM Research and Develop.*, vol. 2, pp. 100–108, 1958.

33. Post, E. L.: Recursively Enumerable Sets of Positive Integers and Their Decision Problems, *Bull. Am. Math. Soc.*, vol. 50, pp. 284–316, 1944.

34. Reed, I. S.: "The Symbolic Design of Digital Computers," MIT Lincoln Laboratory, Lexington, Mass., Tech. Mem. 23, 1953.

35. Reed, I. S.: "Symbolic Design Techniques Applied to a Generalized Computer," MIT Lincoln Laboratory, Lexington, Mass., Rept. 141, Jan. 3, 1957.

36. Reed, I. S.: Symbolic Synthesis of Digital Computers, *Proc. 1952 Meeting, Assoc. for Computing Machinery*, pp. 90–94, Richard Rimbach Associates, Pittsburgh, Pa.

37. Robin, M. O.: Two-way Finite Automata, *Proc. Summer Inst. Symbolic Logic*, pp. 366–369, Cornell University Press, Ithaca, N.Y., 1957.

38. Robin, M. O., and D. Scott: Finite Automata and Their Decision Problems, *IBM J. Research and Develop.*, vol. 3, 1959.

39. Rogers, H., Jr.: Gödel Numberings of Partial Recursive Functions, *J. Symbolic Logic*, vol. 23, 1958.

40. Shannon, C. E.: A Universal Turing Machine with Two Internal States, "Auto-

mata Studies," Annals of Mathematics Studies, no. 34, Princeton University Press, Princeton, N.J., 1956.

41. Simon, J. M.: Some Aspects of the Network Analysis of Sequence Transducers, *J. Franklin Inst.*, vol. 65, pp. 439–450, June, 1958.
42. Simon, J. M.: A Note on the Memory Aspects of Sequence Transducers, *IRE Trans. on Circuit Theory*, vol. CT-6, pp. 26–30, March, 1959.
43. Turing, A.: On Computable Numbers with Applications to the Entscheidungsproblem, *Proc. London Math. Soc.*, series 2, vol. 42, pp. 230–265, 1936; vol. 43, pp. 544–546, 1937.
44. Turing, A.: Computing Machinery and Intelligence, *Mind*, vol. 59, pp. 433–460, 1950.
45. Turing, A.: Solvable and Unsolvable Problems, *Sci. News*, no. 31, pp. 7–23, Penguin Books, Inc., Baltimore, 1954.
46. Turing, A.: Computability and λ-definability, *J. Symbolic Logic*, vol. 2, pp. 153–163, 1937.
47. Unger, S. H., and M. C. Paull: Minimizing the Number of States in Incompletely Specified Sequential Switching Functions, *IRE Trans. on Electronic Computers*, vol. EC-8, pp. 356–366, September, 1956.
48. Von Neumann, J.: Probabilistic Logic and the Synthesis of Reliable Organism from Unreliable Components, "Automata Studies," Annals of Mathematics Studies, no. 34, Princeton University Press, Princeton, N.J., 1956.
49. Von Neumann, J.: The General and Logical Theory of Automata, from "Cerebral Mechanisms in Behavior," John Wiley & Sons, Inc., New York, 1951.
50. Wiener, N.: "Cybernetics," John Wiley & Sons, Inc., New York, 1948.

Appendix

Number Representation Systems

In order for a digital machine to perform numerical calculations, it is first necessary to represent numerical data by means of the states of physical devices. The devices which have appeared most desirable for storing and processing information in digital form have been devices which appear to be most effectively operated in a *bistable* manner; that is, these devices can be operated at high speeds with considerable reliability when used as two-state devices. Most of the early electronic machines operated the physical devices used at extreme operating points: magnetic cores were either saturated in one direction or the other, vacuum tubes and transistors were either saturated or cut off. More recently, designers have tended to place the operating points of these devices away from the extremes, but still to restrict the operating points to within two regions, and hence still to operate the devices in a binary manner.

Since the devices in present-day machines are used in a binary manner, machine designers have been prone to use the most natural and efficient number representation system subject to this constraint, the binary number representation system. The second most prevalent class of machines represents numbers in a binary-coded decimal system, which is less efficient from a purely machine viewpoint, but which is more natural from the viewpoint of some machine users. (The decimal number system is, indeed, a more efficient system for manual computation, record keeping, etc.)

In actual practice the machine user rarely writes his instructions to the machine in anything resembling actual machine language. Every present-day general-purpose computer of any magnitude is provided with a set of conversion programs which automatically convert the symbols written by the user to machine language. Concern over the machine's internal number system must therefore be directed toward a consideration of the process of making programs operational and toward the actual maintenance of the machine. Indeed, most modern machines are provided with automatic program systems which far exceed the minimal set of programs needed to convert decimal numbers to binary, assign internal machine memory addresses to instructions and data, convert mnemonic instruction codes to the correct machine representations, and so on.

The extent of present-day program systems is staggering, and the facilities offered by such programs are expanded daily. The user of a binary-number general-purpose machine can now write his programs using the English language, common mathematical symbols (somewhat restricted by the characters and programs available), and decimal numbers. The combination of programs and machine will then do the rest, converting the English, decimal, mathematical, logical expressions of the human to machine representation and then performing the necessary information processing. Indeed, some systems will inform the user if he has made certain obvious mistakes in communicating with the machine.

This is not to say that all the work necessary for bridging the natural gap between machine and user has been provided, or even that the problem has been especially well handled, but rather to indicate that the question of natural and efficient machine design versus providing a convenient system for human utilization must take into account the very significant contribution of the programmers who design and prepare the program systems which so greatly facilitate use of the machine. A primary concern of the designer of general-purpose machines must certainly be the providing of instructions and other facilities needed by the utility programs, as well as the requirements of the actual user.

In most cases the binary number representation system has been a rather obvious choice for the designer of special-purpose machines, and certainly the preponderance of special-purpose machines use this most natural system. The next best choice for certain special-purpose machines may well lie in the newer and lesser-known representations based on residue number classes,[6] Mersenne primes,[5] etc., which may eventually provide more efficient algorithms for multiplication, division, and other operations. Several references to such systems are included in the list at the end of this appendix. As yet, the known disadvantages of such systems outweigh the known advantages, and further research is required before these can be seriously considered.

One special-purpose computer has been constructed using Galois field elements for the number system, thus simplifying certain specialized computations; Ref. 4 describes this machine and the number system.

Number Representation Systems. For us, the most familiar number representation system is the decimal system. Before describing this system and others, let us define the terms *radix* and *positional notation*.

Positional notation refers to a technique which is used to represent numbers symbolically, where each number is represented by a sequence of symbols, and the value of a given symbol is determined not only by the symbol itself, but by its position in the sequence.

For instance, in our familiar decimal system the over-all value of the symbol 2 in the sequence 323 is different from its value in the sequence 20,000. The sequence 323 is really a shorthand notation for the polynomial $3 \times 10^2 + 2 \times 10^1 + 3 \times 10^0$, where only the coefficients of the polynomial are written. In general, the sequence of symbols $c_1 c_2 \cdots c_n$ is an abbreviation for the sum $\sum_{i=1}^{n} c_i 10^{n-i}$ whereby we write the coefficients of the polynomial in their natural order.

When a positional notation system is used, the value of each symbol in a sequence is determined by the symbol, the position of the symbol in the sequence, and the *base* or *radix* of the number representation system. The radix or base is defined as the number of different symbols which can occur in a given position in the sequence. The radix is equal to 10 in the decimal system. Commonly each position in a sequence representing a number is to the same base.

Other number representation systems using positional notation have other radices. For instance, the binary number representation system, with a radix 2, has only two symbols which can occur in each position in the sequence, and we use the symbols 0 and 1 for these. The radix 3 or *ternary* number representation system has three symbols per position; commonly 0, 1, and 2 are used. The *octal* number system, with a radix of 8, has eight possible symbols for each position; we use 0, 1, 2, . . . , 7. For the *sexadecimal* system, which has sixteen symbols per position, the symbols $a, b, c, d, e,$ and f are commonly appended to the symbols 0, 1, 2, . . . , 9. In the octal system the sequence 624 represents the sum $6 \times 8^2 + 2 \times 8^1 + 4 \times 8^0$ which is equal to the sequence 404 in the decimal system. In general the sequence $c_1 c_2 \cdots c_n$ is an abbreviation for the sum $\sum_{i=1}^{n} c_i r^{n-i}$, where r is the radix of the system.

Thus far, we have used the term number as an undefined or primitive term. We shall leave it so. (There are three schools of thought concerning the definition of number: the logistic school, headed by Russell,[20] Whitehead,[27] and more recently Quine;[17] the set theoretic group, initiated by Cantor and with such present day exponents as Halmos,[8] Suppes,[22] and perhaps Kleene;[13] and the intuitionist school, which was founded by Brouer. We recommend Heyting[10] for a good description of the latter. The definitions of number proposed by these doctrines would differ, but all would agree that the sequence 348 is a representation of a number.) Let us therefore assume that we know what numbers are and agree that the set of *natural numbers* or *nonnegative integers* can be used in an intelligent manner. Let us also assume that we can use the *proof by induction* and can define a function using recursion. We therefore define addition, multiplication, and exponentiation as follows, using 0 as the first natural number and 1 as the successor to 0:

$$0 + x = x$$
$$x + (y + 1) = (x + y) + 1$$
$$x \times 0 = 0$$
$$x \times (y + 1) = (x \times y) + x$$
$$x^0 = 1$$
$$x^{y+1} = x^y \times x$$

These rules are sufficient to permit the calculation of any sum, product, or power formed of nonnegative integers.

One of the features of our positional notation system is that it permits calculations to be formed in a very simple way. We commonly perform addition and multiplication using only the coefficients of the numbers represented (exponentiation is performed by repeated multiplication). Let $A = a_1, a_2, \ldots, a_n$ and $B = b_1, b_2, \ldots, b_m$, both to the radix r. Then

$$A + B = \sum_{i=1}^{n} a_i r^{n-i} + \sum_{j=1}^{m} b_j r^{m-i}$$

or, if $m = n$,

$$A + B = \sum_{i=1}^{n} (a_i + b_i) r^{n-i}$$

$$A \cdot B = \left(\sum_{i=1}^{n} a_i r^{n-i} \right) \left(\sum_{j=1}^{m} b_j r^{m-j} \right) = \sum_{i=1}^{n} \sum_{j=1}^{m} (a_i \cdot b_j) r^{m+n-i-j}$$

By first calculating the addition and multiplication table for the symbols used to represent digits, we can form sums and products using simple pencil-and-paper techniques. The following are ternary addition and multiplication tables and samples of the algorithms we now use.

Ternary System

Addition table

	0	1	2
0	0	1	2
1	1	2	10
2	2	10	11

$$\begin{array}{r} 10212 \\ + \ 21120 \\ \hline 102102 \end{array}$$

Multiplication table

	0	1	2
0	0	0	0
1	0	1	2
2	0	2	11

$$\begin{array}{r} 122 \\ \times \ 210 \\ \hline 000 \\ 122 \\ 1021 \\ \hline 111020 \end{array}$$

We can check the sum and product above in the decimal system by first converting the ternary numbers to decimal and then performing the calculations. It is common practice when using several radices to append the particular radix associated with a given number as a subscript; thus 1012_3 is ternary 1012, 1129_{10} is decimal 1129, 10110_2 is binary 10110.

$$10212_3 = 1 \times 3^4 + 0 \times 3^3 + 2 \times 3^2 + 1 \times 3 + 2 \times 1$$
$$= 104_{10}$$
$$21120_3 = 2 \times 3^4 + 1 \times 3^3 + 1 \times 3^2 + 2 \times 3 + 0 = 204_{10}$$
$$10212_3 + 21120_3 = 102102_3$$
$$104_{10} + 204_{10} = 308_{10}$$
$$102102_3 = 1 \times 3^5 + 0 \times 3^4 + 2 \times 3^3 + 1 \times 3^2 + 0 \times 3^1$$
$$+ 2 \times 3^0 = 308_{10}$$
$$122_3 = 1 \times 3^2 + 2 \times 3 + 2 = 17_{10}$$
$$210_3 = 2 \times 3^2 + 1 \times 3 + 0 = 21_{10}$$
$$122_3 \times 210_3 = 111020_3$$
$$17_{10} \times 21_{10} = 357$$
$$111020_3 = 357_{10}$$

Binary-coded Decimal. Another system for representing numbers in machines is the *binary-coded decimal* system. In this system sets of binary digits are handled in a pseudodecimal manner. A large number of the earlier machines used binary-coded decimal, and many general-purpose business machines use this system.

Basically, a binary-coded decimal system can be established by simply associating a unique m-digit binary number with each of the ten decimal digits, where $m \geq 4$. The most used binary-coded decimal system is the most obvious; it is the 8,4,2,1 code, which associates the binary number 0000_2 with the decimal digit 0_{10}, 0001_2 with 1_{10}, 0010_2 with 2_{10}, . . . , and 1001_2 with 9_{10}. Using this, the decimal number 243 would be represented in the binary-coded decimal 8,4,2,1 code as 0010 0100 0011.

In binary-coded decimal, each decimal digit must be represented by at least four binary digits, and there are a total of 16!/6! different ways to code four binary digits into the ten decimal digits. Of these approximately 3×10^9 different encodings, only a few have been actually used, and always for rather special reasons. Generally binary-coded decimal codes are divided into two classes: *weighted* codes and *nonweighted* codes. The 8,4,2,1 code is a weighted code, for each position in a code group of four binary digits carries a specific assigned weight, and each decimal digit represented is equal to the sum of the weights corresponding to 1's in the binary code group. A weighted binary-coded decimal system is therefore one where each decimal digit corresponds to a sequence $b_1 b_2 b_3 b_4$ with value $\sum_{i=1}^{4} b_i w_i$, where the b_i are binary digits, and the w_i the weights assigned. Weeg[25] has shown there are only 17 different four-binary-digit weighted codes when only positive integers are assigned to the w_i. As examples, the 2,4,2,1 and 7,4,2,1 codes follow.

	w_1 w_2 w_3 w_4 2 4 2 1										w_1 w_2 w_3 w_4 7 4 2 1								
0	0	0	0	0							0	0	0	0					
1	0	0	0	1							0	0	0	1					
2	1	0	0	0	or	0	0	1	0		0	0	1	0					
3	0	0	1	1	or	1	0	0	1		0	0	1	1					
4	0	1	0	0	or	1	0	1	0		0	1	0	0					
5	1	0	1	1	or	0	1	0	1		0	1	0	1					
6	1	1	0	0	or	0	1	1	0		0	1	1	0					
7	0	1	1	1	or	1	1	0	1		1	0	0	0	or	0	1	1	1
8	1	1	1	0							1	0	0	1					
9	1	1	1	1							1	0	1	0					

It is interesting to note that only the 8,4,2,1 code represents each decimal digit uniquely; all other positive-weighted codes have choices in certain positions, as shown above.

It is also possible to introduce negative weights, thus increasing the number of weighted codes to 88, for there are 71 weighted codes containing both negative and positive weights.[25] One of these is the 8, 6, −4, 1 code, which follows.

	w_1 w_2 w_3 w_4 8 6 −4 1			
0	0	0	0	0
1	0	0	0	1
2	0	1	1	0
3	0	1	1	1
4	1	0	1	0
5	1	0	1	1
6	0	1	0	0
7	0	1	0	1
8	1	0	0	0
9	1	0	0	1

This is another example of a unique representation code, and 21 of the 71 codes with mixed positive and negative weights have unique representations.

Many other binary-coded decimal systems have been used and suggested. A *biquinary* code in which each decimal digit is represented by two binary code groups, one to the base two and the other to the base five, was used in an early Bell Telephone Labs machine.[26] A *two-out-of-five* code has also been used in digital systems concerned with communications; this code is formed by assigning code groups of five binary digits, each containing only two 1's to each decimal digit. Then, if a code group containing more or fewer than two 1's is detected, it is clear that an error has occurred.   For this reason the code is known as an *error-detecting* code.   There are many other error-detecting and correcting codes.   The most popular are doubtless the *parity check* codes in which additional digits are appended to each group of binary digits, so that the number of 1's among certain sets of digits is always even (odd in some systems).   References 3, 7, 15, 16, and 18 describe such codes in more detail.   (See also Prob. 10-10.)

Fractions and Negative Numbers.   The representation of fractions follows directly in a positional notation system.   A symbol called a *radix point* is used in a sequence representing fractions or mixed numbers, where the radix point is defined as the point separating the members of the sequence associated with zero and positive powers of the radix from the members associated with negative powers.   The sequence

$$c_1 c_2 \cdots c_n . c_{n+1} c_{n+2} \cdots c_{n+m}$$

represents the sum $\sum_{i=1}^{m+n} c_i r^{n-i}$, where the c_i are digits, the . indicates the radix point, and r is the radix of the system.   (The radix point in the decimal system is often called a decimal point, and the radix point in the binary system is called a *binary point.*)

We now consider the problem of representing negative numbers in a machine.   The information as to whether a given number is negative or positive may be stored in a single binary digit, so let us assign the symbol c_0 to the leftmost position in each sequence representing a number, and agree that we shall let a 0 in this position signify that the sequence represents a positive number, and a 1 signify that the sequence represents a negative number.   We shall often call c_0 the *sign digit.*

We are now left with several choices as to the over-all meaning of the sequence, but the most obvious choice is to let the sequence $c_0 c_1 c_2 \cdots c_n$ represent the sum $(1 - 2c_0) \sum_{i=1}^{n} c_i r^{n-i}$.

In this case c_0 indicates whether the number is positive or negative, and the remaining symbols in the sequence, $c_1 c_2 \cdots c_n$, represent the magnitude of the number.   The sequence 01010_2 therefore represents $+1010_2$ or 10_{10}, and the sequence 11010_2 represents -10_{10}.   Likewise the sequence 19424_{10} represents -9424_{10} and 09424_{10} represents $+9424_{10}$.

Two other representation systems are often used.   One is called the *true complement* system, and the other the *radix-minus-one complement* system.   These are both used to facilitate addition and subtraction, as will be explained.

The *true complement* of a number $c = c_1 c_2 \cdots c_n$ is defined as the number $r^n - c$, where r is the radix of the system.   The true complement of $c_1 c_2 \cdots c_n$ can be formed by simply subtracting each c_i from $r - 1$, thus forming a $\bar{c}_i$, and then adding 1 to the resulting number $\bar{c} = \bar{c}_1 \bar{c}_2 \cdots \bar{c}_n$. For instance, the true complement of 2413_8 is defined as $8^4 - 2413_8 = 5365_8$ and can be formed in the following way:

$$(7 - 2) \times 8^3 + (7 - 4) \times 8^2 + (7 - 1) \times 8^1 + (7 - 3) \times 8^0 + 1$$
$$= 5364 + 1 = 5365_8$$

For example, the true complement of 7414_{10} is 2586_{10}, and the true complement of 1010_2 is 0110_2.

The radix-minus-one complement of the number $c = c_1 c_2 \cdots c_n$ is defined as the number $r^n - c - 1$, where r is the radix. The radix-minus-one complement of a given number is formed by simply subtracting each digit c_i in the sequence representing the number from $r - 1$. For instance, the radix-minus-one complement of 2417_{10} is 7582_{10}, and the radix-minus-one complement of 1010_2 is 0101_2.

The true complement of a number in a system with radix r is generally called the r's complement; for instance, the true complement of a decimal number is called the 10's complement, and the true complement of a binary number the 2's complement. Similarly, the radix-minus-one complement of a decimal number is called the 9's complement and the radix-minus-one complement of a binary number the 1's complement.

The true and radix-minus-one complements are generally used to represent negative numbers only, so, given the number $+2417_{10}$, the negative of this would be represented by -7582_{10} in the 9's complement system and -7583_{10} in the 10's complement system.

The virtue of these systems lies in the fact that these complements are rather easily formed, and if complements are used, the subtraction and addition operations can be performed using only addition and a few simple rules.

Let us assume that negative numbers are stored in 9's-complement form, and that all numbers are to be represented by sequences of length 5 consisting of $c_0 c_1 c_2 c_3 c_4$, where c_0 (the sign digit) indicates whether the number is positive or negative and the digits following represent the value of the number. For instance, the sequence 00019_{10} represents the number $+19_{10}$ and the sequence 10114_{10} represents the number -9885_{10}. We can add the numbers represented by simply adding the sequences,

$$
\begin{array}{rr}
19 & 00019 \\
-9885 & 10114 \\
\hline
-9866 & 10133 \\
\end{array}
$$

for 10133 represents the number -9866 in this system. Care must, of course, be taken to see that the sum or difference is not greater or less than the numbers which can be represented. We cannot, for instance, represent the sum of 9484 and 5966 in this system, nor the sum of -5966 and -5976. One other rule is required: c_0 *is to be interpreted as a binary digit and added using the binary addition table regardless of the radix.* If a carry is generated from the sign digits in a radix-minus-one system, then a 1 is added to the c_n position in the sum as follows:

$$
\begin{array}{rl}
-27 & \quad 19972 \\
-54 & \quad 19945 \\
\hline
-81 & \quad 1\ 19917 \\
& \quad \hookrightarrow 1 \qquad \text{end-around carry} \\
& \quad \overline{19918}
\end{array}
$$

or

$$
\begin{array}{rl}
-496 & \quad 19503 \\
-501 & \quad 19498 \\
\hline
& \quad 1\ 19001 \\
-997 & \quad \hookrightarrow 1 \qquad \text{end-around carry} \\
& \quad \overline{19002}
\end{array}
$$

Most machines represent negative numbers in either a true or radix-minus-one complement form and represent numbers in fractional form. Assume a 1's-complement system, let numbers be represented by sequences of five binary digits $c_0 c_1 c_2 c_3 c_4$, where c_0 is the sign digit and $c_1 c_2 c_3 c_4$ represents a binary fraction with binary point to the left of c_1, and let negative numbers be represented in 1's-complement binary form. Then -0.0101_2 is represented as 11010_2 and $+0.0101_2$ as 00101_2. The sum of these is 0 or 11111_2, which is negative 0 in this system. The system has another representation for 0; this is 00000, often called positive 0.

In the corresponding 2's-complement system, the number $+0.001_2$ is represented as 0001, the number -0.001_2 as 1111. Now in the 2's-complement system, or in any true-complement system, any carry which occurs out of the c_0 position during an addition is dropped as follows:

$$
\begin{array}{rl}
-10 & \quad 11110 \\
+10 & \quad 00010 \\
\hline
00 & \quad 00000 \\
& \quad 1 \quad \text{is dropped}
\end{array}
$$

So 00000 is 0 and there is only one 0 in this system. We can define the value of a sequence $c = c_0 c_1 c_2 \cdots c_n$ in the fractional representation 2's-complement binary number system as follows:

$$
\delta_2(c) = -c_0 + \sum_{i=1}^{n} c_i 2^{-i}
$$

This particular 2's-complement binary fraction system will represent numbers in the interval $-1 \leq \delta_2 \leq 1 - 2^{-n}$.

The 1's-complement binary fraction representation system of a number $c = c_0c_1c_2 \cdots c_n$ can be defined thusly:

$$\delta_1(c) = c_0(2^{-n} - 1) + \sum_{i=1}^{n} c_i 2^{-i}$$

Similarly, the true-complement and radix-minus-one complement systems can be used with systems based on other radices.   We can define the value of a sequence $c_0c_1c_2 \cdots c_n$ to the radix r in a true-complement fractional representation system as

$$\delta_r(c) = -c_0 + \sum_{i=1}^{n} c_i r^{-i}$$

and the value in a radix-minus-one fractional representation system as

$$\delta_{r-1}(c) = c_0(r^{-n} - 1) + \sum_{i=1}^{n} c_i r^{-i}$$

In a true-complement fractional system numbers can be represented in the interval $-1 \leq \delta_r \leq 1 - r^{-n}$, and in a radix-minus-one complement fractional system numbers can be represented in the interval $-1 + r^{-n} \leq \delta_{r-1} \leq 1 - r^{-n}$.

Since the only system other than binary which has been used to any extent is the binary-coded decimal system, some of the characteristics of such systems will be described. Systems using other radices have similar characteristics.

The negative of a number represented in a 9's-complement fractional form can be easily formed.   The sign digit c_0 is changed in value and each of the other digits c_i is subtracted from 9.   Thus the negative of 04454 is 15545 and the negative of 12426 is 07573.   Addition and subtraction can therefore be performed quite easily in such a system using only an addition operation; to subtract we merely form the 9's complement and add. Two things must be considered: the forming of the 9's complement and the end-around carry which sometimes results, and which must then be added to the least-order digit in the sum.

Let us consider the forming of the 9's complement first.   If decimal digits are represented in a machine using some binary-coded decimal system, construction is simplified if the 9's complement of a given code group representing a decimal digit can be easily formed.   The most direct way to form a 9's complement would be simply to complement each binary digit in the code group.   But notice that if this is done in the

8,4,2,1 system, the 9's complement will not result. For instance, when complemented digit by digit, the code group 1000_2, which represents 8_{10}, becomes 0111, which represents 7_{10}, not 1_{10}. The 2,4,2,1 code described previously will work, however: $1111 = 9_{10}$, complemented, it is $0000 = 0_{10}$; $0001 = 1_{10}$ complemented is $1110 = 8_{10}$; and a check of the table will indicate that the 9's complement of a given code group can be formed by simply complementing each binary digit in the group.

Another binary-coded decimal system which form 9's complements rather easily is the *excess-3 code* which was used in the early Harvard machines and which represents a decimal digit with a four-digit binary code group formed by adding $3_{10} = 0011$ to the 8,4,2,1 code group for the digit. For instance, the excess-3 code represents 3_{10} as 0110, 4_{10} as 0111, 5_{10} as 1000, etc. A check will indicate that the 9's complement of a given code group can be formed by complementing each binary digit in the group. For instance, $0111 = 4_{10}$, complemented, it is $1000 = 5_{10}$; $0100 = 1_{10}$, complemented, it is $1011 = 8_{10}$; etc.

The excess-3 code is not a weighted code, and difficulty arises during calculations involving both positive and negative numbers. (The excess-3 carried in both the addend and augend becomes an excess 6 in the sum if both addend and augend are positive, an excess 0 if one is negative and the other positive. Interested readers may find details in Ref. 9.)

The 9's complement of a digit stored in an 8,4,2,1 system can of course be formed by a combinational network. Many systems represent decimal numbers in a serial parallel manner, where each code group of four binary digits arrives in parallel, and as many sets of digits arrive as there are decimal digits represented, plus a set containing the sign digit. (If a word contains sign plus 10 binary-coded decimal digits, 11 groups of four binary digits will be transmitted serially; only one binary digit of the four representing the sign will have sign meaning, and the remaining three may contain other information.) When this system is used, only one 9's-complement circuit and only one adder need be used, for each four-binary-digit code group can be transmitted to these circuits in serial (see Refs. 3, 16, and 21 for details). When the 9's-complement system is used, any end-around carry must be added in when it arises, and this requires an additional circulation of the sum plus carry through the adder.

The 10's complement of a number stored in binary-coded decimal is not so easily formed. It is not possible to form the 10's complement by simply forming a complement of each of the binary code groups representing the decimal digits individually, for the digits interact. It is possible to form the 10's complement in a serial-parallel or serial system by means of a sequential circuit or by forming the 9's complement of each code group (digit) and then adding 1 to the least-order digit (this can be done during an actual subtraction or addition). In parallel systems the parallel full adders may be used in a similar manner.

When a 10's-complement system is used, any carry generated while adding the sign digits is simply discarded, so no additional passes through the adder are required for serial or serial-parallel systems. This somewhat compensates for the additional complexity which arises in forming the 10's complement.

REFERENCES

1. Alt, F. L.: "Electronic Digital Computers," Academic Press, Inc., New York, 1958.
2. Alt, F. L.: "Advances in Computers," Academic Press, Inc., New York, 1960.
3. Bartee, T. C.: "Digital Computer Fundamentals," McGraw-Hill Book Company, Inc., New York, 1960.
4. Bartee, T. C., and D. I. Schneider: An Electronic Decoder for Bose-Chaudhuri-Hoquenghem Error-correcting Codes, *IRE Trans. on Information Theory*, vol. IT-8, no. 5, Proceedings of International Conference on Information Theory, Brussels, Belgium, September, 1962.
5. Fraenkel, A. S.: The Use of Index Calculus and Mersenne Primes for the Design of a High-speed Digital Multiplier, *J. Assoc. for Computing Machinery*, vol. 8, no. 1, pp. 87–97, January, 1961.
6. Garner, H. L.: The Residue Number System, *IRE Trans. on Electronic Computers*, vol. EC-8, no. 2, pp. 140–147, June, 1959.
7. Garner, H. L.: Generalized Parity Checking, *IRE Trans. on Electronic Computers*, vol. EC-7, no. 3, pp. 207–213, September, 1958.
8. Halmos, P. R.: "Naive Set Theory," D. Van Nostrand Company, Inc., Princeton, N.J., 1960.
9. Harvard University Computation Laboratory: "Investigations for Design of Digital Calculating Machinery," Progress Reports 1 through 10.
10. Heyting, A.: "Intuitionism—An Introduction," North Holland Publishing Co., Amsterdam, 1956.
11. Hollingdale, S. H.: "High Speed Computing," The Macmillan Company, New York, 1959.
12. Kautz, W. H.: Optimized Data Encoding for Digital Computers, *IRE National Convention Record*, part 4, pp. 47–57, 1954.
13. Kleene, S. C.: "Introduction to Metamathematics," D. Van Nostrand Company, Inc., Princeton, N.J., 1952.
14. Lehman, M.: High Speed Multiplication, *IRE Trans. on Electronic Computers*, vol. EC-6, pp. 204–205, 1957.
15. Peterson, W. W.: "Error-correcting Codes," The M.I.T. Press, Cambridge, Mass., 1960.
16. Phister, Montgomery: "Logical Design of Digital Computers," John Wiley & Sons, Inc., New York, 1958.
17. Quine, W. V.: "Mathematical Logic," Harvard University Press, Cambridge, Mass., 1951.
18. Richards, R. K.: "Arithmetic Operations in Digital Computers," D. Van Nostrand Company, Inc., Princeton, N.J., 1955.

19. Robertson, J. E.: A New Class of Division Methods, *IRE Trans. on Electronic Computers*, vol. EC-7, pp. 218–222, 1958.
20. Russell, B.: "Introduction to Mathematical Philosophy," George Allen and Unwin, Ltd., London, 1919.
21. Scott, N. R.: "Analog and Digital Computer Technology," McGraw-Hill Book Company, Inc., 1960.
22. Suppes, P.: "Axiomatic Set Theory," D. Van Nostrand Company, Inc., Princeton, N.J., 1960
23. Tocher, K. D.: Techniques of Multiplication and Division for Automatic Binary Computers, *Quart. J. Mechanics and Appl. Math.*, vol. 11, part 3, pp. 364–384, 1958.
24. Wadley, W. G.: Floating-point Arithmetics, *J. Assoc. for Computing Machinery*, vol. 7, no. 2, pp. 129–140, April, 1960.
25. Weeg, G. P.: Uniqueness of Weighted Code Representation, *IRE Trans. on Electronic Computers*, vol. EC-9, no. 4, pp. 487–490, December, 1960.
26. White, G. S.: Coded Decimal Number Systems for Digital Computers, *Proc. IRE*, vol. 41, no. 10, pp. 1450–1452, October, 1953.
27. Whitehead, A. N., and B. Russell: "Principia Mathematica," Cambridge University Press, New York, 1925.

Index